High Definition Metrology Based Surface Quality Control and Applications

AF348753

Shichang Du · Lifeng Xi

High Definition Metrology Based Surface Quality Control and Applications

 Springer

Shichang Du
School of Mechanical Engineering
Shanghai Jiao Tong University
Shanghai, China

Lifeng Xi
School of Mechanical Engineering
Shanghai Jiao Tong University
Shanghai, China

ISBN 978-981-15-0281-1   ISBN 978-981-15-0279-8   (eBook)
https://doi.org/10.1007/978-981-15-0279-8

© Springer Nature Singapore Pte Ltd. 2019
This work is subject to copyright. All rights are reserved by the Publisher, whether the whole or part of the material is concerned, specifically the rights of translation, reprinting, reuse of illustrations, recitation, broadcasting, reproduction on microfilms or in any other physical way, and transmission or information storage and retrieval, electronic adaptation, computer software, or by similar or dissimilar methodology now known or hereafter developed.
The use of general descriptive names, registered names, trademarks, service marks, etc. in this publication does not imply, even in the absence of a specific statement, that such names are exempt from the relevant protective laws and regulations and therefore free for general use.
The publisher, the authors and the editors are safe to assume that the advice and information in this book are believed to be true and accurate at the date of publication. Neither the publisher nor the authors or the editors give a warranty, expressed or implied, with respect to the material contained herein or for any errors or omissions that may have been made. The publisher remains neutral with regard to jurisdictional claims in published maps and institutional affiliations.

This Springer imprint is published by the registered company Springer Nature Singapore Pte Ltd.
The registered company address is: 152 Beach Road, #21-01/04 Gateway East, Singapore 189721, Singapore

Foreword

Surface quality control is an essential technology in advanced manufacturing with substantial significance in the context of the global market. This is especially important in the automotive and aerospace industries, in which surface quality has critical impacts on functionality, safety, and performance. However, due to the limitations of measurement techniques in early years, surface topography could only be assessed by visual inspections and touch detections. In recent years, novel platforms, namely, high-definition metrology (HDM), have emerged in which high-density point cloud data is measured for reflection of the entire surface contour or profile. HDM provides great capabilities and opportunities for developing new advanced surface quality control methodologies. And, with the continuous emergence of novel measurement instruments and measuring methods, surface quality control methods continue to evolve and improve.

I am enthused to learn that Profs. Shichang Du and Lifeng Xi are writing a book on surface quality control and the applications based on HDM. Professors Du and Xi are leading scholars in quality control engineering of complex manufacturing processes with more than 200 peer-reviewed papers published in international journals and conferences. The results of their research have been implemented in industrial systems in Shanghai Automotive Industry Corporation General Motors Wuling company (SGMW) with significant economic impacts.

This book, which is based on the authors' extensive research and implemented experience, covers a comprehensive understanding of HDM. The key topics include surface evaluation methods, surface filtering methods, surface classification methods, surface monitoring methods, surface prediction methods, and online compensation methods based on HDM. Both methodological discussions and real case studies are presented in the book. The book will not only help readers to master the advanced quality control methods based on HDM but also to train their capabilities in their implementations in practice.

Dr. Jianjun (Jan) Shi
Carolyn J. Stewart Chair and Professor
H. Milton Stewart School of Industrial and Systems Engineering
Georgia Institute of Technology
Atlanta, GA, USA

Preface

The purpose of this book is to provide insight into surface quality control techniques and applications based on high-definition metrology (HDM). It is intended to serve as a reference for engineers who routinely use a variety of quality control methods and are interested in understanding the data processing from HDM data to final control actions. This book can also be used as a text for advanced courses in engineering quality control applications for students who are already familiar with quality control methods and practices.

Most of the material is presented with the objective of enabling the reader to not only assimilate the quality control methods involved but also to be able to quickly implement the techniques in practical engineering problems. For that reason, a number of case studies are included that highlight the implementation of the methods using measured HDM data of surface features. The book employs MATLAB extensively for these case studies. Familiarity with MATLAB will aid in understanding the material.

Also, it should be pointed at the outset that this book assumes a general familiarity with surface quality control methods, for readers unfamiliar with some of the quality control techniques, several excellent textbooks are available, such as statistical method from the viewpoint of quality control (Walter Andrew Shewhart and William Edwards Deming, 1939), statistical quality control (Richard S. Leavenworth and Eugene L. Grant, 2000), etc.

Shanghai, China

Shichang Du
Lifeng Xi

Acknowledgements

We are extremely grateful for National Natural Science Foundation of China for providing Key Program (Grant No. 51535007) and General Program (Grant No. 51775343). Numerous colleagues and students also volunteered to review sections of the text, provided insight into some of the techniques discussed here, and also supplied several excellent references. We express our sincere gratitude to the following people: Zhongqin Lin, Xinmin Lai, Sun Jin, Juntong Xi, Zhenqiang Yao, Min Xu, Xueping Zhang, Zhuoqi Wu, Sen Wang, Zhiyuan Yuan, Xiaobo Chen, Ershun Pan, Yixi Zhao, Yanting Li, Lu Chen, Guolong Wang, Tangbin Xia, Xiaojun Zhou, Meng Wang, Delin Huang, Yiping Shao, Guilong Li, Xufeng Yao, Changping Liu, Lan Fei, Rui Xu, Rui Zhang, Tao Liu, Ke Song, Ziren Zhao, Heng Wang, Tingting Song, Kun Wang, Yaxiang Yin, Yafei Deng, and Chen Zhao. The above people all come from Shanghai Jiao Tong University. We are grateful for Jianjun Shi from Georgia Institute of Technology, Jun Ni from University of Michigan, Hui Wang from Florida State University, and Zhenhua Huang and Guodong Jiao from Coherix company. Special acknowledgement goes to two individuals Yiping Shao and Delin Huang who directly helped in editing and reviewing the entire manuscript draft of the book.

We are also grateful for Shanghai Automotive Industry Corporation General Motors Wuling company (SGMW) in Liuzhou, China for offering an extraordinary environment for the engineering practice of surface quality control. The company's experienced engineers, state-of-the-art laboratories, and its high-precision machine tools offer students and faculty unique opportunities for research and teaching. This text has benefited greatly from the company's resources and the energy and enthusiasm of its managers, engineers, and workers. We thank them for their support.

Contents

Chapter 1
Introduction

1.1 History and Current Status of Surface Topography

With the rapid development of the automotive and aircraft industries, the effect of surface quality on the functional behavior of a product has always been a focus of research. As an important feature of surface quality, surface topography, also known as surface texture, plays an essential role in the rapidly developing manufacturing industry. At the beginning of the twentieth century, due to the limitations of measurement technique, surface topography could only be assessed by visual inspections and touch detections. Subsequently, with the continuous emergence of novel measurement instruments and measuring methods, more and more surface parameters and characterization methods emerged. In the early 1970s, the Motif parameters were proposed in the French automobile industry, and the fractal method was developed by Sayles and Thomas in the late 1970s. By the 1980s, a number of parameters that could only be adapted to specific functional requirements were more and more proposed, which led to "parametric explosions". In the 1990s, the international standard of surface topography was on the historical stage, and more and more manufacturing and measurement information were added to the specifications. When the twenty-first century comes, a series of standards for surface topography under the framework of the new generation of GPS (Geometrical Product Specification) have been issued gradually. The release of ISO 1302:2002 indicates that the specification of surface topography has entered an era of integral expression. Different from the ISO standard, ASME B46.1-2009 is widely used in the United States instead, which contains all the information of the surface topography specification.

As the two-dimensional (2D) characterization matures, scholars and engineers gradually realize its limitations. The 2D profile cannot reflect the shape, direction, and various elements (such as peaks, valleys, and saddle regions) of the entire surface. Therefore, the study of three-dimensional (3D) characterization methods comes into being. In the early 1990s, the project team of Birmingham University

© Springer Nature Singapore Pte Ltd. 2019
S. Du and L. Xi, *High Definition Metrology Based Surface Quality Control and Applications*, https://doi.org/10.1007/978-981-15-0279-8_1

proposed the famous "Birmingham 14" parameters. In 1998, Brunel and Huddersfield University proposed the definition of element parameters in the AUTOSURF and SURFSTAND projects. Subsequently, ISO/TC 213 founded the task force WG16 in June 2002 to conduct a series of studies on 3D surface topography. In 2012, WG 16 issued the new international standard ISO 25178-2 for the definition of 3D surface topography parameters. So far, ISO 25178 has been basically completed, and it also shows that the study of 3D surface topography gradually matures.

Surface topography can be considered as the linear superposition of flaws, form error, waviness, and roughness. It is a crucial link between a component, the manufacturing process that generates it, and the functionality that is expected of it. The presence of surface topographies with different properties can radically alter the part function. In engineering applications, surface topography affects many functions of the part, such as wear, friction, lubrication, fatigue, sealing, fitting, vibration, and optical properties. The primary reason for developing this study is to clarify the relationship between surface topography, part functionality, and manufacturing process parameters. Therefore, achieving the reasonable and accurate characterization, evaluation, filtering, classification, monitoring, and prediction of surface topography is not only an important way to improve manufacture precision but also an important guarantee for the performance of parts.

1.2 Scope and Objectives

This book describes the subjects that almost contain all aspects of surface topography. The full text begins with surface measurements, gradually extends to surface characterization and feature extraction (filtration), then explores mapping relationships between surface topography and functional behaviors of parts (classification, monitoring, and prediction), and finally lists some practical engineering applications (compensation manufacturing). The framework of the book is shown in Fig. 1.1, and the procedure involves the following aspects:

- **High-Definition Metrology**. With the development of high-definition metrology (HDM) technologies, high-density point cloud data is measured for reflection of the entire surface contour, which provides great opportunities for surface characterization and evaluation, surface filtering, surface classification, surface monitoring, surface prediction, etc.
- **Surface Characterization and Evaluation**. Characterization and evaluation of surface features through the measured surface data is the first step to study surface topography. From the view of the dimensions, surface topography characterization and evaluation can be applied in two aspects: profile (2D surface) and areal surface topography (3D surface). We focus on 3D surface in this book.

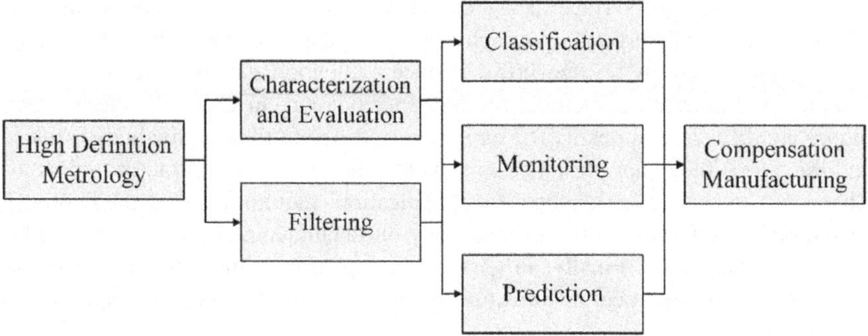

Fig. 1.1   The framework of the book

- **Surface Filtering**. Filtration is a process to partition a surface into different components and extract the surface features of interest. It has always been important in surface metrology, and it has a significant aspect of surface topography analysis.
- **Surface Classification**. The surface appearance is sensitive to changes in the manufacturing process and is one of the most important product quality characteristics. The classification of workpiece surface patterns is critical for quality control, because it can provide feedback on the manufacturing process.
- **Surface Monitoring**. The surface quality is one of the most important factors that can directly influence the performance of the final product. Therefore, surface monitoring is necessary since it can help improve surface quality by special cause identification, removal, and variation reduction.
- **Surface Prediction**. Satisfied surface topography is important to achieve the function of a part, thereby prediction of surface topography in advance is essential since it can help optimize process parameters and promote quality control of the manufacturing processes.
- **Online Compensation Manufacturing**. The manufacturing error is a very important factor that influences the quality of workpieces, which will obviously reduce the manufacturing accuracy of the workpieces. Online compensation is a method of real-time monitoring of the machining condition of a machine tool in the machining process, and compensating in real time when deviations occur.

1.3   Organization

We have organized the whole book into eight chapters. The remaining parts are high-definition metrology, surface evaluation, surface filtering, surface classification, surface monitoring, surface prediction, and online compensation manufacturing, respectively. In Chap. 2, a new noncontact laser measurement instrument is introduced to achieve surface measurement. Chapter 3 describes current standards of profile and areal surface topography, which is used to realize the surface characterization. Furthermore, two novel surface evaluation methods are proposed to complete the current geometrical product specification. In Chap. 4, filtering techniques are discussed in detail. The evolution of different filtering methods is traced, and three new filters are presented to separate and extract the surface components. Chapters 5, 6, and 7 are devoted to classification, monitoring, and prediction. In each case, the effects of surface topography on manufacturing process and product quality are illustrated. Finally, in Chap. 8, compensation manufacturing methods are applied to the practical manufacturing process, which benefits the improvement of surface quality.

Chapter 2
High-Definition Metrology

2.1 A Brief History of Measurement Technology

2.1.1 Contact Measurement

The contact measurement method mainly refers to the measurement method based on mechanical probe [1–4]. The probe is in contact with the measured surface and moves linearly along the measured surface. The upper and lower displacements of the probe are transmitted to the measuring sensor through the lever mechanism, and the output value of the sensor is proportional to the measured variable. The relationship between the vertical displacement of the probe (the measured variable, z_s) and linear displacement (x_s) is represented by the function $z_s = f(x_s)$. Thus, values of the form error, waviness, and roughness of the measured surface can be obtained. The probe is usually made of diamond, and the shape of the probe is conical. The radius of the probe tip generally ranges from 0.1 to 10 μm, and the cone angle is generally 60° or 90°. This probe-based method can be applied for vertical resolution in both microscale and nanoscale surface measurements. If a material with a low coefficient of thermal expansion and a very sharp probe are used, the vertical resolution can be close to 0.05 nm and the lateral resolution is more than 0.1 μm.

Coordinate measuring machine (CMM) is a typical representative of contact measurements. As an efficient measuring instrument, CMM can measure geometric parameters such as size, shape, and position. CMM extracts several coordinate values of the measured feature through the measurement of spatial three-dimensional coordinate points. The measured coordinate values are mathematically calculated and fitted to obtain the estimated value of the feature to be measured, and digital measurement of the geometric parameters is achieved. CMM can automatically make measurements according to the program codes or computer-aided design (CAD) drawings. CMM can also automatically select

© Springer Nature Singapore Pte Ltd. 2019
S. Du and L. Xi, *High Definition Metrology Based Surface Quality Control
and Applications*, https://doi.org/10.1007/978-981-15-0279-8_2

measurement methods and measurement paths according to common sense and rules of measurement. Many researches have been conducted on CMM [5–7].

The contact measurement methods have the advantages of stable and reliable measurement results, good repeatability, large range, and high precision. Besides, contact measurement methods require low measurement environment and simple instrument operation. These methods can be applied to measuring not only metal surfaces but also nonmetal surfaces. However, there are several disadvantages of contact measurement methods. (1) The probe diameter has a certain size, which will affect the surface measurement accuracy of workpieces with deep grooves, small holes, etc. (2) The pressure between probe and the measured surface may cause the probe to scratch the measured surface and a measurement error will be generated. (3) The 3D measurement of the surface takes a lot of time, which is the most important drawback of the contact measurement methods.

2.1.2 *Noncontact Measurement*

In noncontact measurement, the detection component of the measuring device does not directly contact with the measured surface, so the measured surface is neither scratched nor damaged by the measuring device, and the measurement error introduced due to the contact is avoided. The vast majority of noncontact measurement methods use optical measurement technology, which combines traditional optical metrology and information processing techniques. Optical measurement methods have the following main advantages. (1) All data in the image plane can be obtained at the same time. (2) No damage is caused to the measured surface since there is no contact with the measured surface during the measurement. (3) Measurement speed is relatively fast. (4) Surfaces of almost all materials can be measured.

Optical measurement is an important component of noncontact measurement [8]. Noncontact optical measurement is divided into active measurement and passive measurement [9]. Active measurement systems have specialized instruments to generate measurement light sources or sound sources, while passive measurements do not. The most commonly used active measurements are laser triangulation [10–12] and grating methods [13–15]. Since the measurement process is performed by optical methods, there are certain requirements on the surface of the measured object. Reflective or dark objects are not suitable for optical measurement, or more complicated optical techniques are required when such objects are encountered or certain treatments are required to ensure the measurement. In the passive measurement methods, the 3D information of the object is obtained by directly using the image produced by natural light, instead of emitting a controllable light beam to a measured object. The accuracy of the depth data obtained by this method is low, and only fuzzy concepts of the relative distance of the scene are obtained.

Active measurements are based on the technique of transmitting energy (such as laser, ultrasonic, X-ray, etc.) to the scene and using the structural information provided by a special light source to obtain depth information. The active measurement methods have high ranging accuracy, strong anti-interference performance, strong real-time performance, and a wide range of applications, and they are relatively easy to directly obtain the 3D contour information of the measured object when compared to passive measurement methods.

2.2   High-Definition Metrology

2.2.1   Definitions and Measurement Principles

The laser triangulation measurement used the HeNe laser as the light source in the early days and was bulky. In recent years, with the development of semiconductor lasers, photoelectric position detector (PSD), and charge-coupled device (CCD), the laser triangulation measurement has been widely used in the measurement of displacement and surface topography. A single-point laser triangulation method is used as an example to describe the principle of laser triangulation measurement. Two structures of direct-type and oblique-type are usually used by the single-point laser triangulation measurement. The principle of direct-type laser triangulation measurement is shown in Fig. 2.1a. The light emitted by the laser is focused on the surface of the measured object after being focused by the converging lens. If the object moves or the surface changes, the incident light spot will move along the incident light axis. The receiving lens receives scattered light from the incident light spot and images it on the sensitive surface of the photoelectric position detector (such as PSD, computed tomography (CT)). If the displacement of the light spot on the imaging surface is x', the displacement of the measured surface can be obtained by Eq. (2.1).

$$x = \frac{ax'}{b \sin \theta - x' \cos \theta} \tag{2.1}$$

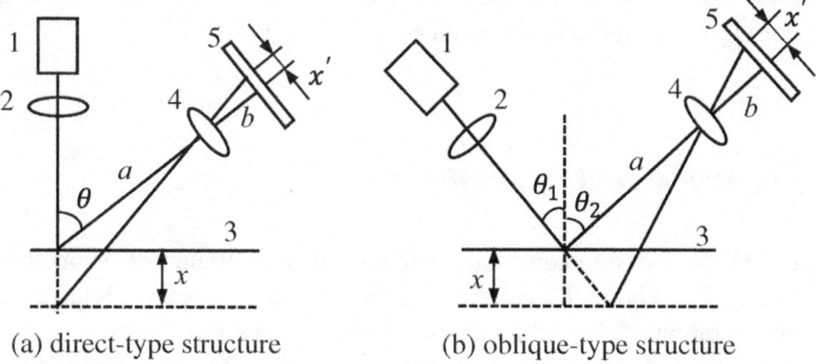

(a) direct-type structure   (b) oblique-type structure

Fig. 2.1 The principle of laser triangulation measurement

where a is the distance from the intersection point of the laser beam axis and optical axis of the receiving lens to the front surface of the receiving lens, b is the distance from the back surface of the receiving lens to the center of the imaging surface, and θ is the angle between the optical axis of the laser beam and the optical axis of the receiving lens.

The principle of oblique-type laser triangulation measurement is shown in Fig. 2.1b. The light emitted by the laser is incident on the measured surface at an angle from the normal of the measured surface. A receiving lens is also used to receive scattered light or reflected light from the light spot on the measured surface. If the spot image moves x' on the detector sensitive surface, the distance of the measured surface along the normal direction is calculated by

$$x = \frac{ax' \cos \theta_1}{b \sin(\theta_1 + \theta_2) - x'(\theta_1 + \theta_2)} \tag{2.2}$$

where θ_1 is the angle between the laser beam axis and the normal of the measured surface, and θ_2 is the angle between the optical axis of the imaging lens and the normal of the measured surface.

The characteristics of direct-type and oblique-type laser triangulation measurement are shown as follows. More details about laser triangulation measurement can be referred in Chap. 10 in [16].

(1) The oblique-type laser triangulation measurement can receive the reflected light from the measured surface. When the measured surface is a mirror surface, the output signal of the photodetector will not be too small due to the weak scattering light, which makes the measurement impossible. Due to the characteristics of receiving scattered light, the direct-type laser triangulation measurement is suitable for measuring surfaces with good scattering properties.
(2) When the displacement x' (shown in Fig. 2.1) occurs on the measured surface, the oblique incident light spot shines on different points of the surface. It is different to detect the displacement of a certain point of the measured surface, while the direct-type measurement can detect the displacement of the point.
(3) The sensor of oblique-type laser triangulation measurement has a higher resolution than the direct-type laser triangulation measurement, but it has a small measuring range and a large volume.

2.2.2   Examples and Applications

An online HDM measurement equipment using laser triangulation metrology is shown in Fig. 2.2. Figure 2.3 exhibits the measurement process. Figure 2.4 shows the actual operation of the online measurement equipment.

The field view of the HDM system is within 75×56 mm^2, and the depth of field view is 15 mm. Accuracy in X (translation direction) is ± 1 μm, accuracy in

(a) Measurement component   (b) Industrial personal computer

Fig. 2.2 Online measurement equipment based on HDM system

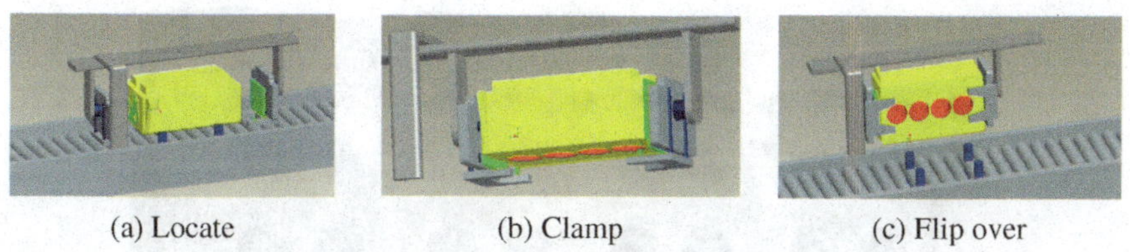

(a) Locate   (b) Clamp   (c) Flip over

Fig. 2.3 The measurement process

Y (direction of line laser) is ± 10 μm, and accuracy in Z is ± 20 μm. Resolution of the system is 0.02 mm^3. Repeatability is 0.02 mL. The HDM system is developed to measure the volumes of chambers. The outputs of the HDM system consist of the volume values and point clouds of the chamber. The design volume of the chambers is 24.4 ± 0.4 mL, which indicates that the measurement error should be within 0.1 mL. Since the repeatability of the HDM system is 0.02 mL, the measurement error of the HDM system is allowable in the production process.

The online measurement equipment based on HDM system has been applied to the measurement of cylinder heads (shown in Fig. 2.5). The moving speed of the guide rail is set at 10 mm/s, and the acquisition frame rate of the 3D measurement sensors is 110 f/s. The scan time of the cylinder head depends on the length of the cylinder head and speed of guide rail. The scan time of the cylinder head is 32 s. Besides, the measurement time also includes time of grab (20 s), drop (15 s), and leave (10 s). Therefore, the total measurement time is 77 s, which is less than the cycle time 87.5 s of manufacturing a cylinder head, and the online measurement is

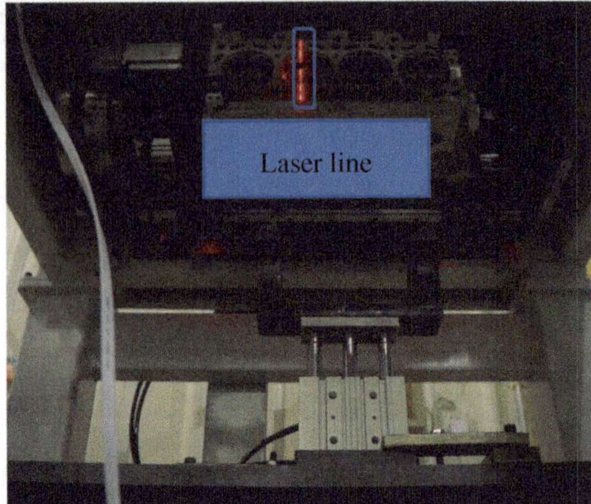

(a) On-line measurement   (b) An example of measurement

Fig. 2.4 Actual operation of the online measurement equipment

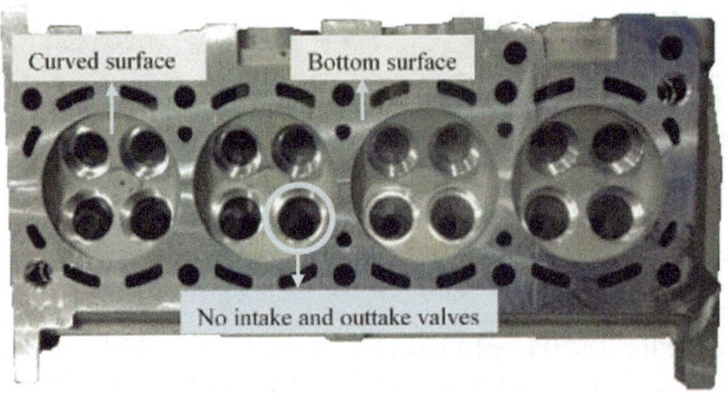

Fig. 2.5 An engine cylinder head with four combustion chambers

easily implemented. The measurement time of the system is adjustable, and faster measurement can be achieved by increasing the speed of guide rail and acquisition speed of the system. 640×1280 is the number of points that are collected from a cylinder head by the HDM system. It includes the bottom surface and all chambers of a cylinder head. There are 640 points on a laser line, and the total measurement contains 1280 laser lines. During the measurement, coordinates of the missing points are represented by specific coordinates and these specific coordinates are

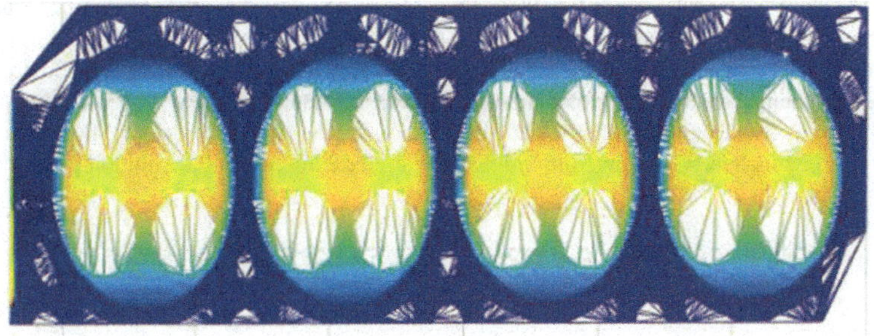

Fig. 2.6 Point cloud of an engine cylinder head

same. The points with same coordinates are regarded as missing points since coordinates of effective points are different. The point cloud data of an engine cylinder head is shown in Fig. 2.6.

References

1. Garratt JD (1982) A new stylus instrument with a wide dynamic range for use in surface metrology. Precis Eng 4(3):145–151
2. Badami VG, Smith ST, Raja J, Hocken RJ (1996) A portable three-dimensional stylus profile measuring instrument. Precis Eng 18(2–3):147–156
3. Chetwynd DG, Liu X, Smith ST (1996) A controlled-force stylus displacement probe. Precis Eng 19(2–3):105–111
4. Clark SR, Greivenkamp JE (2002) Ball tip–stylus tilt correction for a stylus profilometer. Precis Eng 26(4):405–411
5. Vermeulen M, Rosielle P, Schellekens PHJ (1998) Design of a high-precision 3D-coordinate measuring machine. CIRP Ann Manuf Technol 47(1):447–450
6. Umetsu K, Furutnan R, Osawa S, Takatsuji T, Kurosawa T (2005) Geometric calibration of a coordinate measuring machine using a laser tracking system. Meas Sci Technol 16(12):2466–2472
7. Park J, Kwon K, Cho N (2006) Development of a coordinate measuring machine (CMM) touch probe using a multi-axis force sensor. Meas Sci Technol 17(9):2380–2386
8. Chen F, Brown GM, Song M (2000) Overview of 3-D shape measurement using optical methods. Opt Eng 39(1):10–23
9. Bidanda B, Hosni YA (1994) Reverse engineering and its relevance to industrial engineering: a critical review. Comput Ind Eng 26(2):343–348
10. Dorsch RG, Häusler G, Herrmann JM (1994) Laser triangulation: fundamental uncertainty in distance measurement. Appl Opt 33(7):1306–1314
11. Wang G, Zheng B, Li X, Houkes Z, Regtien PP (2002) Modelling and calibration of the laser beam-scanning triangulation measurement system. Robot Auton Syst 40(4):267–277
12. Molleda J, Usamentiaga R, García DF, Bulnes FG (2010) Real-time flatness inspection of rolled products based on optical laser triangulation and three-dimensional surface reconstruction. J Electron Imaging 19(3):1–14
13. Eichler HJ, Klein U, Langhans D (1980) Coherence time measurement of picosecond pulses by a light-induced grating method. Appl Phys 21(3):215–219

14. Andresen K, Kamp B, Ritter R (1992) Three-dimensional surface deformation measurement by a grating method applied to crack tips. Opt Eng 31(7):1499–1505
15. Liu W, Wang Z, Mu G, Fang Z (2000) Color-coded projection grating method for shape measurement with a single exposure. Appl Opt 39(20):3504–3508
16. MVTec Software GmbH (2010) Halcon solution guide III-C 3D vision. MVTec Software GmbH, Munchen, Germany. http://download.mvtec.com/halcon-9.0-solution-guide-iii-c-3d-vision.pdf

Chapter 3
Surface Characterization and Evaluation

3.1 A Brief History of Surface Evaluation

Most manufactured parts depend on their surface quality, and surface topography has become the dominant functional features of a part. In order to clarify the relationships between these features and the functional behavior of the part, the first step is to characterize and evaluate these surface features through the measured surface data. Research into surface topography of characterization and evaluation has been carried out for several decades and is still very active. From the view of the dimensions, surface topography characterization and evaluation can be applied in two aspects: profile (2D surface) and areal surface topography (3D surface).

In scientific research, profile surface topography has been proposed for nearly half a century. With the maturation of profile characterization system, it has been standardized for some time. The terms, definitions, and parameters are defined in ISO 4287. During the last decade, due to the rapid development of measurement techniques, the evaluation of surface geometric qualities gradually changes from profile to areal characterization. Although profile surface topography can show the manufacturing process change effectively, much more functional information about the surface can be obtained by analyzing the areal surface topography. Comparing with the profile surface, the areal surface is a closer representation of the "real surface". It is easier to determine the exact nature of a topographic feature and used for problem diagnostics and function prediction. In summary, profile characterization is still the most utilized method for surface characterization in manufacturing industry, especially for process and quality control purposes. But, areal methods of analysis are becoming more commonplace as the development of manufacturing industry. Up to now, areal characterization is still under development but the standards are progressively being published.

Two methods are described in detail in Sects. 3.2 and 3.3.

© Springer Nature Singapore Pte Ltd. 2019
S. Du and L. Xi, *High Definition Metrology Based Surface Quality Control
and Applications*, https://doi.org/10.1007/978-981-15-0279-8_3

3.2 3D Surface Form Error Evaluation

3.2.1 Introduction

During the last decade, the evaluation of surface geometric qualities (surface texture and surface form) gradually changes from 2D profile to 3D areal characterization [1]. For example, the evaluation of surface texture has extended from roughness and waviness to 3D surface texture, from 2D-Motif to 3D-Motif, and from 2D material ratio curve to 3D material ratio curve. However, compared with the remarkable evaluation methods on 3D surface texture, there is limited research on 3D surface form error. In general, surface form error is evaluated using surface flatness measured by coordinate measuring machines (CMMs). As CMM measures only a few scattered points or profiles due to economic constraints [2], it cannot sample high-density points describing 3D surface form error in industrial application. Recently, noncontact high-definition metrology (HDM) has been adopted for its 3D inspection of the entire surface as HDM can generate a surface height map of millions of data points within seconds [3]. HDM provides possibility to evaluate 3D surface form error in many aspects besides surface flatness. Therefore, the purpose of this research is to develop a proper method to evaluate the 3D surface form error using HDM data.

To begin with, the scope of 3D surface form error is first discussed. In the authors' point of view, 3D surface form error describes relatively long-wavelength deviations from a 3D areal characterization of an entire machined surface. According to the authors' definition, there are differences and similarities between 3D surface form error and 2D surface texture, 3D surface texture, as well as flatness measured by CMM. Four types of surface geometric qualities are classified by a 2D coordinate system (Fig. 3.1). The horizontal coordinate is the lateral resolution of the corresponding measurement techniques, and the vertical coordinate is the metrology dimension. Therefore, 3D surface form error differs from flatness measured by CMM in metrology dimension and 3D surface texture in lateral resolution. As for similarities, 3D surface form error has the same level of lateral resolution with surface flatness and has the same metrology dimension with 3D surface texture.

As both 3D surface texture and 3D surface form error are obtained based on 3D areal measurement, the evaluation methods for 3D surface texture can be adopted to evaluate 3D surface form error to a certain degree. 3D surface texture is typically measured by areal-topography methods (ATM) such as phase-shifting interferometry, coherence scanning interferometry, and atomic force microscopy [4]. A series of successful studies such as "Birmingham 14 parameters" [5–8] and international standards on 3D surface texture [9] were achieved. Among these parameters, certain 3D surface texture parameters such as height parameters in ISO 25178-Part 2 [9] can be extended to evaluate HDM data. However, other 3D surface texture parameters are not fit to evaluate HDM data. There are two reasons. The first reason is the measurement continuity. ATM usually measures a simple square area that has

Fig. 3.1 Classification of surface geometrical qualities. Reprinted from Ref. [27], copyright 2014, with permission from Elsevier

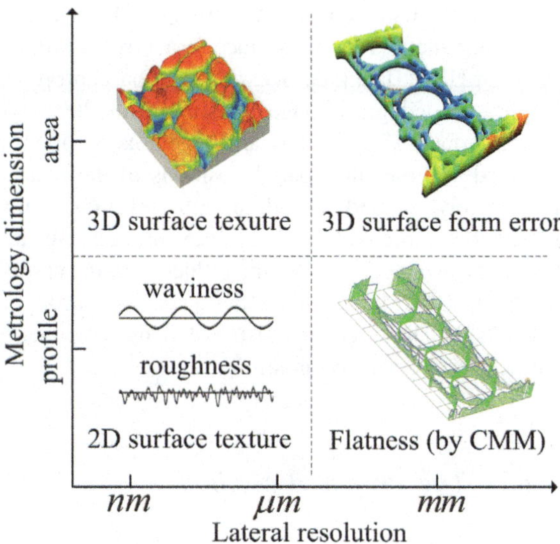

no holes or empty zones. Therefore, ATM data is continuously sampled, so the 3D surface texture parameters can be calculated directly. On the contrary, HDM measures the entire surface probably with many holes and empty zones. For instance, engine block faces measured by HDM in Fig. 3.1 have cylinder holes, bolt holes, and cooling holes. Therefore, HDM data is not continuously measured, so the 3D surface form error cannot be analyzed directly by using 3D surface texture methods such as 3D frequency filtering. The second reason is the lateral resolution or deviation wavelength between HDM and ATM. ATM measures typically a selected area with an edge of several millimeters with lateral resolution at the scale of micrometer or even nanometer. However, HDM measures the entire surface with an edge of several hundred millimeters at sub-millimeter resolution. Therefore, 3D surface texture parameters, which focus on the short-wavelength, high-frequency deviation of local regions, are not designed for evaluating the long-wavelength, low-frequency deviation of the entire surface. The above two reasons also explain why HDM measurement is assigned to the scope of 3D surface form error instead of 3D surface texture.

Recently published research has adopted some characteristic extraction techniques using HDM data for various applications [10–13]. However, the extracted characteristics are inadequate for evaluating 3D surface form error. Because characteristics for surface form error evaluation must have engineering meaning whereas those for surface classification and surface partition do not have to. In a short summary, 3D surface form error information has not been mined and analyzed sufficiently using HDM data.

The evaluation of 3D surface form error should focus on two problems. The first problem is to process raw HDM data into a suitable data type for evaluating the entire surface that has holes or empty zones. The second problem is to extract meaningful characteristics describing 3D surface form error. In this book, an evaluation method for 3D surface form error is proposed to tackle the two problems. First, an HDM data preprocessing method is proposed to convert HMD data into a height-encoded gray image containing all the height and spatial information of entire surface. Gray values and positions of pixels in the image are assigned to height deviations and spatial positions of the measurement points of the surface, respectively. Second, a modified gray-level co-occurrence matrix (GLCM) method is developed to evaluate 3D surface form error of machined surface with complex geometry, which eliminates the influence of the empty zone of the surface. From the modified GLCM, characteristics such as entropy, contrast, and correlation are calculated. These characteristics having engineering meaning complement 3D surface form error evaluation.

3.2.2   The Proposed Method

The proposed method for 3D surface form error evaluation includes two parts: an HDM data preprocessing method and a modified GLCM method, which are illustrated in Sects. 3.2.2.1 and 3.2.2.2, respectively.

3.2.2.1   HDM Data Preprocessing Method

In this subsection, an HDM data preprocessing method is proposed to convert HDM data into a height-encoded and position-maintained gray image. The height information (Z coordinates of HDM data) is converted to pixel gray intensities, and the spatial information (X-and Y-coordinates) is converted to pixel index. The converted gray image reduces the data size to about one-third of the size of raw HDM data and serves as a suitable data type for further analysis. Figure 3.2 shows the four steps of the HDM data preprocessing method, including HDM data alignment, grid generation, grayscale converting, and empty zone removing.

The first step is to obtain the orthogonal height deviations from raw HDM data $[x_i\, y_i\, z_i]$, where $i = 1, 2, \ldots, N$ is the number of the measurement points. The raw HDM data is relative to the reference plane of HDM. Unless the measured surface is perfectly aligned with the reference plane, raw data will be titled with respect to the measured surface. A least square plane P is fitted to acquire the nominal measured surface:

$$P : ax + by + c - z = 0 \tag{3.1}$$

The original coordinates $[x_i\, y_i\, z_i]$ can be transformed into new coordinates $[X_i\, Y_i\, Z_i]$ on the fitted plane P. The orthogonal height deviation Z_i is calculated by

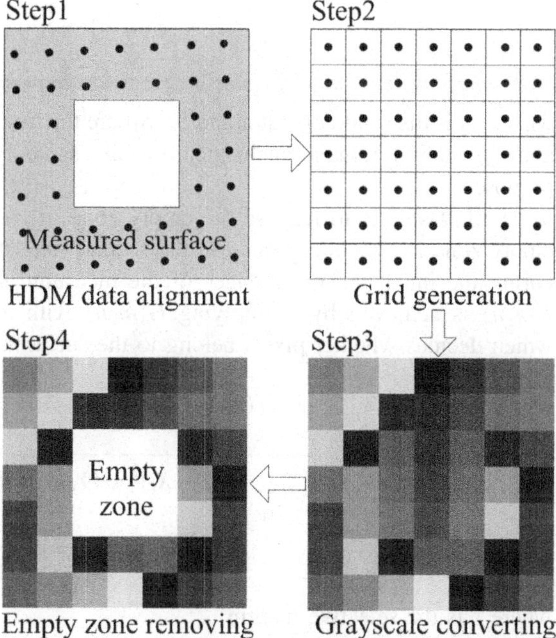

Fig. 3.2 HDM data preprocessing method. Reprinted from Ref. [27], copyright 2014, with permission from Elsevier

$$Z_i = \frac{z_i - ax_i - by_i - c}{\sqrt{1 + a^2 + b^2}} \tag{3.2}$$

The second step is to generate a regular grid of height deviation Z_i. A continuous surface is interpolated using coordinates $[X_i\,Y_i\,Z_i]$ by Delaunay triangulation [14]. From the interpolated surface, a 2D grid matrix $Z(m, n)$ is resampled. The index m and n are calculated as follows, given the resampling interval l.

$$\begin{aligned} m &= \frac{X_i - X_{\min}}{l} \\ n &= \frac{Y_i - Y_{\min}}{l} \end{aligned} \tag{3.3}$$

The grid matrix $Z(m, n)$ stores both height information and spatial information of the entire surface: the value of $Z(m, n)$ contains vertical deviation Z_i and the index m and n represent the spatial information X_i and Y_i.

The third step is to convert the grid matrix $Z(m, n)$ into a gray image $G(m, n)$. Each element of the grid matrix is assigned a pixel of the image. The spatial indexes m and n in the grid matrix are correlated to the positions of pixels in the image. The vertical deviations in the grid matrix $Z(m, n)$ are scaled to pixel gray intensity $G(m, n)$ with gray interval from 0 (black) to 255 (white). Gray intensity $G(m, n)$ is calculated as

$$G(m,n) = \left[\frac{Z(m,n) - S_L}{S_U - S_L} \times 255 \right] \tag{3.4}$$

where $[*]$ is the round operator and S_L, S_U are the lower and upper specifications. S_L and S_U should remain unchanged for the same type of parts for comparison purpose.

The last step is to remove the empty zone of the gray image $G(m,n)$. As the empty zone is also interpolated in the second step, it should be removed by comparing the inner and outer edges of the measured surface. The final gray image $I(m,n)$ is achieved by multiplying $G(m,n)$ with a selective parameter $\delta(m,n)$, which decides whether pixels belong to the surface:

$$I(m,n) = G(m,n) \cdot \delta(m,n) \tag{3.5}$$

$$\delta(m,n) = \begin{cases} 0, & if \sqrt{(X_{\min} + ml - X_i)^2 + (Y_{\min} + nl - Y_i)^2} > \varepsilon, \forall i = 1,\ldots,N \\ 1, & otherwise \end{cases}$$

$$\tag{3.6}$$

where ε is the selective margin.

Through the above four steps, HDM data is converted to a gray image by establishing correspondence relationship between coordinates $[x_i \, y_i \, z_i]$ and pixels in gray image $I(m,n)$. Then the gray image $I(m,n)$ that contains all the height and spatial information of HDM data can be analyzed by image analysis techniques directly for 3D surface form error evaluation purpose.

3.2.2.2 Modified GLCM Method

In this subsection, a modified GLCM is adopted to evaluate 3D surface form error using the converted gray image. GLCM describes how frequent one gray level appears in a specified relationship to another gray level of the image [15], which is a successful image analysis technique to study textural features. Given the gray image $I(x,y)$, the normalized co-occurrence frequencies $p(i,j)$ between gray level i and j are defined as

$$p(i,j) = P(i,j) / \sum_{i=1}^{N_g} \sum_{j=1}^{N_g} P(i,j) \tag{3.7}$$

$$P(i,j) = \#\{I(x_1,y_1) = i, I(x_2,y_2) = j, x_2 = x_1 + d\cos\theta, y_2 = y_1 + d\sin\theta\}, i \neq 0, j \neq 0$$

$$\tag{3.8}$$

where N_g is the number of gray levels, i and j are corresponding gray levels of pixels $I(x_1, y_1)$ and $I(x_2, y_2)$ with a distance d and an angle θ, and $\#$ denotes the number of pixel pairs satisfying the conditions. As computing GLCM for all distances d and angles θ would require a lot of calculations, GLCM calculated for angles quantized to 45° and distance of one or two pixels is suggested [15]. The dimension of $p(i, j)$ is equal to the gray levels N_g in the image. For example, if the gray levels are between 0 and 255, the size of the $p(i, j)$ will be 256 by 256. Note that the calculated GLCMs vary for different numbers of gray level N_g. Along with the decrease of gray levels from 256, the spatial and height information will be lost. Therefore, gray levels equal to 256 are suggested to make most of the information from the converted gray image.

Unlike the traditional definition of GLCM, a further condition $i \neq 0, j \neq 0$ is added when calculating GLCM of machined surfaces. This condition ensures that pixel pairs with either one pixel equal to zero are not counted in calculating GLCM. Because the pixel values of the measured surface are usually at the middle range of the gray interval $[0, 255]$, only the empty zone can have zero values. Therefore, pixel pairs with one zero pixel represent the edge of the surface and those with two zero pixels represent the empty zone, which should be eliminated when evaluating 3D surface form information.

Figure 3.3 shows an example of how to calculate the modified GLCM of a 4-by-6 gray image. The image with zero pixels in the middle is a simple illustration of a machined surface with a hole (Fig. 3.3a). The hole looks like a smooth circle in the image provided enough pixels, especially millions of HDM data points. As the gray level of the image is 4r, the dimension of corresponding GLCM is 4-by-4. Figure 3.3b shows the general form of the modified GLCM, where $\#$ denotes the number of pixel pairs satisfying the conditions. Considering the additional condition $i \neq 0, j \neq 0$, the pixel pairs $(0, 0)$, $(0, j)$, and $(i, 0)$ are not counted in the GLCM to eliminate the information of the hole. Four GLCMs for a distance equal to one and an angle equal to $0°, 45°, 90°, 135°$ are shown in Fig. 3.3c. More specifically, the value of element $(1, 2)$ in the GLCM of $d = 1, \theta = 0°$ is 2 because horizontally adjacent pixels with the values 1 and 2 appear twice. Continuing this procedure will fill in the values in GLCM.

From the modified GLCM, three uncorrelated features, entropy, contrast, and correlation, are calculated as follows:

$$entropy = -\sum_i \sum_j p(i,j) \log(p(i,j)) \tag{3.9}$$

$$contrast = \sum_{n=0}^{N_g-1} (i-j)^2 \left\{ \sum_{i=1}^{N_g} \sum_{j=1}^{N_g} p(i,j) \right\} \tag{3.10}$$

$$correlation = \frac{\sum_i \sum_j (ij)p(i,j) - \mu_x\mu_y}{\sigma_x\sigma_y} \tag{3.11}$$

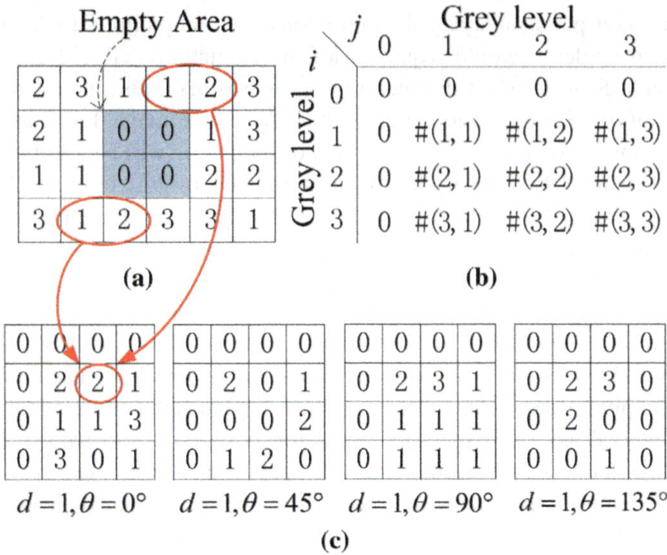

Fig. 3.3 Calculating modified GLCM: **a** a gray image of machined surface, **b** the general form of GLCM with condition $i \neq 0, j \neq 0$, **c** four GLCMs of $d = 1$ and $\theta = 0°, 45°, 90°, 135°$, Reprinted from Ref. [27], copyright 2014, with permission from Elsevier

where p_x and p_y are marginal probability of $p(i,j)$, and μ_x, μ_y, σ_x, and σ_y are the mean and standard deviations of p_x and p_y.

Entropy evaluates the randomness of the surface height distribution of the measured points. If the height of the measured points is regularly distributed within a narrow range, entropy will be small. If the height of the measured points is randomly distributed, entropy will be large. Specifically, entropy will be zero for a surface of which the measured points have the same height, and entropy will reach its maximum for a surface with complete random height distribution.

Contrast measures the amount of local surface height difference between a pixel and its neighbor over the whole image. Since the height difference $i - j$ is magnified to $(i - j)^2$, contrast is sensitive to the local height difference. Therefore, contrast can be used to distinguish whether surface flatness is caused by global variation or local variation. For example, suppose two surfaces A (Fig. 3.4a) and B (Fig. 3.4b) with same flatness but caused by local variations and global trend, respectively. As the local height difference of surface A is greater than that of surface B, contrast of surface A is greater than contrast of surface B. In a similar perspective, low flatness is a sufficient, but not necessary condition to keep low contrast [16].

Correlation reflects the degree of how correlated a pixel to another pixel in certain direction and distance. It indicates directionality of surface height distribution in the row or column. High correlation implies significant directionality of the surface in that direction and vice versa. Correlation could be used to distinguish surfaces machined by different processing techniques.

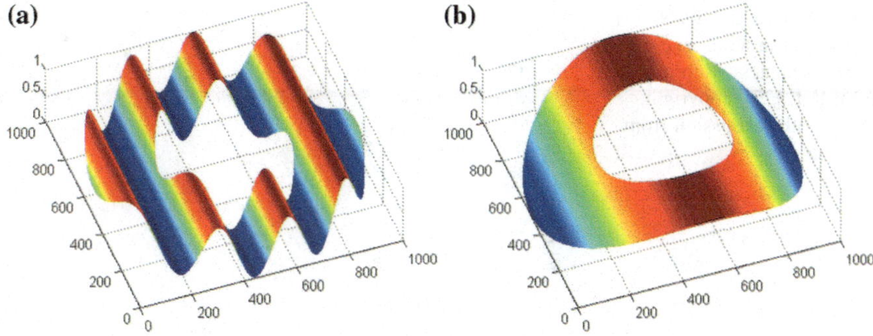

Fig. 3.4 Simulated plates with the same flatness: **a** contrast = 5.11, entropy = 6.35, **b** contrast = 0.54, entropy = 6.30. Reprinted from Ref. [27], copyright 2014, with permission from Elsevier

These characteristics, entropy, contrast, and correlation, can be used with flatness together to evaluate 3D surface form error measured by HDM. Through 3D surface form error evaluation, engineers can better understand quality of machined surface from various angles and making adjustment of the machining process. Moreover, these characteristics also provide possibility to evaluate certain 3D form error-related properties, for example, the mechanical sealing property for large sealing surfaces.

3.2.3   Case Study

In this subsection, a case study on engine block faces is presented to illustrate the evaluation of 3D surface form error using HDM data. The extracted 3D form error characteristics are compared with the flatness, and the usefulness of the proposed method is discussed.

3.2.3.1   Experimental Conditions

The experimental parts are engine blocks, the face of which is a major sealing surface in automotive power train. The material of the engine block is Cast Iron FC250. The milling process is carried out in an EX-CELL-O machining center using a CBN milling cutter with a diameter of 200 mm. Quaker 370 KLG cutting fluid is used. The cutting speed is 816.4 m/min, the depth of cut is 0.5 mm, and feed rate is 3360 mm/min. Ten engine blocks are sampled from different production lines and measured by a 3D laser holographic system (ShaPix® Surface Detective™, Ann Arbor, MI) with lateral resolution 0.15 mm. About 0.8 million points on an area of 320 mm × 160 mm are acquired for each engine block. The

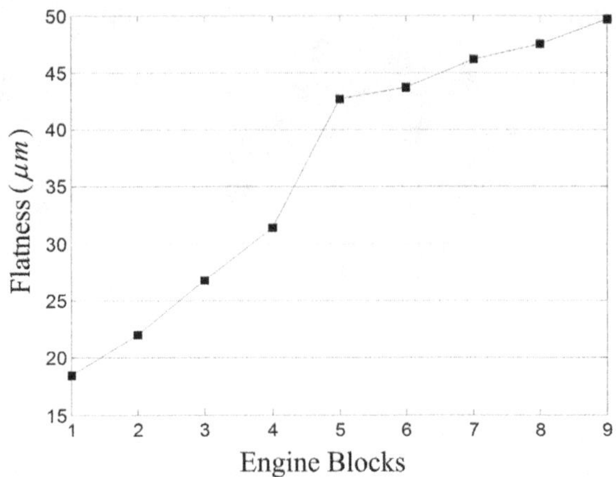

Fig. 3.5 Engine blocks rearranged in an ascending order of flatness. Reprinted from Ref. [27], copyright 2014, with permission from Elsevier

flatness of the ten engine blocks is calculated using the least square method. The numbers of engine blocks are rearranged in an ascending order of flatness: the engine block with smallest flatness is marked as Block1 and that with largest flatness is marked as Block 10. However, Block 10 is removed as its flatness is out of flatness tolerance (50 mm). Flatness of the rest nine blocks is shown in Fig. 3.5.

3.2.3.2 Experimental Results

The HDM data preprocessing method in Sect. 2.1 is applied to convert HDM data into gray images. The interpolation interval l is 0.2 mm. The lower and upper specifications S_L and S_U are assigned from -30 mm to 30 mm to 256 gray levels of the image. Figure 3.6 shows a converted grayscale image of the engine block face.

The GMCMs of the nine blocks are calculated using the modified GLCM method in Sect. 2.2 with pixel distance equal to 1, 2, 3, 4, and 5 and the angles equal to $0°$, $45°$, $90°$, and $135°$. Figure 3.7 shows GLCMs of Block 1 to Block 9 with pixel distance of 1. It is found that the nonzero elements of the GLCMs are centered near the matrix diagonal; the length of matrix diagonal is longer for the surface with larger flatness. Other researches pointed out that the surface flatness is inversely propisitional to the width of the matrix diagonal [17, 18], which is also correct for large pixel distance.

Entropy, contrast, and correlation are calculated as an average of GLCM in four directions for pixel distance d equal to 1, 2, 3, 4, and 5 (Fig. 3.8). For the same engine block, entropy and contrast increase and correlation decreases with the increase of pixel distance. The reason is that the distribution of two concurrence pixels for larger pixel distance becomes more disordered and less correlated. However, for all the nine engine blocks, entropy, contrast, and correlation show a similar trend with various pixel distances. Therefore, pixel distance equal to one is suggested for surface face milling, when the spatial and height information of two

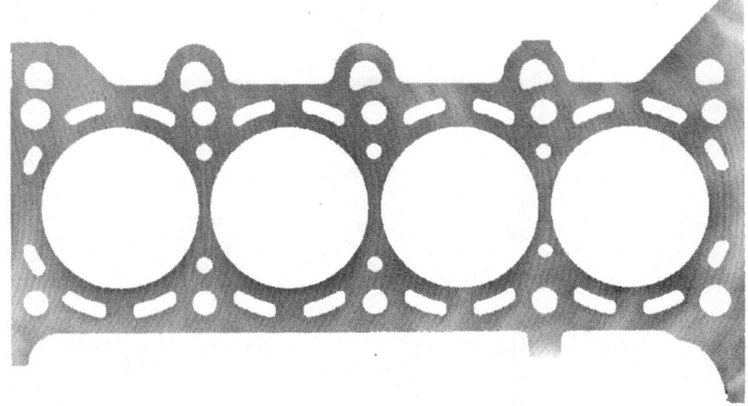

Fig. 3.6 A gray image of engine block face. Reprinted from Ref. [27], copyright 2014, with permission from Elsevier

adjacent pixels in all the four directions is of most concern. Moreover, if the machined surface has highly periodic feed marks, such as turned surfaces, a power spectral density technique could be used to choose the proper pixel distance [19].

A more detailed analysis of the results shows that entropy increased with flatness for the nine blocks except Block 6 and Block 8. For those surfaces with flatness more than 40 mm (Blocks 5 to Block 9), Block 8 has the least random height distribution. Thus, it can be predicted that Block 8 has different functional properties in certain areas with respect to Block 5, 6, 7, and 9. Contrast also shows a correlation with flatness except Block 4 and Block 8. The relatively lower contrast of Block 4 and Block 8 implies that surface form error of Block 4 and Block 8 is more likely caused by global deviations of the entire surface than local deviations. This implies that engineers can make systematic adjustment to the corresponding production line that manufactured Block 4 and Block 8. Correlation shows weak correlation with flatness but could be used to distinguish surface processing methods such as milling, turning, and grinding. Correlation of all the nine face-milled parts is high because each pixel is neighbored by another pixel with a similar gray level with pixel distance equal to one.

3.2.4   Conclusions

This subsection has developed an evaluation method for 3D surface form error using high-definition metrology. An HDM data preprocessing method is proposed to convert millions of three-dimensional data points into a height-encoded and position-maintained gray image. The height information (Z-coordinates) is converted to pixel gray intensities, and the spatial information (X- and Y-coordinates) is

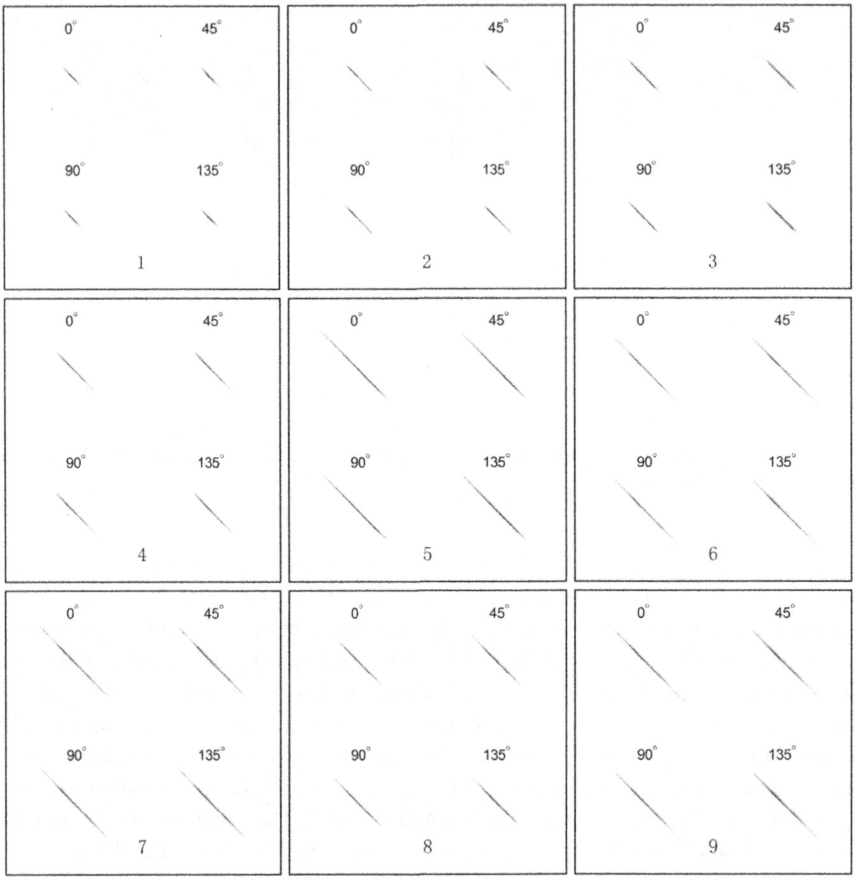

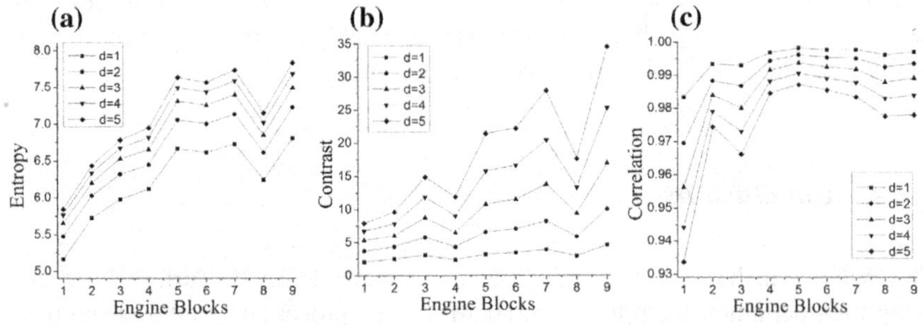

Fig. 3.7 GLCMs of Block 1 to Block 9 with pixel distance of 1. Reprinted from Ref. [27], copyright 2014, with permission from Elsevier

Fig. 3.8 GLCM characteristics of Block 1 to Block 9 with pixel distance equal to 1, 2, 3, 4, and 5: **a** entropy, **b** contrast, and **c** correlation. Reprinted from Ref. [27], copyright 2014, with permission from Elsevier

converted to the pixel index. Then a modified GLCM is adopted to evaluate 3D surface form error using the converted gray image. Gray level of 256 and pixel distance of one are suggested when calculating GLCMs for machined surface. Entropy, contrast, and correlation are calculated as an average of GLCM in four directions $0°$, $45°$, $90°$, and $135°$. Entropy indicates randomness of the surface height distribution and it is proportional to flatness. Contrast measures the degree of surface local deviation, which is the amount of local surface height difference between a pixel and its neighbor over the image. Correlation reflects the degree of how correlated a pixel to another pixel in certain direction and distance. These characteristics, including entropy, contrast, and correlation, are complementary to surface flatness and provide a comprehensive evaluation of 3D surface form error. Moreover, these characteristics help engineers for better understanding of the surface quality from various angles and making due adjustment of the process.

Starting from 3D surface form error evaluation, three directions of further research are outlined. First, latest image analysis methods should be studied to evaluate 3D form error using HDM data. Second, 3D surface form error can be extended to machining process diagnosis, such as the wear of wiper insert in face milling and erroneous fixation of tool or part. Third, the relation of 3D surface form error and surface functional properties should be studied. For example, flatness is required to prevent the potential leakage of combustion gas, coolant, and lubricant in the mating surfaces of engine blocks and engine heads. However, sometimes leakage may still happen with qualified flatness. Therefore, 3D surface form error characteristics could have more potentials than flatness to describe the sealing performance.

3.3   Co-kriging Method for Form Error Estimation Incorporating Condition Variable Measurements

3.3.1   Introduction

Surface form is predominantly considered as one of the most important features of practical product surfaces due to its crucial influence on the functional behavior of a machined part [20–27]. The form error estimation under various machining conditions is an essential step in the assessment of part surface quality generated in machining processes [28]. In order to estimate form error, the general approach is to fit a set of discrete measurement points to an underlying model to obtain an artificial surface and calculate the magnitude of form error. Coordinate measuring machines (CMMs) are widely used as an excellent measurement equipment due to its accuracy and versatility in measuring complicated geometries [29, 30]. Considering measurement cost, only a few measurement points are collected offline using a CMM for a part surface in mass production. On the other hand, with the development of new advanced machining condition measurement equipments, a lot of machining condition data can be available online. Therefore, fitting an artificial surface with a more accurate representation of the actual surface using the finite offline sample points and lots of online measurement data is of great importance under certain machining conditions.

For a nonstationary process, the form error is composed of two components: (1) deterministic error representing a deterministic trend surface that accounts for large-scale variation and is considered to be spatially dependent and (2) random error representing small-scale variation on the trend surface and is considered to be residuals and stochastic in nature and spatially independent. Incorporating the deterministic error into assessment procedure is crucial in form error estimation for geometric features [31]. Numerous methods have been used to model deterministic errors. Yan and Menq [32] explored a two-step method to estimate the deterministic error. The measurements are used to construct an artificial surface and this surface is fitted to the nominal surface. Then, the deterministic error is estimated using the orthogonal deviation from the nominal surface. Furthermore, a polynomial of different orders [33], B-spline function [34], and Fourier analysis [35–37] are presented to model the systematic error. These methods require a large amount of CMM data in order to estimate the relatively large number of parameters involved in the polynomial or B-spline function approaches or to allow a clean separation of the frequencies in the Fourier analysis approach [38].

Correlation arises frequently in data observed within certain intervals of space, and the data indeed exhibits a significant amount of positive autocorrelation [39]. Measurements are often spatially correlated because they are obtained in similar machining conditions during the machining process and related to similar (local) properties of the machined material. Spatial correlation is different from temporal correlation, which is usually represented via time series models [40]. In fact, spatial correlation models allow one to represent contiguity in space rather than in time.

Sayles and Thomas [41] use structure function to model the spatial correlation in product surface quality studies. Colosimo and Semeraro [42] use a spatial autoregressive regression (SARX) model (i.e., a regression model with spatial autoregression errors) to characterize the roundness of a manufactured product. Xia et al. [38] present a Gaussian process (GP) method for modeling the deterministic errors using a Gaussian correlation function. The analysis result shows that the Gaussian process method generally gives a less biased estimate of the form error than the traditional minimum zone (MZ) and least squares (LS) methods [43–45]. Suriano et al. propose a method for efficiently measuring and monitoring surface spatial variations by fusing multi-resolution measurements and process information [46]. Furthermore, inverse distance weighted interpolation (IDW) and triangulated irregular network (TIN) [47] are also widely used methods.

Little research on the spatial statistics-based methods to model deterministic errors of part form has been conducted and the literature on this topic is sparse. Spatial statistics, e.g., Kriging method, is one of the most important meta-models for interpolation in random field [48, 49]. Detailed descriptions of existing research on Kriging meta-modeling for interpolation are provided in a review [50]. Yang and Jackman [51] apply one univariate spatial statistic method, i.e., ordinary Kriging (OK), to predict machining part form error. The spatial correlation of the height values (Z-coordinates) is modeled as an explicit function of the other two coordinates (X-, Y-coordinates). The general prediction method is proposed to estimate the trend of a surface component and to subtract the trend from the sample data to obtain the residuals component. The residuals are treated as stationary and a variogram is fitted to the residuals. Finally, the estimated residuals are combined with the trend surface to obtain estimates of the actual surface. The results show that fitted surface obtained through OK can provide more accurate estimates of form error.

Yang and Jackman [52] only consider spatial correlation on height values obtained offline. With the development of the measurement equipments, more online measurement data is acquired to monitor the conditions of the machining processes. As well known, the machining conditions (cutting forces, feed rate, cutting tool vibration, etc.) are ubiquitous in machining processes and highly influence the final machining part form errors [52–54]. Chen [55] points out that cutting conditions, tool wear, the material properties of tool, and workpiece, as well as cutting process parameters (including cutting speed, depth of cut, feed rate, tool geometry, etc.) significantly influence the form error of finished surface on machined parts. Roth et al. [56] think that a machining condition has a direct effect on part quality and proposed online sensor-based approaches for tool condition monitoring that provide a means to assess the underlying tool condition during the cutting process. Benardos and Vosniakos [57] review the approaches that predict the surface form error and roughness under various machining conditions.

However, no research has combined spatial correlation with the machining conditions to analyze the surface form error for a machined part. Thus, in order to obtain a more accurate artificial surface for further improving the form error estimation, it is necessary to consider both spatial correlation on measured height values and the influence of the machining conditions. The spatial prediction for surface form error considering multiple machining conditions is a multivariate spatial estimation problem. The goal of this section is to estimate form error using multivariate spatial statistics method.

3.3.2   Problem Description and Univariate Spatial Method for Form Error Estimation

A set of points measured on the part surface can be denoted by

$$\{Z(s), s \in D\} \tag{3.12}$$

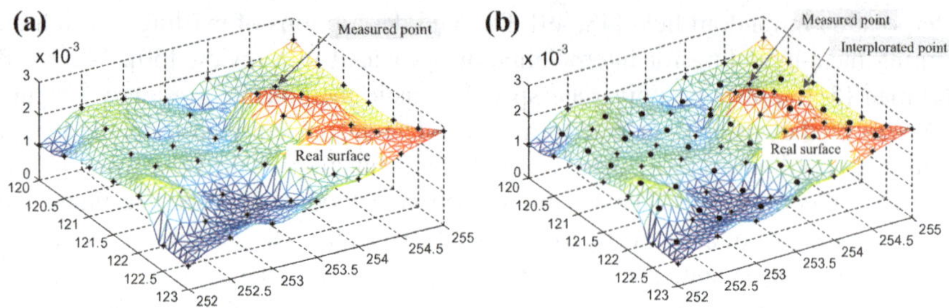

Fig. 3.9 Methods of form error estimation. Reprinted from Ref. [77], copyright 2015, with permission from ASME. **a** By directly measured points, **b** By interpolated and measured points

where s is a spatial location vector in R^2 on the surface and the index set D defines a finite region on the surface.

Through a CMM, a set of height value measurements, $\{z(s_1),\ldots,z(s_n)\}$, can be collected from locations $\{s_1,\ldots,s_n\}$ located within this finite region D. Traditionally, the collected points $\{z(s_1),\ldots,z(s_n)\}$ are directly used to estimate surface form error with LS or MZ methods (see Fig. 3.9a, unit: mm). More accurate form can be characterized by directly obtaining more data from the manufactured surface. However, collecting a large number of measured points requires high cost and decreases production efficiency. Univariate spatial statistics (e.g., Kriging) makes it possible to interpolate some points and to reconstruct artificial surfaces for form error estimation. Univariate spatial statistics method outperforms the direct method in some cases [58] (see Fig. 3.9b).

Univariate spatial statistics method supposes that the height values on the surface $\{Z(s), s \in D\}$ are real-valued intrinsically stationary random fields, which for all $s, h \in R$ satisfies [59]

$$E[Z(s) - Z(s+h)] = 0 \tag{3.13}$$

$$2\gamma(h) = Var[Z(s) - Z(s+h)] = E[Z(s) - Z(s+h)]^2 \tag{3.14}$$

where $E(\cdot)$ is the mathematic expectation, $Var(\cdot)$ is the variance of the increment between two points on the surface with the distance h, $2\gamma(h)$ called variogram in spatial statistics, which denotes the degree of spatial correlation in a spatial stationary process. In addition, the correlation at various locations depends only on the distance of two locations $h = \|s_i - s_j\|$.

The semi-variogram (called variogram for short) $\gamma(h)$ is a measurement for the spatial correlation structure at a given distance h, which is denoted by [60]:

$$\hat{\gamma}(h) = \frac{1}{2N_n(h)} \sum_{i,j \in N_n(h)} [Z(s_i) - Z(s_j)]^2, \, h \in R \tag{3.15}$$

where n is the total number of measured points, $N_n(h) = \{(i,j) : i,j \in [1,n], \|s_i - s_j\| = h\}$ is the number of pairs of measurements with the given distance h, and s_i and s_j are the locations of the measured points. The summation is conducted over all i, j pairs in the distance class.

When n surface data points, $\{z(s_1), z(s_2), \ldots, z(s_n)\}$, are measured, the height value $Z(s_0)$ at a new point location s_0 can be estimated through Kriging with the help of variogram. In particular, Kriging estimation $Z(s_0)$ is determined by a linear combination of the data values $Z^*(s_0) = \sum_{i=1}^{n} \lambda_i [Z(s_i)]$ where λ_i are weights chosen on the condition that the estimation is unbiased and the estimation error variance is smallest.

OK uses sparse scatting measured points to interpolate more certain points on the part surface and to reconstruct a more accurate artificial surface for form error estimation, which excels the form error estimation only with the limited direct measurements.

However, the height values of some interpolated points are not exactly equal to that on the real surface (see Fig. 3.10, unit: mm). There is a deviation e between the estimation and the real surface height value, which leads to the inaccuracy in artificial surface construction and form error estimation. The reason of this phenomenon is that Kriging predicting the height values only uses known height value information on measured locations considering the spatial correlation. It may not meet the accuracy demand of form error estimation based on univariate spatial correlation. How to decrease this deviation e to improve the precision of form error estimation is one of the crucial problems.

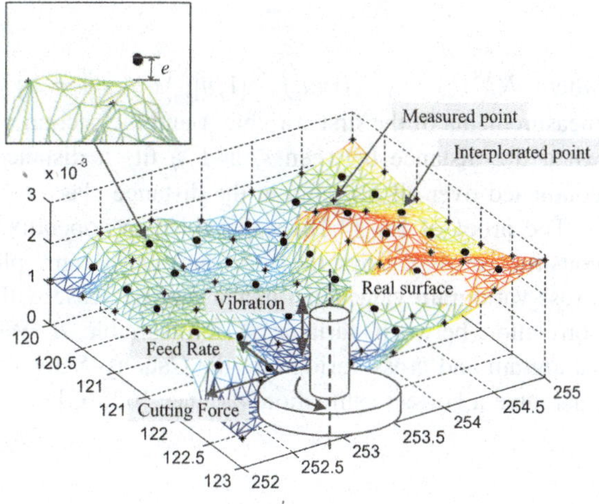

Fig. 3.10 Influence of machining conditions. Reprinted from Ref. [77], copyright 2015, with permission from ASME

In fact, the form error estimation of a part surface is greatly influenced by the machining conditions, such as the cutting forces, feed rate, and tool vibration. To establish a more accurate artificial surface, it is necessary to combine spatial correlation both on the height values and on the machining conditions.

3.3.3 The Proposed Multivariate Spatial Statistics Method

3.3.3.1 Variogram and Cross-Variogram

Usually, the real values of the machining conditions at different locations can be online obtained by sensors. The values of the machining conditions are denoted by

$$\{V_t(s), s \in D\} \tag{3.16}$$

where s is a spatial location vector in the finite region D on the surface and t is the index of the machining condition types.

Incorporated with the machining conditions V_t, the spatial correlation takes place in the process of $\{Z(s), s \in D\}$ and $\{V_t(s), s \in D\}$. Therefore, the variogram has to be extended to cross-variogram, which characterizes the multivariate spatial correlation:

$$2\gamma_{AB}(h) = Var[A(s) - B(s+h)] = E[A(s) - B(s+h)]^2 \tag{3.17}$$

The sample cross-variogram by n measured points conducted from auto-variogram is calculated as

$$\hat{\gamma}_{AB}(h) = \frac{1}{2N_n^{AB}(h)} \sum_{i,j \in N_n^{AB}(h)} [A(s_i) - A(s_j)][B(s_i) - B(s_j)], \ h \in R^d, A, B \in \{Z, V_t\}$$

$$\tag{3.18}$$

where $N_n^{AB}(h) = \{(i,j) : i,j \in [1,n], \|s_i - s_j\| = h\}$ is the number of pairs of measurements of the first variable A and the second variable B at locations s_i and s_j when the distance between s_i and s_j fits a distance class h. The summation is conducted over all i, j pairs in the distance class.

The proof of Eq. (3.18) is provided in appendix. For more details on how to construct cross-variogram from auto-variogram, please refer to the Ref. [61]. Cross-variogram values can increase or decrease with distance h depending on the correlation between variable A and variable B. When models are fitted to the variogram and cross-variogram, the Cauchy–Schwarz relation must be checked to guarantee a correct estimation variance [62, 63]:

$$|\gamma_{AB}(h)| \leq |\gamma_{AA}(h) * \gamma_{BB}(h)|^{\frac{1}{2}}, A, B \in \{Z, V_t\} \qquad (3.19)$$

The Cauchy–Schwarz inequality is considered in variogram modeling process to avoid facing with negative definite matrices and insoluble equation system in co-Kriging algorithm [64]. Therefore, the Cauchy–Schwarz inequality should be satisfied whether using the cross-covariances or using the cross-variograms [65]. Many efforts on constructing valid cross-covariance structure in multivariate spatial process can be found in the literature of design and analysis of computer experiments, where a covariance representation (instead of variogram) of the Kriging method is often used [66–69]. It can be shown that the Cauchy–Schwarz inequality is necessary but not sufficient for joint nonnegative definiteness of the fitted cross-covariance models. There is a controversy over the applicability of the Cauchy–Schwarz inequality to the cross-variogram matrices [70]. In order to generate nonnegative definite functions for cross-variogram, Rehmanand Shapiro [71] and Armstrong and Diamond [72] proposed the Fourier transform to follow the result which is a natural generalization of the "sufficient" part of Bochner's theorem.

Cross-variogram $\gamma_{AB}(h)$, as well as variogram $\gamma(h)$, can be modeled to a theoretical variogram through a least square fitting process. As the distance h gets larger, the variogram values increases, which indicates that as two locations get farther apart, the expected difference of the measured values between these two locations increases as well.

The theoretical variogram model has three main parameters: (1) Nugget (C_0) is the semivariance at zero distance due to measurement error and micro-scale variation. (2) Sill ($C_0 + C_1$) is the semivariance value of the plateau, which is the maximum height of the variogram curve. As distance h gets large, the correlation between the two points becomes negligible and the value of variogram tends to be stationary. (3) Correlation length (a) is the distance at which the semivariance achieves the plateau, which means that pairs of points larger than this distance apart are negligibly correlated.

There exist various models for fitting diverse empirical variograms in practice. Some popular theoretical fitting models include

$$\text{Spherical model}: \quad \gamma(h) = \begin{cases} C_0 + C_1\left[1.5\left(\frac{h}{a}\right) - 0.5\left(\frac{h}{a}\right)^3\right], & 0 \leq h \leq a \\ C_0 + C_1, & h > a \end{cases} \qquad (3.20)$$

$$\text{Exponential model}: \quad \gamma(h) = C_0 + C_1\left[1 - \exp\left(-\frac{3h}{a}\right)\right] \qquad (3.21)$$

$$\text{Periodicity model}: \quad \gamma(h) = W\left\{1 - \frac{1}{\theta}\sin(\theta)\right\}, \theta = \frac{2\pi h}{\omega} \qquad (3.22)$$

where W is the amplitude and ω is the wavelength of periodicity variogram.

The differences among the theoretical variogram models vary in the fitting degree, and thus it is important to choose the best theoretical variogram model by appropriate fitting standards. Among them, the ratio of the structural effects is frequently used, which is the ratio of nugget to sill and sill $\frac{C_1}{C_0 + C_1}$. It characterizes the degree of spatial heterogeneity, which means the spatial variability caused by the autocorrelation occupies the proportion of the total variation of the system. The larger the ratio is, the stronger the spatial correlation is. The stronger spatial correlation for the theoretical variogram, the more precise estimation for the predictive modeling interpolation by spatial statistics method, and thus it makes form error estimation more accurate.

3.3.3.2 Multivariate Spatial Method

Co-Kriging (CK) is a multivariate spatial interpolation method based on the multivariate spatial correlation structure [73]. Compared to Kriging, it organizes not only a sparse set of data (height value, denoted Z) but also other datasets that are dense and can be sampled more frequently and regularly (machining condition, denoted V_t). Figure 3.11 shows the different variables and data information used in Kriging and CK to interpolate the unmeasured points. CK requires more information which is designated by machining conditions V_1, V_2. In Fig. 3.11, black dots $Z(s_i)$, $i \in \{1, 2, 3, 4\}$ are measurements of height values at locations s_i, and black squares are unmeasured points s_0 that will be interpolated. $V_1(s_a), a \in \{2, 3, 4, 6, 7\}$ and $V_2(s_b), b \in \{1, 2, 4, 5, 6\}$ are measurements of machining conditions at locations s_a and s_b, locations s_5, s_6, s_7 on surface in Fig. 3.11b are new measured points required for CK.

The height value Z is the target variable and the machining condition V_t gives the supportive information for the target variables. Both autocorrelation and cross-correlations between Z and all other machining condition variable V_t are used to make better predictions. In fact, at some location s*, if there is only the measurement of machining condition variable V_t (as the measurement $A(s^*)$), and there

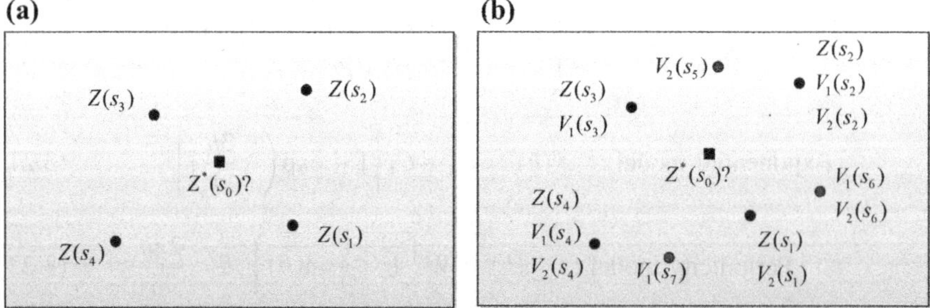

Fig. 3.11 Comparison of Kriging and CK variables, Reprinted from Ref. [77], copyright 2015, with permission from ASME. **a** Kriging, **b** CK

is not any measurement of height value Z (as the measurement $B(s^*)$), the data of $A(s^*)$ is still useful for the estimator. Even if there is no cross-correlation between Z and V_t, or even if no measurement of Z at some locations, it can still fall back on the autocorrelation for height value. So CK can theoretically do no worse than Kriging. At the same time, the deviation between the real surface and the estimated height value in the univariate spatial method can more or less decrease. Therefore, it is possible to improve the precision of form error estimation using CK.

Incorporated with machining conditions, the height value $Z(s_0)$ at the unmeasured point location s_0 can also be determined by a linear combination of the measured CMM data. The interpolated height value by CK can be calculated by [74, 75].

$$Z_{CK}^*(s_0) = \sum_{i_1=1}^{N_1} \lambda_{1i_1} V_1(s_{1i_1}) + \sum_{i_2=1}^{N_2} \lambda_{2i_2} V_2(s_{2i_2}) + \cdots + \sum_{i_T=1}^{N_T} \lambda_{Ti_T} V_T(s_{Ti_T}) + \sum_{k=1}^{N_0} \lambda_{0k} Z(s_{0k})$$

$$(3.23)$$

where $V_t(t = 1, 2, \ldots, T)$ are T machining conditions, Z is the main variable, height value, on measured points, $s_{ti_t}(t = 1, \ldots, T)$ are measured locations of T machining conditions, and s_{0k} is the measured location of height value, $N_t(t = 1, 2, \ldots, T)$ are the total numbers of the measured locations for T machining conditions, N_0 is the total number of measured locations for height value, $\lambda_{ti_t}(t = 1, \ldots, T)$ are weights characterized by the impacts of machining conditions on the height value of the estimated point s_0, and λ_{0k} characterizes the impact of the height values from known measured locations.

Since CK weights have to satisfy the unbiased constraint $E[Z_{CK}^*(s_0) - Z(s_0)] = 0$, calculated by

$$E[Z_{CK}^*(s_0) - Z(s_0)]$$

$$= E\left\{ \left[\sum_{i_1=1}^{N_1} \lambda_{1i_1} V_1(s_{1i_1}) + \sum_{i_2=1}^{N_2} \lambda_{2i_2} V_2(s_{2i_2}) + \cdots + \sum_{i_T=1}^{N_T} \lambda_{Ti_T} V_T(s_{Ti_T}) + \sum_{k=1}^{N_0} \lambda_{0k} Z(s_{0k}) \right] - Z(s_0) \right\}$$

$$= E\{Z(s_0)\}\left[\sum_{k=1}^{N_0} \lambda_{0k} - 1\right] + E\{V_1(s_{1i_1})\} \sum_{i_1=1}^{N_1} \lambda_{1i_1} + E\{V_2(s_{2i_2})\} \sum_{i_2=1}^{N_2} \lambda_{2i_2} + \cdots + E\{V_T(s_{Ti_T})\} \sum_{i_T=1}^{N_T} \lambda_{Ti_T}$$

$$= 0$$

$$(3.24)$$

Then we obtain

$$\sum_{k=1}^{N_0} \lambda_{0k} = 1, \quad \sum_{i_1=1}^{N_1} \lambda_{1i_1} = 0, \quad \sum_{i_2=1}^{N_2} \lambda_{2i_2} = 0 \ , \cdots, \quad \sum_{i_T=1}^{N_T} \lambda_{Ti_T} = 0 \qquad (3.25)$$

Minimizing CK error variance $Var[Z_{CK}^*(s_0) - Z(s_0)]$, calculated by

$$\min\{Var[Z_{CK}^*(s_0) - Z(s_0)]\}$$

$$= E\left\{\left\{[Z_{CK}^*(s_0) - Z(s_0)] - E[Z_{CK}^*(s_0)] - Z(s_0)\right\}^2\right\}$$

$$= E\left\{[Z_{CK}^*(s_0) - Z(s_0)]^2\right\}$$

$$= E\left\{\left\{\left[\sum_{i_1=1}^{N_1} \lambda_{1i_1} V_1(s_{1i_1}) + \sum_{i_2=1}^{N_2} \lambda_{2i_2} V_2(s_{2i_2}) + \cdots + \sum_{i_T=1}^{N_T} \lambda_{Ti_T} V_T(s_{Ti_T}) + \sum_{k=1}^{N_0} \lambda_{0k} Z(s_{0k})\right] - Z(x_0)\right\}^2\right\}$$

$$(3.26)$$

Lagrange multipliers are usually used to adjoin constraint Eq. (3.26) to an objective function. CK error variance is minimized. Then we obtain

$$\begin{cases}
\sum_{i_1=1}^{N_1} \lambda_{1i_1}\gamma_{11}(s_{1i} - s_{l_1}) + \sum_{i_2=1}^{N_2} \lambda_{2i_2}\gamma_{12}(s_{2i_2} - s_{l_2}) + \cdots + \sum_{i_T=1}^{N_T} \lambda_{Ti_T}\gamma_{1T}(s_{Ti_T} - s_{l_T}) + \sum_{k=1}^{N_0} \lambda_{0k}\gamma_{10}(s_{0k} - s_K) + \mu_1 = \gamma_{10}(s_0 - s_{l_1}) \\
\sum_{i_1=1}^{N_1} \lambda_{1i_1}\gamma_{21}(s_{1i} - s_{l_1}) + \sum_{i_2=1}^{N_2} \lambda_{2i_2}\gamma_{22}(s_{2i_2} - s_{l_2}) + \cdots + \sum_{i_T=1}^{N_T} \lambda_{Ti_T}\gamma_{2T}(s_{Ti_T} - s_{l_T}) + \sum_{k=1}^{N_0} \lambda_{0k}\gamma_{20}(s_{0k} - s_K) + \mu_2 = \gamma_{20}(s_0 - s_{l_2}) \\
\cdots \\
\sum_{i_1=1}^{N_1} \lambda_{1i_1}\gamma_{T1}(s_{1i} - s_{l_1}) + \sum_{i_2=1}^{N_2} \lambda_{2i_2}\gamma_{T2}(s_{2i_2} - s_{l_2}) + \cdots + \sum_{i_T=1}^{N_T} \lambda_{Ti_T}\gamma_{TT}(s_{Ti_T} - s_{l_T}) + \sum_{k=1}^{N_0} \lambda_{0k}\gamma_{T0}(s_{0k} - s_K) + \mu_T = \gamma_{T0}(s_0 - s_{l_T}) \\
\sum_{i_1=1}^{N_1} \lambda_{1i_1}\gamma_{01}(s_{1i} - s_{l_1}) + \sum_{i_2=1}^{N_2} \lambda_{2i_2}\gamma_{02}(s_{2i_2} - s_{l_2}) + \cdots + \sum_{i_T=1}^{N_T} \lambda_{Ti_T}\gamma_{0T}(s_{Ti_T} - s_{l_T}) + \sum_{k=1}^{N_0} \lambda_{0k}\gamma_{00}(s_{0k} - s_K) + \mu_0 = \gamma_{00}(s_0 - s_K)
\end{cases}$$

$$(3.27)$$

where the addition parameters $\mu_i, i = 0, 1, 2, \ldots, T$ are Lagrange multipliers used in minimization of CK error variance σ_{CK}^2 to satisfy the unbiased condition and $\gamma_{pq}(s_i - s_j)$, $p, q = 0, 1, 2, \ldots, T$ is the cross-variogram $(p \neq q)$ or auto-variogram $(p = q)$ value $\gamma_{pq}(h)$ between the height values and machining conditions for the distance $h = \|s_i - s_j\|$ calculated in spatial correlation.

CK weights can be calculated by the two constraints Eqs. (3.25) and (3.27). Combined the interpolated points with the measured points, the artificial surface can be reconstructed closer to the real surface, which makes it more precise for form error estimation.

3.3.3.3   An Illustrated Example

In order to illustrate the proposed multivariate spatial method, a sample surface in the 30×30 square with four measured points $P_1(-30, 30), P_2(0, -30), P_3(30, 0), P_4(0, 30)$ is presented (see Fig. 3.12).

At each point, there are three variables (two independent machining conditions V_1, V_2 and one dependent height value Z) that have the spatial correlations. The machining condition V_1 at point P_1, P_2, P_3, P_4 is known as $V_{11} = 8, V_{12} = 3, V_{13} = 10, V_{14} = 7$, the machining condition V_2 at point P_1, P_2 is known as $V_{21} = 4, V_{22} = 11$, and the height value Z at point P_1, P_2 is known as $Z_1 = 9, Z_2 = 12$. The goal is to estimate the value of the variable Z at point $P_0(0, 0)$. The auto-variogram of the variables V_1, V_2, Z is calculated as (exponential model)

Fig. 3.12 Distribution of measured and estimated points. Reprinted from Ref. [77], copyright 2015, with permission from ASME

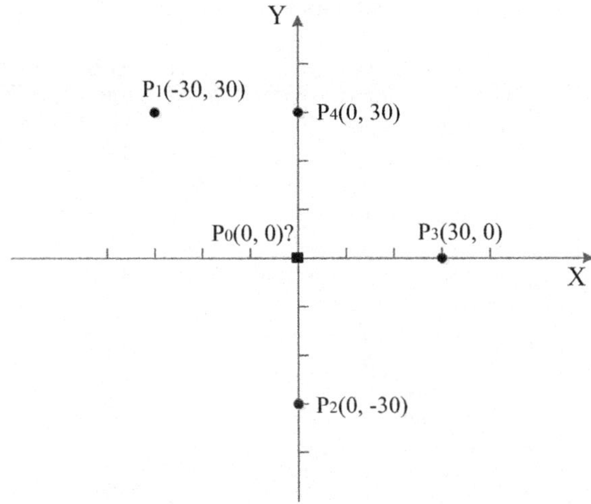

$$\gamma_{11}(h) = 0.06 + 0.7(1 - \exp(-h/300))(V_1)$$

$$\gamma_{22}(h) = 0.14 + 0.8(1 - \exp(-h/200))(V_2)$$

$$\gamma_{00}(h) = 0.22 + 0.9(1 - \exp(-h/100))(Z)$$

The cross-variogram between variables is calculated by spherical model as

$$\gamma_{12}^{+}(h) = \gamma_{20}^{+}(h) = \gamma_{10}^{+}(h) = \begin{cases} 0.4 + 0.32(1.5(h/100) - 0.5(h/100)^3), h \leq 100 \\ 0.72, h > 100 \end{cases}$$

$$\gamma_{12}(h) = 0.5[\gamma_{12}^{+}(h) - \gamma_{11}(h) - \gamma_{22}(h)]$$

$$\gamma_{20}(h) = 0.5[\gamma_{20}^{+}(h) - \gamma_{22}(h) - \gamma_{00}(h)]$$

$$\gamma_{10}(h) = 0.5[\gamma_{10}^{+}(h) - \gamma_{11}(h) - \gamma_{00}(h)]$$

According to Eqs. (3.25)–(3.27), the constraints are given by

$$\sum_{k=1}^{N_0} \lambda_{0k} = 1, \quad \sum_{i=1}^{N_1} \lambda_{1i} = 0, \quad \sum_{j=1}^{N_2} \lambda_{2j} = 0 \, (N_1 = 4, \quad N_2 = 2, \quad N_0 = 2)$$

$$\sum_{i-1}^{N_1} \lambda_{1i}\gamma_{11}(s_{1i} - s_I) + \sum_{j-1}^{N_2} \lambda_{2j}\gamma_{12}(s_{2j} - s_J) + \sum_{k-1}^{N_0} \lambda_{0k}\gamma_{10}(s_{0k} - s_K) + \mu_1 = \gamma_{10}(s_0 - s_I)$$

$$\sum_{i-1}^{N_1} \lambda_{1i}\gamma_{21}(s_{1i} - s_I) + \sum_{j-1}^{N_2} \lambda_{2j}\gamma_{22}(s_{2j} - s_J) + \sum_{k-1}^{N_0} \lambda_{0k}\gamma_{20}(s_{0k} - s_K) + \mu_2 = \gamma_{20}(s_0 - s_J)$$

$$\sum_{i-1}^{N_1} \lambda_{1i}\gamma_{01}(s_{1i} - s_I) + \sum_{j-1}^{N_2} \lambda_{2j}\gamma_{02}(s_{2j} - s_J) + \sum_{k-1}^{N_0} \lambda_{0k}\gamma_{00}(s_{0k} - s_K) + \mu_0 = \gamma_{00}(s_0 - s_K)$$

Convert the constraints to the matrix, and the weights are achieved by

$$
\begin{bmatrix} \lambda_{11} \\ \lambda_{12} \\ \lambda_{13} \\ \lambda_{14} \\ \lambda_{21} \\ \lambda_{22} \\ \lambda_{01} \\ \lambda_{02} \\ \mu_1 \\ \mu_2 \\ \mu_0 \end{bmatrix}
=
\begin{bmatrix}
0.0600 & 0.2003 & 0.2003 & 0.1266 & 0.1000 & 0.0527 & 0.0600 & -0.0932 & 1 & 0 & 0 \\
0.2003 & 0.0600 & 0.1523 & 0.1896 & 0.0527 & 0.1000 & -0.0932 & 0.0600 & 1 & 0 & 0 \\
0.2003 & 0.1523 & 0.0600 & 0.1523 & 0.0527 & 0.0731 & -0.0932 & -0.0460 & 1 & 0 & 0 \\
0.1266 & 0.1869 & 0.1523 & 0.0600 & 0.0808 & 0.0596 & -0.0201 & -0.0798 & 1 & 0 & 0 \\
0.1000 & 0.0527 & 0.0527 & 0.0808 & 0.1400 & 0.3680 & 0.0200 & -0.1771 & 0 & 1 & 0 \\
0.0527 & 0.1000 & 0.0731 & 0.0596 & 0.3680 & 0.1400 & -0.1771 & 0.0200 & 0 & 1 & 0 \\
0.0600 & -0.0932 & -0.0932 & -0.0201 & 0.0200 & -0.1771 & 0.2200 & 0.6598 & 0 & 0 & 1 \\
-0.0932 & 0.0600 & -0.0460 & -0.0798 & -0.1771 & 0.0200 & 0.6598 & 0.2200 & 0 & 0 & 1 \\
1 & 1 & 1 & 1 & 0 & 0 & 0 & 0 & 0 & 0 & 0 \\
0 & 0 & 0 & 0 & 1 & 1 & 0 & 0 & 0 & 0 & 0 \\
0 & 0 & 0 & 0 & 0 & 0 & 1 & 1 & 0 & 0 & 0
\end{bmatrix}^{-1}
\times
\begin{bmatrix}
-0.046 \\ -0.0201 \\ -0.0201 \\ -0.0201 \\ -0.1163 \\ -0.0825 \\ 0.5312 \\ 0.4533 \\ 0 \\ 0 \\ 1
\end{bmatrix}
$$

Through the MATLAB programming procedure, the values of the weights are obtained as

$$\lambda_{11} = 0.3703; \quad \lambda_{12} = 0.2212; \quad \lambda_{13} = -0.2783; \quad \lambda_{14} = -0.3132; \quad \sum \lambda_{1i} = 0$$

$$\lambda_{21} = 0.0456; \quad \lambda_{22} = -0.0456; \quad \sum \lambda_{2j} = 0$$

$$\lambda_{31} = 0.4515; \quad \lambda_{32} = 0.5485; \quad \sum \lambda_{0k} = 1$$

$$\mu_1 = 0.0047; \quad \mu_2 = -0.0266; \quad \mu_0 = 0.0271$$

The estimated height value on $P_0(0,0)$ using CK is obtained as

$$Z^*_{CK}(0,0) = \sum_{i=1}^{N_1} \lambda_{1i} V_1(s_{1i}) + \sum_{j=1}^{N_2} \lambda_{2j} V_2(s_{2j}) + \sum_{k=1}^{N_0} \lambda_{0k} Z(s_{0k}) = 8.9773$$

The height value of the unmeasured point $P_0(0,0)$ is achieved by CK using the machining conditions. The cross-variogram function is simply calculated by known auto-variogram, and all these variograms characterize the spatial correlation on this exampled surface.

3.3.4   Comparison of Univariate and Multivariate Spatial Statistics Methods Using Simulated Data

3.3.4.1   Simulation Comparison Procedure

In order to compare CK considering machining conditions with OK without considering machining conditions for form error estimation, the simulation experiment has been conducted. Below is the simulation procedure:

Step 1: Simulate a geometric feature on the surface with an area of 10×10 mm^2.

Step 2: Generate a dense enough set of measurements every 0.1 mm, a total of N = 10000 points on the selected geometric feature so that the measurements closely represent the actual surface. In an OK simulation, only the height values are generated, while in CK simulation, the height values and one of the machining conditions—machine tool vibration value—are generated at the same time corresponding to locations.

Step 3: Select M = 1000 data points from the set of dense measurements. The M locations and their corresponding observations are chosen using a random sampling approach. This method enforces M samples are evenly spread over the surface.

Step 4: Calculate the variogram for the height, vibration value, and the cross-variogram between the height and the vibration using M selected sample data points.

Step 5: Take (M − m) locations as samples to estimate the height value of m = 50, 100, 150, 200, 250, 300, which are as training samples using OK. Randomly remove half of the M height values, but insert their corresponding vibration value, and then take (M − m) locations as testing samples to estimate the height value of m training samples using CK.

Step 6: For m = 50, 100, 150, 200, 250, 300, the height value estimations determine the flatness error and denoted by f_{CK} (mm) when using CK, and by f_{OK} (mm) when using OK. Determine the form error f^* (mm) from the real height value using the individual points and treat it as the "true" flatness error.

Step 7: Calculate the form error estimation assessment ratios, f_{CK}/f^* and f_{OK}/f^*. A ratio closer to one indicates a less biased estimation.

Step 8: Repeat steps (5) to (7) 50 times for each m and generate a box–whisker plot of the flatness error estimation assessment.

Table 3.1 Manufacturing scenarios for form error estimation

Scenarios		L	R	A_x	A_y	λ_x	λ_y	A_{vibx}	A_{viby}	λ_{vibx}	λ_{viby}	σ_ε
Case 1	OK	20	0.015	0.05	0.03	2	8	–	–	–	–	0.02
	CK	20	0.015	0.05	0.03	2	8	0.025	0.015	1	4	0.02
Case 2	OK	20	0.025	0.006	0.007	5	4	–	–	–	–	0.008
	CK	20	0.025	0.006	0.007	5	4	0.003	0.0035	2.5	2	0.008
Case 3	OK	20	0.25	0.008	0.008	5	8	–	–	–	–	0.008
	CK	20	0.25	0.008	0.008	5	8	0.004	0.004	2.5	4	0.008

In this simulation, flatness feature is simulated. The data points from the part surface are generated by the following function modified from Dowling's one dimension function for straight feature [39]:

$$z(x, y) = \frac{64}{L^6} R(x^3 (L - x)^2) + \frac{64}{L^6} R(y^3 (L - y)^2) + A_x \sin(\frac{2\pi}{\lambda_x} x) + A_y \sin(\frac{2\pi}{\lambda_y} y) + \varepsilon$$

$$(3.28)$$

$$\text{vibraion}(x, y) = A_{vibx} \sin(\frac{2\pi}{\lambda_{vibx}} x) + A_{viby} \sin(\frac{2\pi}{\lambda_{viby}} y) \qquad (3.29)$$

where the first two terms of Eq. (3.28) represent the surface deflection, the third and fourth terms are wave patterns, and the last one is the random error, assumed to be $N(0, \sigma_\varepsilon^2)$. L is the length of the straight feature, A is the sine wave amplitude ($A_x, A_y, A_{vibx}, A_{viby}$, respectively, represent the sine wave amplitude of height value and vibration value in direction of x-axis and y-axis), λ is the wavelength ($\lambda_x, \lambda_y, \lambda_{vibx}, \lambda_{viby}$, respectively, represent the sine wavelength of height value and vibration value in direction of X-axis and Y-axis), and R is the deflection range.

Table 3.1 shows the parameters in simulated machining scenarios for form error estimation. The simulated surfaces images for different cases are shown in Fig. 3.13.

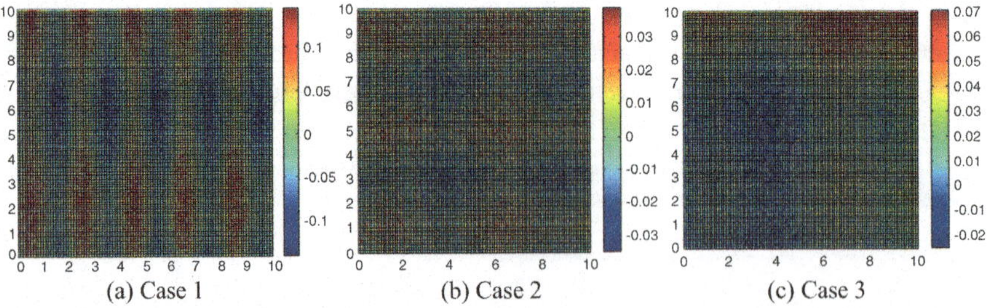

Fig. 3.13 Color-coded images of simulated surfaces. Reprinted from Ref. [77], copyright 2015, with permission from ASME

3.3.4.2 Simulation Results and Discussion

The fitting theoretical variograms of the height and vibration and their cross-variograms for different cases in step 4 in Sect. 3.3.4.1 are shown in Fig. 3.14. The height variograms and cross-variograms for case 3 are modeled by spherical models, and the rest are fitted by periodicity models.

Figure 3.15 shows the ratios of form error estimation for flatness feature. In each box–whisker plot, the locations of the upper limit, the 75% quantile, the median, the 25% quantile, and the lower limit are shown. The crosses outside of the upper and lower limits are usually considered as outliers. The solid line indicates that the estimation of flatness error is the same as the true flatness error (i.e., estimated ratio is equal to one). In other words, the best method is the one that consistently produces box–whisker plots closest to the solid line. From Fig. 3.15, some findings are obtained, which are as follows:

The proposed CK-based method performs significantly better than OK when systematic machining errors exist. For instance, in Case 1 the flatness error estimation assessment ratio is between 0.7 and 0.75 for OK while between 0.85 and 0.95 for CK.

By comparison, OK tends to underestimate the flatness error, even when using a relatively large sample size. The reason is that the flatness error estimation assessment benefits from CK's ability to capture the systematic machining errors, like the influence of vibration on height value and form error. OK only treats the height value as a complete representation of the entire feature.

When the sample size grows larger, CK and OK tend to be unbiased. This is confirmed that when a large sample is used, the distribution of f is centered around the actual flatness error. CK suffers from not having sufficient information when the simulated sample is small. Some authors point out that a small sample size is not sufficient [9, 39]. To a smaller degree, insufficient information from a small sample generally leads to much more uncertainty.

3.3.5 Case Study

3.3.5.1 Case Study I

Experimental Setup

Machining surfaces for this case is the engine cylinder block top surfaces (see Fig. 3.16). The engine block is machined under closely supervised conditions to ensure that no anomalous problems with milling process occur. The material of the engine block is cast iron FC250. The milling process is carried out using an EX-CELL-O machining center and a CBN milling cutter with a diameter of 200 mm. Quaker 370 KLG cutting fluid is used. The cutting speed is 816.4 m/min, the depth of cut is 0.5 mm, and feed rate is 3360 mm/min.

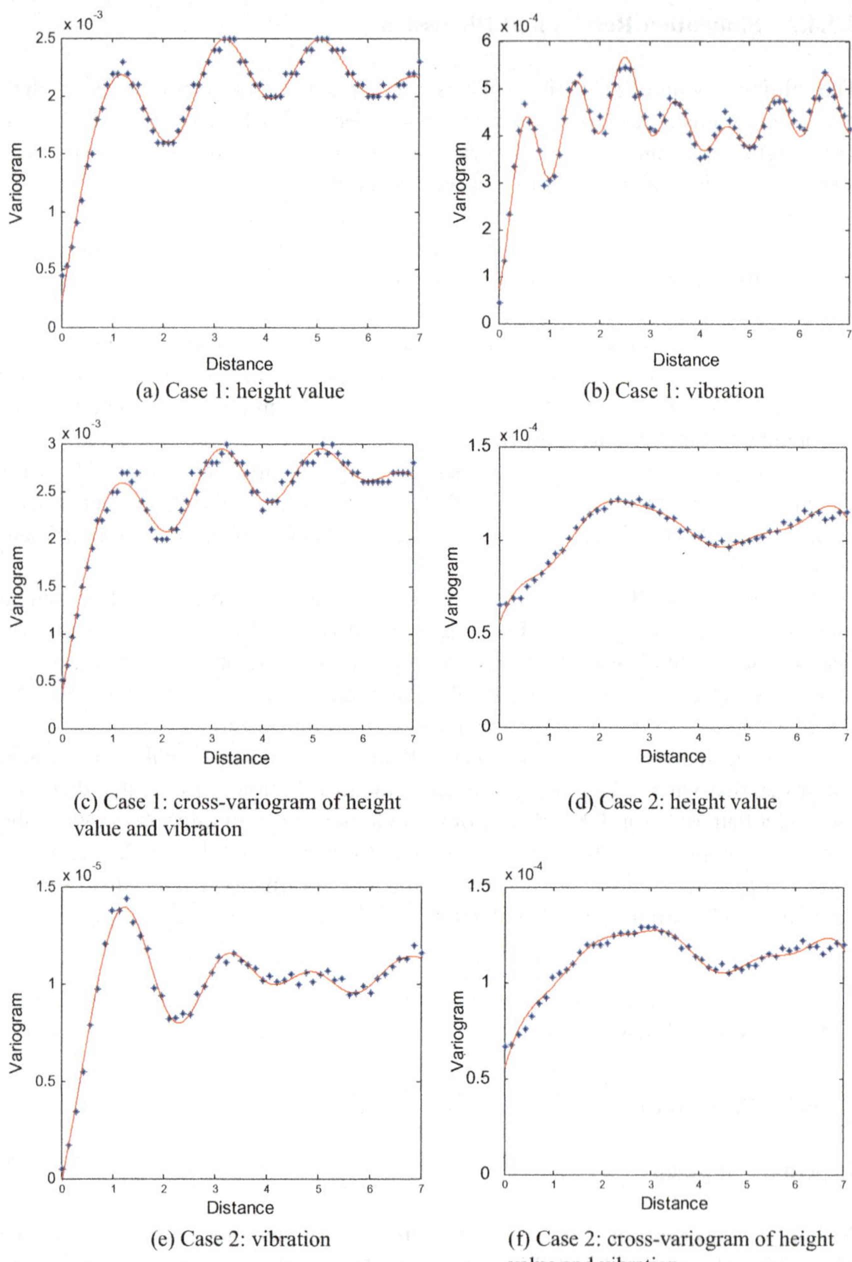

(a) Case 1: height value

(b) Case 1: vibration

(c) Case 1: cross-variogram of height value and vibration

(d) Case 2: height value

(e) Case 2: vibration

(f) Case 2: cross-variogram of height value and vibration

Fig. 3.14 Theoretical variogram of height and vibration and cross-variogram. Reprinted from Ref. [77], copyright 2015, with permission from ASME

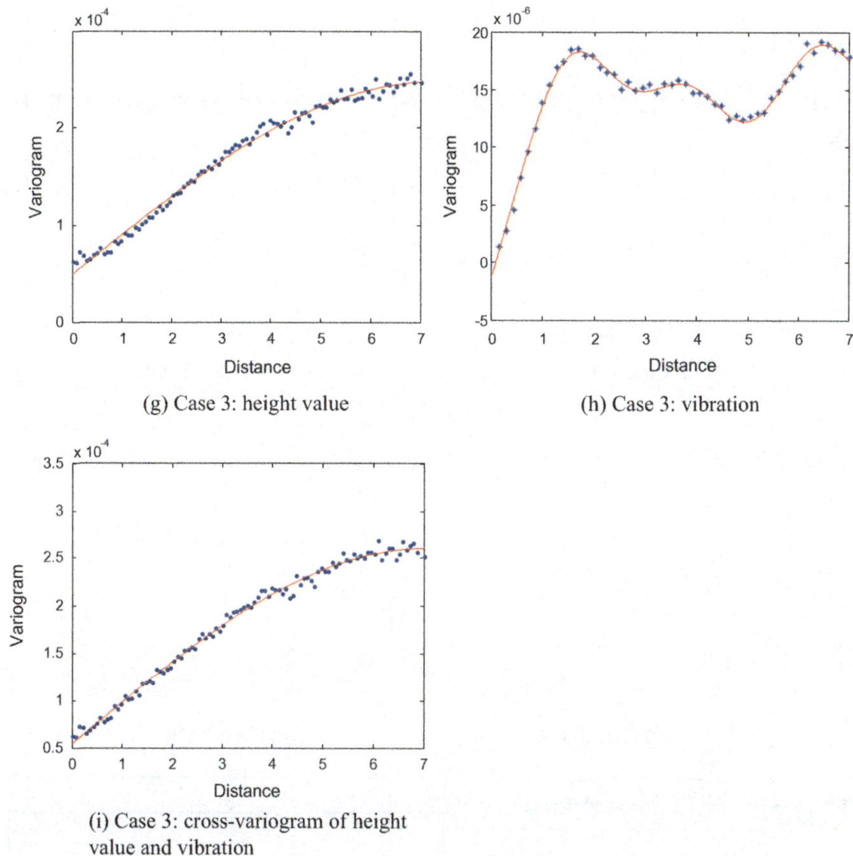

(g) Case 3: height value   (h) Case 3: vibration

(i) Case 3: cross-variogram of height
value and vibration

Fig. 3.14 (continued)

The engine block surface is measured using a CMM and N = 1000 points to an area of 326.5 mm × 174 mm is acquired. The height deviations are from −80 to 80 mm. Figure 3.17 shows artificial surface with sampling data points.

The tool vibration is considered as one of the most important influence factors during the machining process for the surface flatness error estimation. Therefore, vibration is taken into consideration by CK in this case study.

The vibration data collection system is comprised of an accelerometer sensor from which signals are amplified, converted to digital data, and processed using Windows-based software. Figure 3.18 depicts the hardware setup for milling process and the schematic for vibration data collection. The accelerometer sensors are JX20 Eddy current displacement sensors, which are mounted on the shank of the tool holder.

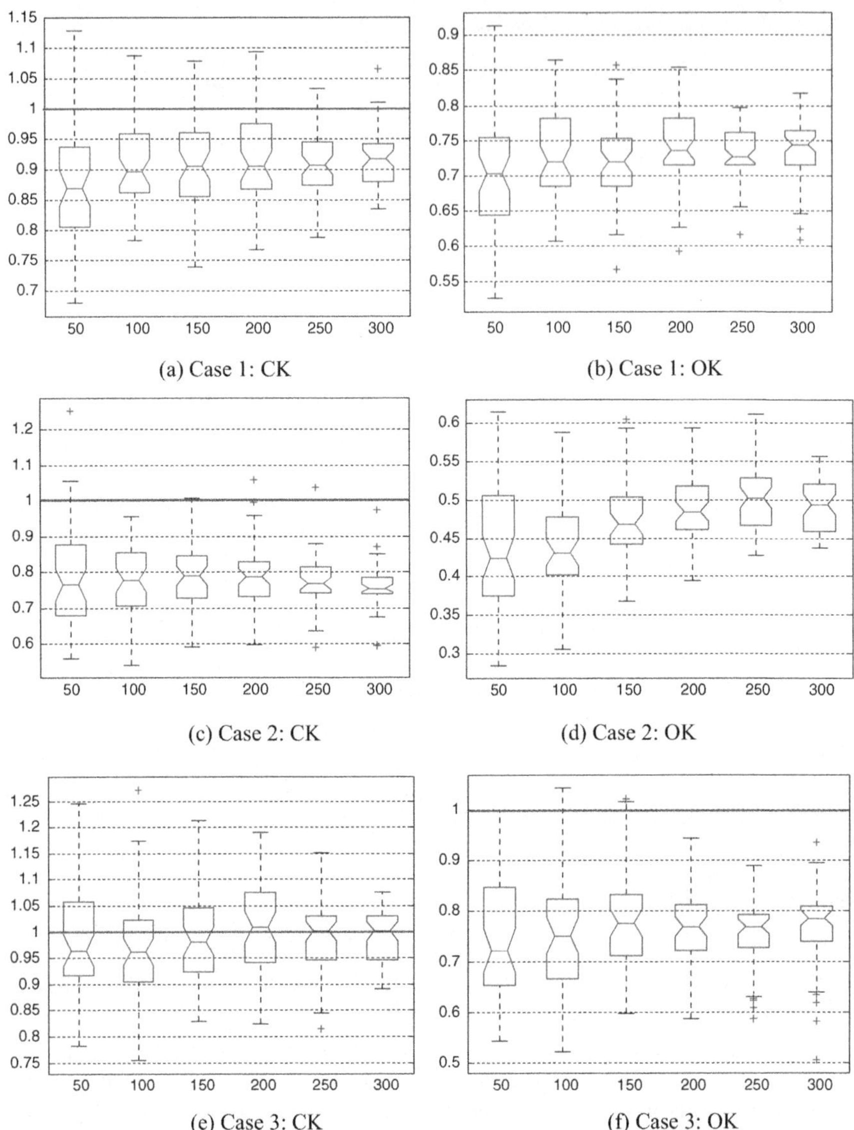

Fig. 3.15 Box–whisker plot of flatness error estimation comparison for different cases. Reprinted from Ref. [77], copyright 2015, with permission from ASME

Data Preprocessing

Basic sample descriptive statistics of surface height values are given in Table 3.2. Since the height and the vibration signal are measured on different scales and units, it is necessary to take these data values into the same measuring scales. Data

Fig. 3.16 Engine cylinder blocks processed by a major domestic car manufacturer. Reprinted from Ref. [77], copyright 2015, with permission from ASME

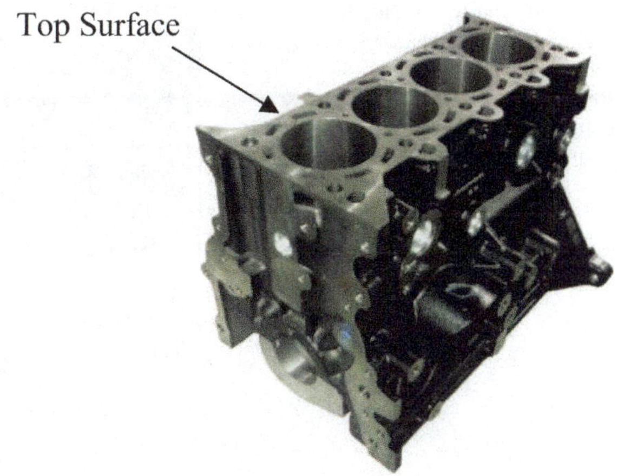

Fig. 3.17 Color-coded image of the engine block face with triangular mesh plot. Reprinted from Ref. [77], copyright 2015, with permission from ASME

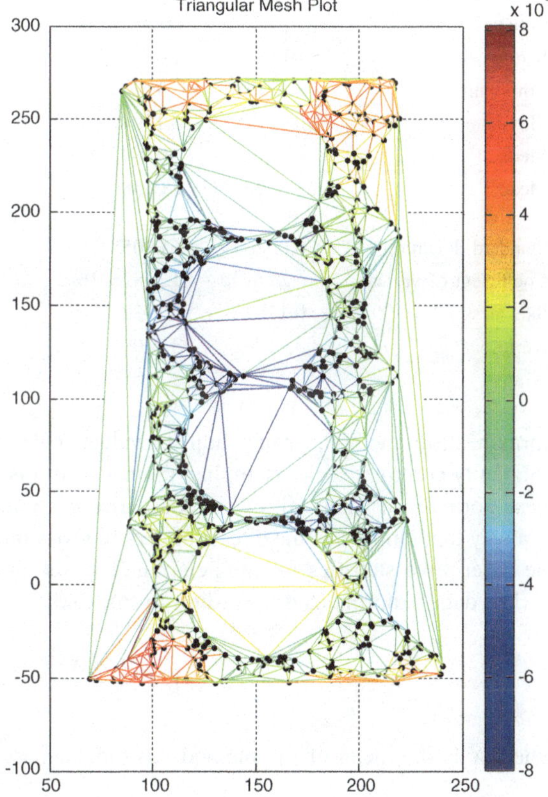

Fig. 3.18 Hardware setup for milling and the schematic for vibration data collection. Reprinted from Ref. [77], copyright 2015, with permission from ASME

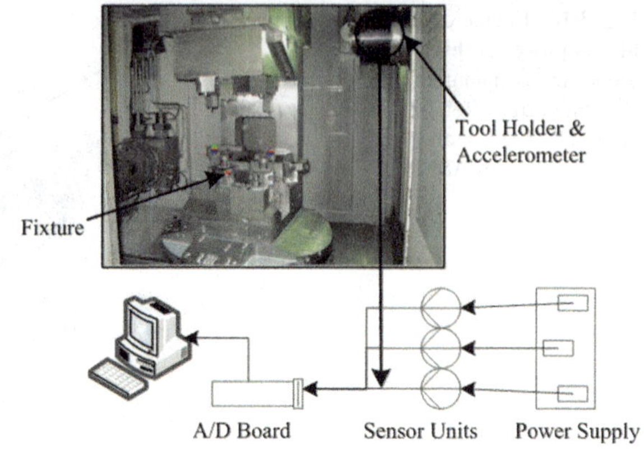

Table 3.2 Basic descriptive statistics for sampling data points

Basic descriptive statistics	Height	Height after normalization	Vibration after normalization
Sample size	1000	1000	1000
Minimum	−0.0080	−3.0311	−1.2653
Maximum	0.0081	3.1394	3.7384
Median	−2e−04	−0.0417	−0.2152
Mean	−9.13e−05	8.0158e−17	4.1178e−15
Standard deviation	0.0026	1.0005	1.0005
Coefficient of variation	−28.5922	1.2482e+16	2.4297e+14
Skewness	0.1782	0.1782	1.0139
Kurtosis	3.1109	3.1109	3.5753

normalization method, which adjusts values measured on different scales to a notionally common scale, is applied to deal with this problem. The intention is that these normalized values allow the comparison of corresponding normalized values in a way that eliminates the effects of certain gross influences. Table 3.2 also shows the descriptive statistics for the height and vibration after normalization.

The data is normalized according to Eq. (3.30):

$$Y = \frac{X - \overline{X}}{s} \tag{3.30}$$

where $\overline{X}$ is the mean of sample and s is the standard deviation of sample.

Results and Discussion

(1)   Comparison analysis of spatial correlation

The spatial correlation theory treats variograms as a measure of variability between measured points instead of the Euclidean distance. The empirical variogram is computed as half the average squared difference on varied lag distance between the data pairs. As shown in Fig. 3.19, the dots characterize the empirical variogram of surface height computed from the 1000 sampled data points.

The fitting process with LS method is used to obtain quantified spatial correlation and theoretical variogram. In Eq. (3.20), the parameters of spherical model are achieved as $[C_0 + C_1, C_0, a] = [1.3839, 0.1644, 197.4504]$. In Eq. (3.21), the parameters of the exponential model are achieved as $[C_0 + C_1, C_0, a] = [2.0492, 0.1330, 495.6395]$. Figure 3.19 also shows the spherical and exponential models' fitting result for the variogram of surface height.

In this case, the spatial variability is assumed to be identical in all directions, and variogram value only increases with the separated distance. It means that two locations close to each other on the surface are more alike, and thus their squared difference is smaller than those that are further apart. From Table 3.3, the variogram of the height value reaches a maximum at 197 mm for the spherical model and at 495 mm for the exponential model before dipping and fluctuating around a sill value. The structural effect ratio in exponential model, compared with the spherical model, has the larger value, which means that the spatial correlation between sample points fitted by exponential model is stronger. Therefore, the predictive value of height by Kriging or CK may theoretically be achieved with optimal minimum error.

The vibration signal value is similarly established using theoretical variogram in spherical and exponential models. Figure 3.20 shows the fitted line for the variograms of the vibration signal value and the cross-variogram of height and

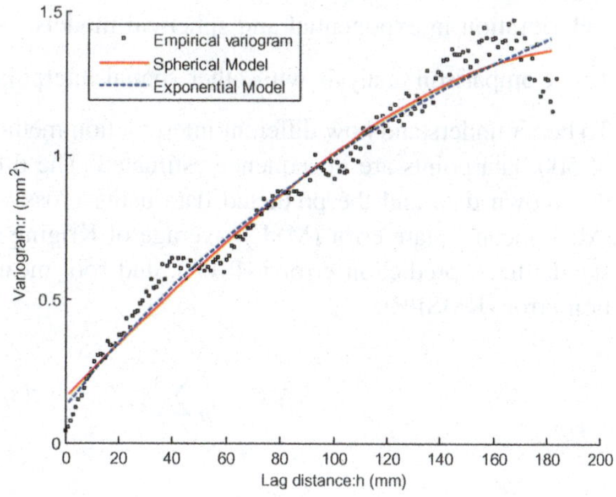

Fig. 3.19 Theoretical variogram of height in spherical model and exponential model. Reprinted from Ref. [77], copyright 2015, with permission from ASME

Table 3.3 Parameters of theoretical variogram and cross-variogram

Type	Variogram	Sill $(C_0 + C_1)$	Nugget (C_0)	Range(a)	Structural effects ratio (%)
Variogram of height	Spherical model	1.3839	0.1644	197.4504	88.12
	Exponential model	2.0492	0.1330	495.6395	93.51
Variogram of vibration	Spherical model	1.1360	0.4449	114.1963	60.84
	Exponential model	1.0198	0.0383	55.6967	96.24
Cross-variogram of height and vibration	Spherical model	2.1535	0.4066	159.9849	81.12
	Exponential model	2.4396	0.2478	240.0733	89.84

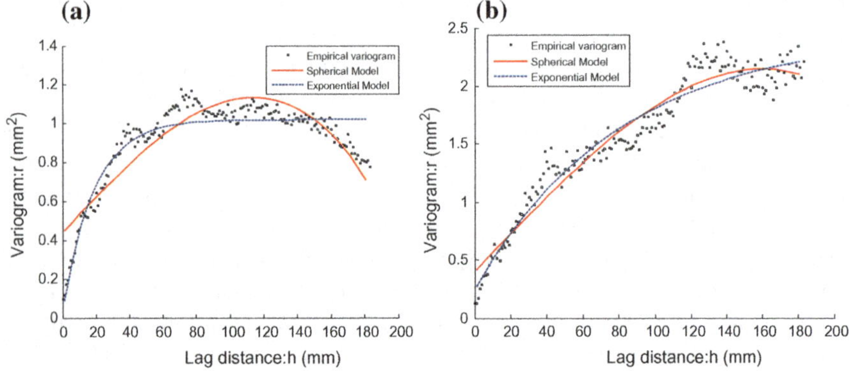

Fig. 3.20 Variogram of vibration and cross-variogram of height and vibration. Reprinted from Ref. [77], copyright 2015, with permission from ASME

vibration. Figure 3.20a is the theoretical variogram of vibration in spherical model and exponential model, and Fig. 3.20b is the theoretical cross-variogram of height and vibration in exponential and spherical models.

(2) Comparison analysis with other spatial interpolation methods

To better understand how different interpolation methods perform, the height values of 500 data points are in sequence estimated. The difference is examined between the known data and the predicted data using cross-validation criteria: Mean error (ME), mean-square error (MSE), average of Kriging standard error (AKSE), mean standardized prediction error (MSPE), and root-mean-square standardized prediction error (RMSPE).

$$ME = \frac{1}{n}\sum_{i=1}^{n}\left[z^*(s_i) - z(s_i)\right] \tag{3.31}$$

$$MSE = \frac{1}{n}\sum_{i=1}^{n}\left[z^{*}(s_i) - z(s_i)\right]^2 \tag{3.32}$$

$$AKSE = \sqrt{\frac{1}{n}\sum_{i=1}^{n}\sigma^2(s_i)} \tag{3.33}$$

$$MSPE = \frac{1}{n}\sum_{i=1}^{n}\frac{Z^{*}(s_i)-Z(s_i)}{\sigma(s_i)} \tag{3.34}$$

$$RMSPE = \sqrt{\frac{1}{n}\sum_{i=1}^{n}\left[\frac{Z^{*}(s_i)-Z(s_i)}{\sigma(s_i)}\right]^2} \tag{3.35}$$

where n is the training sample size, $z(s_i)$ is the true height value, $z^{*}(s_i)$ is the predicted value, and the corresponding Kriging standard error is $\sigma(s_0)$.

The calculated ME is a weak diagnostic for Kriging because it is insensitive to inaccuracies in the variogram. The value of ME also depends on the scale of the data, and is standardized through dividing by the Kriging variance to form the MSPE. An accurate model would have an MSPE close to zero. If the model for the variogram is accurate, then the MSE should equal the Kriging variance, so the RMSPE should be equal to one. If the RMSPE is greater than one, then the variability in the predictions is underestimated and vice versa. AKSE is calculated by the Kriging error variance, which reflects the precision of estimation.

Table 3.4 shows the comparison result after spatial interpolation by 500 estimated points. Since TIN and IDW methods do not have estimation variance error, criterions MSPE and RMSPE cannot be obtained for them. From Table 3.4, some findings are obtained, which are given below:

The unmeasured point can be estimated by nearest measured points using traditional interpolation methods like TIN and IDW, but the result of estimation is less precise than the Kriging methods. The indices of ME and MSE for TIN and IDW are larger than that for the Kriging methods. This is because the number of measured points used for estimation by TIN and IDW is limited and the weights of these measured points are specially fixed.

In comparison of different spatial statistics methods, Kriging method can obtain much more accurate estimation with smaller ME and MSE. Moreover, multivariate

Table 3.4 Comparison result with other spatial interpolation methods

	TIN	IDW	OK	CK
ME	−0.08679	−0.05356	−0.03659	−0.00834
MSE	0.3057	0.2895	0.1889	0.1342
AKSE	–	–	0.4946	0.4215
MSPE	–	–	−0.08148	−0.04880
RMSPE	–	–	0.8186	0.8552

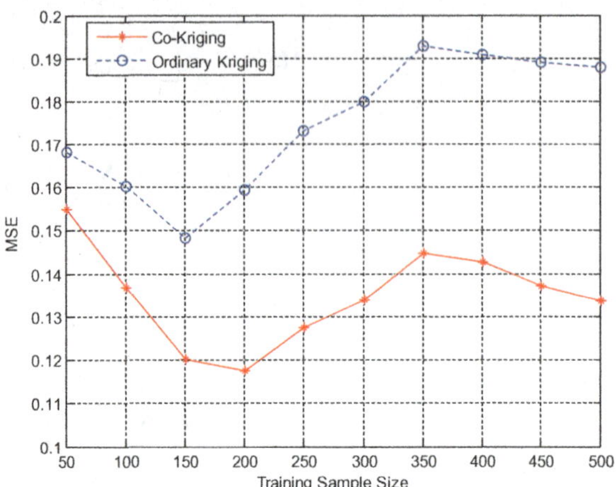

Fig. 3.21 MSE comparisons between CK and OK estimation, Reprinted from Ref. [77], copyright 2015, with permission from ASME

CK method considering the machining conditions has a more accurate estimation by comparing the results of MSPE and RMSPE. According to AKSE, calculated by Kriging error variance, it indicates the uncertainty of the estimation by Kriging methods. Apparently, CK has a much more reliable estimation.

Furthermore, Fig. 3.21 shows the comparison between CK with vibration signal and OK with different training sample sizes. According to the average values of MSE in different numbers of points, CK achieves lower MSE value, which means CK is much more valid for estimating the height value taking vibration into account. On the other hand, with the increasing training sample size, the MSE decreases. When it reaches a specified number of training sample points (150 for OK and 200 for CK), the MSE changes to increase. This phenomenon indicates that appropriate training sample size can reach the most precise estimation by Kriging methods. So Kriging method is also applied to decide the appropriate inspection samples and measure sample size.

(3) Comparison analysis of flatness error estimation with other methods

The flatness errors of the engine block surface are calculated using LS method. The form error of the real height value by 600 measured points is treated as true flatness error, and the general flatness error is calculated using interpolation methods: TIN, IDW, OK, and CK. The inspection process is conducted using 100 sample points from this surface based on a random sampling method, and then interpolation methods are applied to estimate 500 locations in sequence and generate an artificial surface to calculate the flatness error. The 100 sample points are used to calculate the flatness error through the direct method.

The flatness error calculated by CK is 0.01339, which is closer to the true value 0.01336, compared with OK with exponential model (0.01328) and with spherical model (0.01347). Since a rectangular grid is used to represent the predicted surface

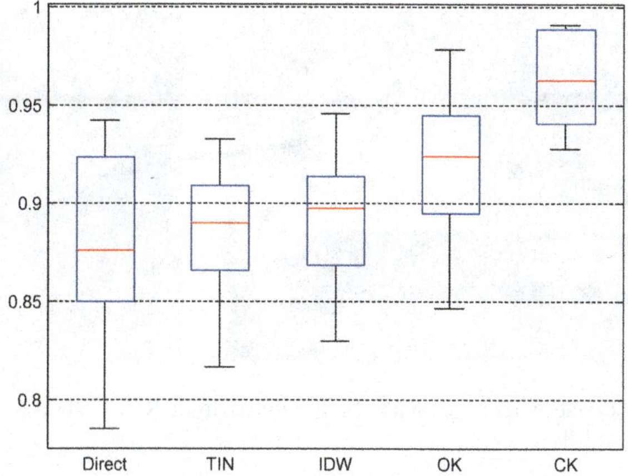

Fig. 3.22 Box–whisker plot of flatness error estimation by different methods. Reprinted from Ref. [77], copyright 2015, with permission from ASME

and calculate the flatness error, the results could be low when the points do not capture the minimum and maximum points. Estimated flatness error is smaller by OK with the exponential model.

Figure 3.22 shows the box–whisker plot of the flatness error estimation by different spatial interpolations methods. 100 samples points are randomly chosen for 50 times to repeat the flatness error estimation by Direct, TIN, IDW, OK, and CK methods. Obviously, CK performs better in flatness error estimation, which is more than 97% of the real value; meanwhile; for OK, the percentage is below 95%.

Figure 3.23 indicates the flatness error estimation by different sample sizes by OK and CK methods. When the sample size is larger, CK and OK tend to be unbiased and CK has a greater convergence speed.

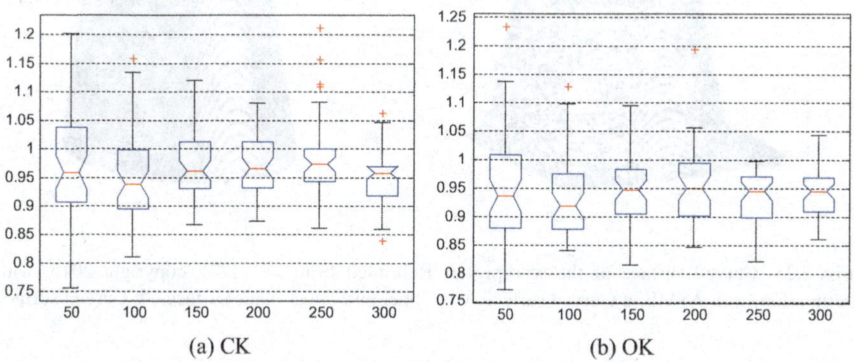

Fig. 3.23 Box–whisker plot of flatness error estimation comparison by different sample sizes. Reprinted from Ref. [77], copyright 2015, with permission from ASME

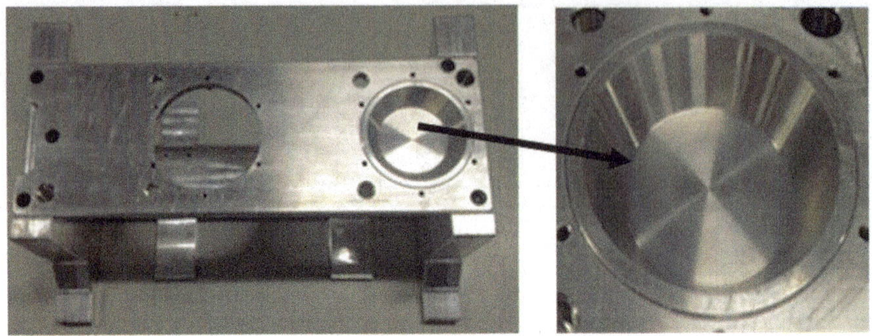

Fig. 3.24 The workpiece for case study II. Reprinted from Ref. [77], copyright 2015, with permission from ASME

3.3.5.2   Case Study II

Experimental Setup

Machining surfaces for this case study are curved surfaces (see Fig. 3.24). Figure 3.25 shows artificial surface with sampling data points.

The height values Z of the curved surface are measured from −40 to 40 mm. 7661 points are measured offline on the curved surface as the true value for the form error estimation. These 7661 measured points are characterized by three dimensions x, y, and z, and the corresponding vibration signal values are obtained by the sensor online during the machining process.

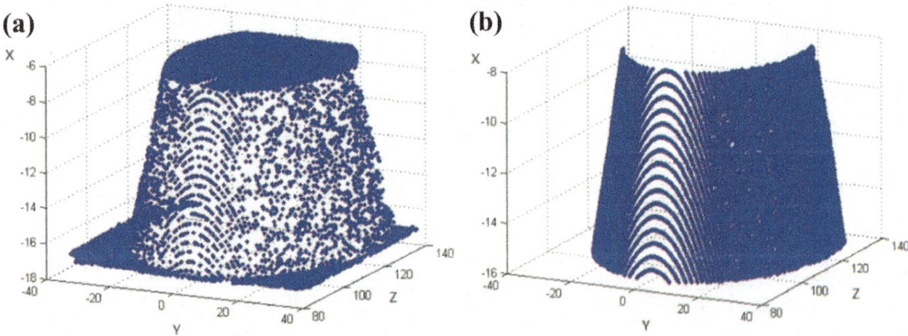

Fig. 3.25 Curved surface of the workpieces, Reprinted from Ref. [77], copyright 2015, with permission from ASME. **a** Curved surface with upper and lower plane surfaces, **b** Curved surface

Results and Discussion

(1) Comparison analysis with other spatial interpolation methods

On the curved surface, 100 points with three dimensions measured offline and with vibration signal measured online are randomly selected as the measured values. OK and CK methods are applied to estimate the height value Z for the resting 7561 locations. The real height values and the estimated height values by OK and CK are calculated by the cross-validation criteria: ME, MSE, AKSE, MSPE, and RMSPE. Table 3.5 shows the result of cross-validation criteria values by OK and CK methods.

Obviously, the CK method considering machining conditions vibration signal can make much more precise estimation. ME, MSE, and MSPE of CK are closer to 0 than those of OK. The uncertainty of Kriging error variance AKSE of CK is less than that of OK, which means the precision of estimation. But the RMSPEs of OK and CK are both less than 1, which means the sample size of initial measured points for estimation is not enough. It needs much more information for Kriging estimation.

Figure 3.26 shows the box–whisker plot result of cross-validation criteria MSE by OK and CK estimations with different initial sample sizes. With the increase of initial sample size for Kriging estimation, the MSE tends to 0. And for CK method, the speed of convergence is much higher than that for OK method.

(2) Comparison analysis of form error estimation with other methods

For the curved surface, the form error estimation is using the MinMax method introduced by Yang and Jackman [76]. It is calculated by the sum of the maximum distance between the upper and lower measured points and the fitting curved surface (see Fig. 3.27), which means the minimum distance between the two enveloping surfaces along the fitting curved surface.

Table 3.6 shows the comparison of form error estimation by direct method, OK and CK methods for the curved surface. The form error estimation of the 7661 measured points is regarded as the true value. The direct method chooses randomly 100, 500, 1000 measured points with three dimensions for the form error estimation. For OK and CK, 100 measured points are randomly chosen for estimation of height values Z in the remaining 7561 locations. The ratio is calculated by the form error of direct, OK, or CK method and that of real measured points.

Table 3.5 Comparison result of cross-validation by OK and CK for case study II

	OK	CK
ME	0.1084	0.0614
MSE	0.3387	0.2587
AKSE	2.4255	1.7615
MSPE	0.0391	0.0165
RMSPE	0.2233	0.3585

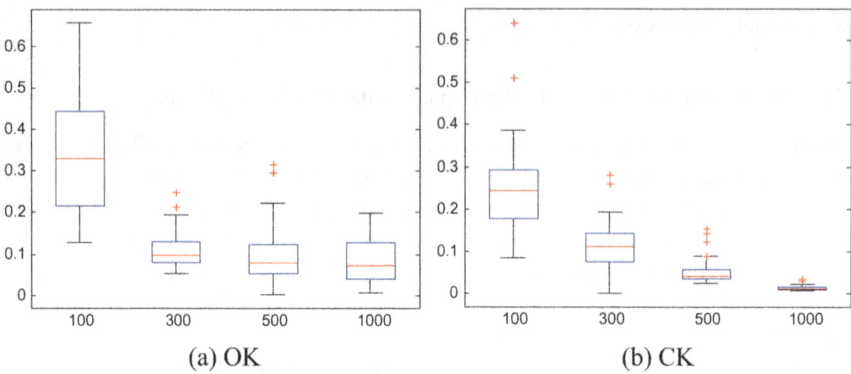

Fig. 3.26 Box–whisker plot of MSE comparison by different sample sizes. Reprinted from Ref. [77], copyright 2015, with permission from ASME

Fig. 3.27 Diagram of form error estimation for curved surface. Reprinted from Ref. [77], copyright 2015, with permission from ASME

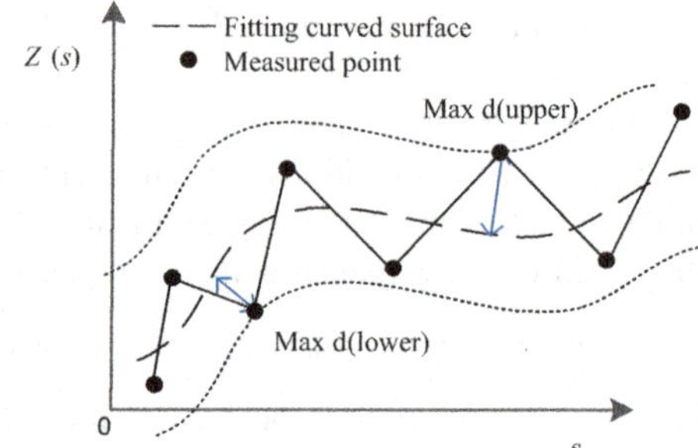

From Table 3.6, it can be found that

(1) With the increase of initial sample size of measured points used to form error estimation, the precision grows from 57 to 62%. But the rate of growth is limited according to the growth of sample size. However, the growth of sample size is at the cost of the measuring cost and efficiency.

(2) The spatial interpolation method can greatly increase the precision of the form error estimation from 57 to 75% even though the number of initial measured points remains low. The improvement for form error estimation is considerable and the measurement cost is reduced.

(3) The proposed multivariate spatial interpolated method still can improve the estimation precision by 2.58% compared to the OK method. This improvement is not large but of great importance for the form error estimation in the manufacturing process.

Table 3.6 Comparison result of form error estimation by OK and CK for case study II

	Real	Direct			OK	CK
Sample size	7661	100	500	1000	100+7561[a]	100+7561[a]
Form error	1.2713e−04	7.3301e−05	7.5525e−05	7.9324e−05	9.6079e−05	9.9355e−05
Ratio (%)	100	57.6583	59.4077	62.3960	75.5754	78.1523

[a]100+7561 indicates 100 initial measured points with 7561 estimated points by OK or CK method

3.3.6   Conclusions

A novel method based on multivariate spatial statistics is developed for more accurate form error estimation. In order to decrease the deviation between the real values and the estimated values, the machining condition data measured online impacting the surface is taken into account. Compared to univariate spatial Kriging method, the multivariate spatial prediction method CK, concerning multiple parameters with different spatial correlations, performs better for form error estimation. The spatial correlation on product surface is characterized by appropriate variogram and cross-variogram.

Through analyzing the simulation cases, CK considering vibration influence could improve the accuracy of flatness error estimation by 10% over OK without considering vibration. For the case studies, the flat surface and curved surface are both illustrated to validate the proposed multivariate spatial statistics CK method. The flatness error estimation on the plan surface by CK also outperforms 10% by direct or TIN/IDW method, and 3–5% by OK. For the form error estimation on the curved surface, CK considering machining conditions can greatly improve the prediction up to 20% compared with the direct method and also performs better than OK.

Appendix

The variance and covariance of the parameters Z_1 and Z_2 at locations s and $s + h$ are described as

$$C_{11}(h) = E[\{Z_1(s) - \mu_1\}\{Z_1(s+h) - \mu_1\}] \tag{A1}$$

$$C_{22}(h) = E[\{Z_2(s) - \mu_2\}\{Z_2(s+h) - \mu_2\}] \tag{A2}$$

$$C_{12}(h) = E[\{Z_1(s) - \mu_1\}\{Z_2(s+h) - \mu_2\}] \tag{A3}$$

where μ_1 and μ_2 are the mean of parameters Z_1 and Z_2.

Therefore, the theoretical calculation of covariance based on limited measured points is described as

$$C_{12}^*(h) = \frac{1}{2N(h)} \sum_{i=1}^{N(h)} \{z_1(s_i)z_2(s_i+h)\} - \frac{1}{2N(h)} \sum_{i=1}^{N(h)} z_1(s_i) \times \frac{1}{2N(h)} \sum_{i=1}^{N(h)} z_2(s_i+h)$$

$$(A4)$$

According to the calculation process of covariance, the variogram and co-variogram are constructed as

$$\gamma_{11}(h) = \frac{1}{2}E\left[\{Z_1(s) - Z_1(s+h)\}^2\right] \tag{A5}$$

$$\gamma_{22}(h) = \frac{1}{2}E\left[\{Z_2(s) - Z_2(s+h)\}^2\right] \tag{A6}$$

$$\gamma_{12}(h) = \frac{1}{2}E[\{Z_1(s) - Z_1(s+h)\}\{Z_2(s) - Z_2(s+h)\}] \tag{A7}$$

And the theoretical co-variogram is calculated as

$$\gamma_{12}^*(h) = \frac{1}{2N(h)} \sum_{i=1}^{N(h)} \{z_1(x_i) - z_1(x_i+h)\}\{z_2(x_i) - z_2(x_i+h)\} \tag{A8}$$

So the co-variogram is constructed as

$$\gamma_{AB}(h) = \frac{1}{2N_n^{AB}(h)} \sum_{i,j \in N_n^{AB}(h)} [A(s_i) - A(s_j)][B(s_i) - B(s_j)], \ h \in R^d, A, B \in \{Z, V_t\}$$

References

1. Jiang X, Scott PJ, Whitehouse DJ, Blunt L (2007) Paradigm shifts in surface metrology. Part II. The current shift. Proc R Soc Math Phys Eng Sci 463(2085):2071–2099
2. Pedonea P, Romanob D (2011) Designing small samples for form error estimation with coordinate measuring machines. Precis Eng 35:262–270
3. Huang Z, Shih AJ, Ni J (2006) Laser interferometry hologram registration for three-dimensional precision measurements. ASME J Manuf Sci Eng 128(4):1006–1013
4. ISO 25178-6 (2010) Geometrical product specifications(GPS)- Surface texture: Areal - Part 6: Classification of methods for measuring surface texture
5. Dong WP, Sullivan PJ, Stout KJ (1992) Comprehensive study of parameters for characterising three-dimensional surface topography I: some inherent properties of parameter variation. Wear 159(2):161–171
6. Dong WP, Sullivan PJ, Stout KJ (1993) Comprehensive study of parameters for characterising three-dimensional surface topography II: statistical properties of parameter variation. Wear 167(1):9–21

7. Dong WP, Sullivan PJ, Stout KJ (1994) Comprehensive study of parameters for characterising three-dimensional surface topography III: parameters for characterising amplitude and some functional properties. Wear 178(1–2):29–43

8. Dong WP, Sullivan PJ, Stout KJ (1994) Comprehensive study of parameters for characterising three-dimensional surface topography IV: parameters for characterising spatial and hybrid properties. Wear 178(1–2):45–60

9. ISO 25178-2 (2012) Geometrical product specifications(GPS)-Surface texture: Areal - Part 2: terms, definitions and surface texture parameters

10. Liao Y, Stephenson DA, Ni J (2010) A multifeature approach to tool wear estimation using 3D workpiece surface texture parameters. ASME J Manuf Sci Eng 132:061008

11. Li YQ, Ni J (2011) B-spline wavelet-based multiresolution analysis of surface texture in end-milling of aluminum. ASME J Manuf Sci Eng 133:011014

12. Tai BL, Stephenson DA, Shih AJ (2011) Improvement of surface flatness in face milling based on 3-D holographic laser metrology. Int J Mach Tools Manuf 51:483–490

13. Suriano S, Wang H, Hu SJ (2012) Sequential monitoring of surface spatial variation in automotive machining processes based on high definition metrology. J Manuf Syst 31:8–14

14. Lee DT, Schachter BJ (1980) Two algorithms for constructing a delaunay triangulation. Int J Parallel Prog 9(3):219–242

15. Haralick RM, Shanmugam K, Dinstein IH (1973) Textural features for image classification. IEEE Trans Syst Man Cybern 3(6):610–621

16. Baraldi A, Parmiggiani F (1995) An investigation of the textural characteristics associated with gray level cooccurrence matrix statistical parameters. IEEE Trans Geosci Remote Sens 33(2):293–304

17. Tien CL, Lyu YR, Jyu SS (2008) Surface flatness of optical thin films evaluated by gray level co-occurrence matrix and entropy. Appl Surf Sci 254(15):4762–4767

18. Tien CL, Lyu YR (2006) Optical surface flatness recognized by discrete wavelet transform and grey level cooccurrence matrix. Meas Sci Technol 17:2299–2305

19. Dutta S, Datta A, Chakladar ND, Pal SK, Mukhopadhyay S, Sen R (2012) Detection of tool condition from the turned surface images using an accurate grey level co-occurrence technique. Precis Eng 36:458–466

20. Du S, Liu C, Huang D (2015) A shearlet-based separation method of 3D engineering surface using high definition metrology. Precis Eng 40:55–73

21. Du S, Liu C, Xi L (2014) A selective multiclass support vector machine ensemble classifier for engineering surface classification using high definition metrology. J Manuf Sci Eng 137 (1):011003

22. Du SC, Huang DL, Wang H (2015) An adaptive support vector machine-based workpiece surface classification system using high-definition metrology. IEEE Trans Instrum Meas 64 (10):2590–2604

23. Guo P, Lu Y, Pei PC, Ehmann KF (2014) Fast generation of micro-channels on cylindrical surfaces by elliptical vibration texturing. J Manuf Sci Eng Trans ASME 136(4)

24. Nguyen HT, Wang H, Hu SJ (2013) Characterization of cutting force induced surface shape variation in face milling using high-definition metrology. J Manuf Sci Eng Trans ASME 135 (4)

25. Rao P, Bukkapatnam S, Beyca O, Kong ZY, Komanduri R (2014) Real-time identification of incipient surface morphology variations in ultraprecision machining process. J Manuf Sci Eng Trans ASME 136(2)

26. Wang M, Ken T, Du S, Xi L (2015) Tool wear monitoring of wiper inserts in multi-insert face milling using three-dimensional surface form indicators. J Manuf Sci Eng 137(3):031006

27. Wang M, Xi LF, Du SC (2014) 3D surface form error evaluation using high definition metrology. Precis Eng J Int Soc Precis Eng Nanotechnol 38(1):230–236

28. Zhu XY, Ding H, Wang MY (2004) Form error evaluation: an iterative reweighted least squares algorithm. J Manuf Sci Eng Trans ASME 126(3):535–541

29. Badar MA, Raman S, Pulat PS (2005) Experimental verification of manufacturing error pattern and its utilization in form tolerance sampling. Int J Mach Tools Manuf 45(1):63–73

30. Raghunandan R, Rao PV (2007) Selection of an optimum sample size for flatness error estimation while using coordinate measuring machine. Int J Mach Tools Manuf 47(3–4):477–482
31. Dowling MM, Griffin PM, Tsui KL, Zhou C (1997) Statistical issues in geometric feature inspection using coordinate measuring machines. Technometrics 39(1):3–17
32. Yan Z, Menq C (1995) Uncertainty analysis for coordinate estimation using discrete measurement data. Proc Manuf Sci Eng 1:595–616
33. Yeh K, Ni J, Hu S (1994) Adaptive sampling and identification of feature deformation for tolerance evaluation using coordinate measuring machines SM Wu, Manufacturing Research Laboratory, Department of Mechanical Engineering and Applied Mechanics, University of Michigan, Ann Arbor, MI, Technical Report
34. Yang B-D, Menq C-H (1993) Compensation for form error of end-milled sculptured surfaces using discrete measurement data. Int J Mach Tools Manuf 33(5):725–740
35. Cho N, Tu J (2001) Roundness modeling of machined parts for tolerance analysis. Precis Eng J Int Soc Precis Eng Nanotechnol 25(1):35–47
36. Desta MT, Feng HY, OuYang DS (2003) Characterization of general systematic form errors for circular features. Int J Mach Tools Manuf 43(11):1069–1078
37. Henke RP, Summerhays KD, Baldwin JM, Cassou RM, Brown CW (1999) Methods for evaluation of systematic geometric deviations in machined parts and their relationships to process variables. Precis Eng J Am Soc Precis Eng 23(4):273–292
38. Xia HF, Ding Y, Wang J (2008) Gaussian process method for form error assessment using coordinate measurements. IIE Trans 40(10):931–946
39. Walker E, Wright SP (2002) Comparing curves using additive models. J Qual Technol 34(1):118–129
40. Whittle P (1954) On stationary processes in the plane. Biometrika, pp 434–449
41. Sayles R, Thomas T (1977) The spatial representation of surface roughness by means of the structure function: a practical alternative to correlation. Wear 42(2):263–276
42. Colosimo BM, Semeraro Q, Pacella M (2008) Statistical process control for geometric specifications: on the monitoring of roundness profiles. J Qual Technol 40(1):1–18
43. Kanada T, Suzuki S (1993) Evaluation of minimum zone flatness by means of nonlinear optimization techniques and its verification. Precis Eng 15(2):93–99
44. Murthy T, Abdin S (1980) Minimum zone evaluation of surfaces. Int J Mach Tool Des Res 20(2):123–136
45. Wang MY (1992) Minimum zone evaluation of form tolerances
46. Suriano S, Wang H, Shao CH, Hu SJ, Sekhar P (2015) Progressive measurement and monitoring for multi-resolution data in surface manufacturing considering spatial and cross correlations. IIE Trans 47(10):1033–1052
47. Li J (2008) A review of spatial interpolation methods for environmental scientists, Record—Geoscience Australia, 137
48. Chen X, Ankenman BE, Nelson BL (2013) Enhancing stochastic Kriging metamodels with gradient estimators. Oper Res 61(2):512–528
49. Kleijnen JPC, Mehdad E (2014) Multivariate versus univariate Kriging metamodels for multi-response simulation models. Eur J Oper Res 236(2):573–582
50. Morimoto Y, Suzuki N, Kaneko Y, Isobe M (2014) Vibration control of relative tool-spindle displacement for computer numerically controlled lathe with pipe frame structure. J Manuf Sci Eng Trans ASME 136(4)
51. Kleijnen JPC (2009) Kriging metamodeling in simulation: a review. Eur J Oper Res 192(3):707–716
52. Yang TH, Jackman J (2000) Form error estimation using spatial statistics. J Manuf Sci Eng Trans ASME 122(1):262–272
53. Weckenmann A, Heinrichowski M, Mordhorst H-J (1991) Design of gauges and multipoint measuring systems using coordinate-measuring-machine data and computer simulation. Precis Eng 13(3):203–207

54. Yan D, Popplewell N, Balkrishnan S, Kaye J (1996) On-line prediction of surface roughness in finish turning. Eng Des Autom 2:115–126
55. Lu C (2008) Study on prediction of surface quality in machining process. J Mater Process Technol 205(1–3):439–450
56. Roth JT, Mears L, Djurdjanovic D, Yang XP, Kurfess T (2007) Quality and inspection of machining operations: review of condition monitoring and CMM inspection techniques 2000 to present. In: Proceedings of the ASME international conference on manufacturing science and engineering, pp 861–872 (2007)
57. Benardos PG, Vosniakos GC (2003) Predicting surface roughness in machining: a review. Int J Mach Tools Manuf 43(8):833–844
58. Dowling MM, Griffin PM, Tsui KL, Zhou C (1995) Comparison of the orthogonal least squares and minimum enclosing zone methods for form error estimation. Manuf Rev 8 (2):120–138
59. Matheron G (1973) The intrinsic random functions and their applications. Adv Appl Probab 5 (3):439–468
60. Cressie N (1992) Statistics for spatial data. Terra Nova 4(5):613–617
61. Zhang RD (2005) Spatial variogram theory and its application. Science and Technology Press, Bei Jing
62. Isaaks EH, Srivastava RM (1989) An introduction to applied geostatistics. Oxford University Press, New York
63. Deutsch CV, Journel AG (1992) Geostatistical software library and user's guide. New York 119:147
64. Goovaerts P (1997) Geostatistics for natural resources evaluation. Oxford University Press, New York
65. Ahmed S, De Marsily G (1987) Comparison of geostatistical methods for estimating transmissivity using data on transmissivity and specific capacity. Water Resour Res 23 (9):1717–1737
66. Gelfand AE, Diggle P, Guttorp P, Fuentes M (2010) Handbook of spatial statistics. CRC Press
67. Kennedy MC, O'Hagan A (2000) Predicting the output from a complex computer code when fast approximations are available. Biometrika 87(1):1–13
68. Qian PZ, Wu CJ (2008) Bayesian hierarchical modeling for integrating low-accuracy and high-accuracy experiments. Technometrics 50(2):192–204
69. Zhou Q, Qian PZG, Zhou SY (2011) A simple approach to emulation for computer models with qualitative and quantitative factors. Technometrics 53(3):266–273
70. Clark I, Basinger KL, Harper WV (1987) A novel approach to co-kriging. In: Geostatistical, sensitivity, and uncertainty methods for ground-water flow and radionuclide transport modeling, San Francisco, pp 473–493
71. Rehman SU, Shapiro A (1996) An integral transform approach to cross-variograms modeling. Comput Stat Data Anal 22(3):213–233
72. Armstrong M, Diamond P (1984) Testing variograms for positive-definiteness. Math Geol 16 (4):407–421
73. Myers DE (1994) Spatial interpolation: an overview. Geoderma 62(1–3):17–28
74. Myers DE (1982) Matrix formulation of co-kriging. J Int Assoc Math Geol 14(3):249–257
75. Myers DE (1983) Estimation of linear combinations and co-kriging. J Int Assoc Math Geol 15 (5):633–637
76. Yang TH, Jackman J (1997) A probabilistic view of problems in form error estimation. J Manuf Sci Eng Trans ASME 119(3):375–382
77. Du S, Fei L (2015) Co-Kriging method for form error estimation incorporating condition variable measurements. J Manuf Sci Eng

Chapter 4
Surface Filtering

4.1 A Brief History of Surface Filtering

4.1.1 M-System and E-System

Engineering surfaces are considered as having long-range deviations called form, superimposed on more general curvature called waviness, and fine texture called roughness. The motivation for such classification comes from the fact that form, waviness, and roughness affect part functional behavior in different ways. Filtration is a process to partition a surface into different components and extract the surface features of interest. It has always been important in surface metrology, and it has a significant aspect of surface topography analysis. There are two competing filter systems: the M-system (linear and robust filters) and the E-system (morphological and segmentation filter).

The M-system began with the basic 2CR filter, which was implemented as a two-stage capacitor–resistance network. The early 2CR filter badly distorted the profile at the ends, and the Gaussian filter was proposed as the standardized phase-corrected profile filter in the M-system subsequently. The French automobile industry developed a purely graphical approach to filtration called roughness and waviness (R&W). They proposed a new concept named "Motif", which means the portion of a profile between local peaks. The approach was to determine the envelope of an operator on the profile. The deviations from the envelope would be the fine texture. This was the first attempt at a simulation of the E-system.

With the advent of many filtering methods, the M-system is the de facto filtering choice and the E-system is applied as part of a larger group of techniques referred to as morphological filters. The M-system consists of linear filter and robust filter, and the E-system consists of morphological filter and segmentation filter. The linear filter, also called the mean line filter, includes Gaussian filter, Spline filter, and Spline wavelet filter. The robust filter includes the robust Gaussian filter and the robust Spline filter. Morphological filter, also called the envelope filter, includes the

© Springer Nature Singapore Pte Ltd. 2019
S. Du and L. Xi, *High Definition Metrology Based Surface Quality Control and Applications*, https://doi.org/10.1007/978-981-15-0279-8_4

disk and horizontal line-segment filters, and scale-space techniques. Segmentation filter is for adopted Motifs to partition a profile into portions according to specific rules.

Three new filtering methods are developed to complete the current geometrical product specification. The proposed filtering methods are described in detail in Sects. 4.2, 4.3, and 4.4.

4.1.2   Current International Standards

In 1996, ISO 11562 is first published to describe the metrological characteristics of phase correct filters for the measurement of surface profiles. In particular, it specifies how to separate the long and short wave content of a surface profile. After long-term use and continuous improvement, it is replaced by ISO 16610 in 2011. In ISO 16610, the filtering approaches can be roughly classified into two categories: profile filter and areal filter. In each category, there are three kinds of filters including linear filter, robust filter, and morphological filter, which is corresponding to the M-system and the E-system.

Up to now, most parts of ISO 16610 have been published which is shown in Table 4.1, under the general title geometrical product specifications (GPSs)—Filtration.

Table 4.1 Published parts of ISO 16610

No.	Title
Part 1	Overview and basic concepts
Part 20	Linear profile filters: Basic concepts
Part 21	Linear profile filters: Gaussian filters
Part 22	Linear profile filters: Spline filters
Part 28	Profile filters: End effects
Part 29	Linear profile filters: Spline wavelets
Part 30	Robust profile filters: Basic concepts
Part 31	Robust profile filters: Gaussian regression filters
Part 32	Robust profile filters: Spline filters
Part 40	Morphological profile filters: Basic concepts
Part 41	Morphological profile filters: Disk and horizontal line-segment filters
Part 49	Morphological profile filters: Scale-space techniques
Part 60	Linear areal filters: Basic concepts
Part 61	Linear areal filters: Gaussian filters
Part 71	Robust areal filters: Gaussian regression filters
Part 85	Morphological areal filters: Segmentation

4.2   A Shearlet-Based Filtering Method for 3D Engineering Surface

4.2.1   Introduction

The analysis of surface texture has gained more and more important in the fields of mechanical manufacturing during recent years, since the functional behavior of a machined part is influenced by its surface texture. Additionally, the analysis of surface texture is considered as an important element in providing feedback on the manufacturing process. The relationship between surface texture, part functionality, and manufacturing process is the main reason for the description and analysis of surface texture.

The surface texture on an engineering surface is defined to have three components from the smaller to the larger scale: roughness, waviness, and form. It is well recognized that different components have different impacts on the functional performance of the parts. To be specific, roughness reveals the irregularities on a surface left after manufacturing and thus can be used to detect errors in the material removal process, and also it has great impact on the part functionality such as friction and wear. Waviness, which may be generated by the vibration of the manufacturing system, has influence on the vibration resistance and tightness of the parts. The form, produced by the poor performance of the machine, can directly affect the assembling of the parts. Therefore, the motivation for classifying the above three components derives from the fact that roughness, waviness, and form have different origins and affect part functionality in different ways. Separating a surface profile into different frequency components is of great significance and an important aspect of surface texture analysis.

Digital filtering is a common practice to realize the above separation process. Filtering of engineering surface has been a hot research topic since it is an important tool for surface texture analysis. Recent advances in filtering techniques are reviewed by Raja et al. [1]. The traditional filtering methods such as 2CR filter and Gaussian filter are first studied, and the Gaussian filter is the most widely used standard filtering technique [2]. However, it is well known that the Gaussian filter causes excessive distortion called "end effect" at the boundaries of the surface profile. Besides, it is not robust against outliers. To solve these problems, on the one hand, some modified methods based on Gaussian filter like regression and robust regression Gaussian filters [3] are proposed; on the other hand, some newer filtering methods like spline filter [4], robust spline filter [5], morphological filter [6], and wavelet-based filters [7, 8] are proposed.

A comprehensive comparison of the above methods has been made by Raja et al. in [9], and wavelet-based filtering is considered to be a very active research topic and has been barely scratched the tip of the iceberg. Different from the previous filtering methods, wavelet-based filters can provide multi-scale analysis, due to the reason that they can divide a given surface profile into different frequency components and study each component with a resolution matched to its scale. Hence, wavelet-based filters are not limited to separate the surface profile into three components (roughness, waviness, and form), and they can separate the profile into fine bandwidths that better reflect the changes of a manufacturing process. In recent years, several researchers have developed wavelet techniques and applied them to analyze engineering surfaces. Fu et al. [10] compared different wavelet bases and recommended Bior6.8 and Coif4 due to their transmission characteristics of the corresponding filters. Jiang et al. [11] developed a lifting wavelet representation for surface characterization, and different frequency components of the surface can be extracted according to the intended requirements of functional analysis. Josso et al. [12] proposed a frequency normalized wavelet transform for surface roughness analysis and characterization.

However, it is worth noting that, in most of the previous work, the engineering surfaces are characterized and analyzed with two-dimensional (2D) parameters since the stylus profilometer is invented in 1930s and widely used since then. Nevertheless, the 2D parameters are inadequate or incapable to characterize the surface texture in three dimensions (3D). With the rapid development of 3D measurement technology, a variety of sophisticated devices and techniques have been developed and applied for 3D measurement of engineering surfaces. For example, a novel measurement technology called noncontact high-definition metrology (HDM) has been adopted for 3D inspection of an entire surface as HDM can generate a surface height map of millions of data points within seconds [13]. In light of this, more appropriate surface characterization methods have been developed to obtain adequate information about 3D surface texture. Blunt and Jiang [14] have made efforts in extending the 2D parameters to characterize 3D surface, and the ASME B46.1 (American Society of Mechanical Engineers 2002) has also included some 3D parameters in its standard [15].

Likewise, it is necessary that the one-dimensional wavelet-based filter techniques are reasonably extended to two-dimensional ones. The discrete wavelet transform (DWT) is a widely used wavelet method in digital signal and image analysis, and it can be easily extended from one dimension to the condition of two dimensions. The application of 2D DWT in analyzing 3D engineering surfaces can be found in [16, 17]. Zeng et al. [18] adopted two-dimensional dual-tree complex wavelet transform (2D DT-CWT) to separate engineering surfaces, which is more superior to 2D DWT in the aspects of shift-invariance and directional selectivity. However, both 2D DWT and 2D DT-CWT are fixed on directions. In other words, they both have only a few directions, which impair their performance in the filtering of 3D surfaces. In order to overcome the limitation of traditional 2D wavelets, the directional sensitivity should be increased. In light of this, a theory called multi-scale geometric analysis (MGA) has been proposed and is suitable for analyzing 3D surfaces. Recently, a new MGA tool called shearlets [19] was proposed. From the view of approximation theory, shearlets form a Parseval frame of well-localized waveforms at various scales and directions, which are the "true" sparse representation for 3D surfaces. The non-subsampled shearlet transform (NSST) proposed by Easley et al. [20] is applied to analyze engineering surfaces. NSST combines the non-subsampled pyramid transform with several different shearing filters, which satisfies the property of shift-invariance. Lacking of shift-invariance means that small shifts in the input signals can induce major variations in the distribution of energy between wavelet coefficients at different scales. NSST possesses several nice properties including fully shift-invariance, multi-scale, and multi-direction, which make it a powerful tool in the analysis of 3D surfaces. It is worth noting that although shearlets have been used in image fusion [21], image denoising [20], and image classification [22], there is little effort made on the applications of using shearlets to filter and characterize 3D engineering surfaces.

4.2.2   The Construction of Shearlets

The shearlet transform is built on the theory of composite wavelets, which is introduced in [19, 20] and provides an effective approach for combining geometry and multi-scale analysis by utilizing the classical theory of affine systems. In two dimensions, the affine systems with composite dilations are defined as follows:

$$\left\{\psi_{j,l,k}(x)\right\} = \left\{|\det A|^{j/2}\psi(B^l A^j x - k) : j, l \in Z, k \in Z^2\right\} \tag{4.1}$$

where $\psi \in L^2(R^2)(L^2(R^2)$ is the two-dimensional Hilbert space) and denotes the basic function, and the anisotropic dilation matrix A and the direction matrix B are both 2×2 invertible matrices and $|\det B| = 1$. j, l, k are scale, direction, and shift parameter, respectively; x denotes the input signal; and Z denotes the set of all integers.

If $\psi_{j,l,k}(x)$ forms a Parseval frame for $L^2(R^2)$, then the elements of this system are called composite wavelets. The dilation matrices A^j refer to scale transformations, while the matrices B^l refer to geometrical transformations, such as rotations and shear. Shearlets are a special example of composite wavelets in $L^2(R^2)$, where $A = A_0$ is the anisotropic dilation matrix and $B = B_0$ is the shear matrix, which are given by

$$A_0 = \begin{pmatrix} 4 & 0 \\ 0 & 2 \end{pmatrix}, \quad B_0 = \begin{pmatrix} 1 & 1 \\ 0 & 1 \end{pmatrix} \tag{4.2}$$

For any $\xi = (\xi_1, \xi_2) \in \hat{R}^2, \xi_1 \neq 0$, let

$$\hat{\psi}^{(0)}(\xi) = \hat{\psi}^{(0)}(\xi_1, \xi_2) = \hat{\psi}_1(\xi_1)\hat{\psi}_2\left(\frac{\xi_2}{\xi_1}\right) \tag{4.3}$$

where $\hat{\psi}_1, \hat{\psi}_2$ are the Fourier transforms of ψ_1, ψ_2, respectively, and $\hat{\psi}_1, \hat{\psi}_2 \in C^\infty(\hat{R})$, supp $\hat{\psi}_1 \subset \left[-\frac{1}{2}, -\frac{1}{16}\right] \cup \left[\frac{1}{16}, \frac{1}{2}\right]$ and supp $\hat{\psi}_2 \subset [-1, 1]$ ("supp" denotes support region).

Equation (4.3) denotes that $\hat{\psi}^{(0)}$ is a compactly supported C^∞ function with support in $\left[-\frac{1}{2}, \frac{1}{2}\right]^2$. In addition, assume that

$$\sum_{j \geq 0} |\hat{\psi}_1(2^{-2j}\omega)|^2 = 1 \text{ for } |\omega| \geq \frac{1}{8} \tag{4.4}$$

And, for each $j \geq 0$,

$$\sum_{l=-2^j}^{2^j-1} |\hat{\psi}_2(2^j\omega - l)|^2 = 1 \text{ for } |\omega| \leq 1 \tag{4.5}$$

The conditions on the support of $\hat{\psi}_1$ and $\hat{\psi}_2$ imply that the functions $\hat{\psi}_{j,l,k}$ have frequency support:

$$\text{Supp } \hat{\psi}_{j,l,k} \subset \left\{(\xi_1, \xi_2) : \xi_1 \in \left[-2^{2j-1}, -2^{2j-4}\right] \cup \left[2^{2j-4}, 2^{2j-1}\right], |\frac{\xi_2}{\xi_1} + l2^{-j}| \leq 2^{-j}\right\} \tag{4.6}$$

Thus, each element $\hat{\psi}_{j,l,k}$ is supported on a pair of trapezoids, contained in a box of approximate size $2^{2j} \times 2^j$, oriented along lines of slope $l2^{-j}$ (As shown in Fig. 4.1a).

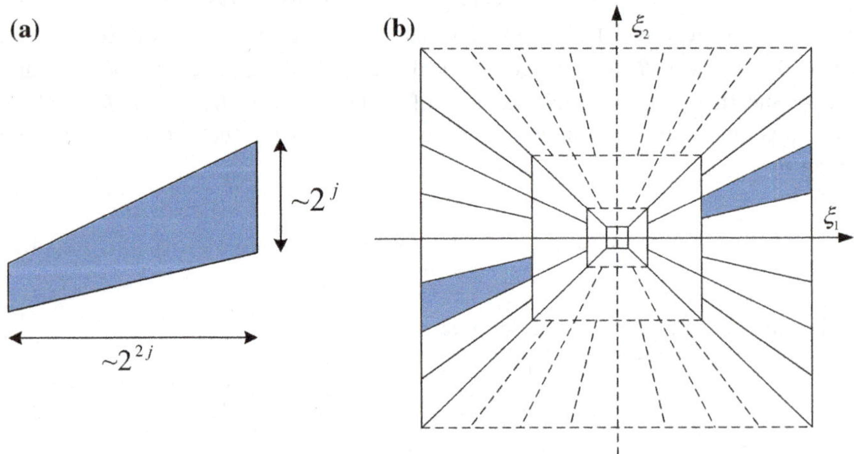

Fig. 4.1 **a** The frequency support of a shearlet **b** The tiling of the frequency plane induced by shearlets. Reprinted from Ref. [33], copyright 2014, with permission from Elsevier

From Eqs. (4.4) and (4.5), it follows that

$$\sum_{j\geq 0}\sum_{l=-2^j}^{2^j-1}|\hat{\psi}^{(0)}(\xi A_0^{-j}B_0^{-l})|^2 = \sum_{j\geq 0}\sum_{l=-2^j}^{2^j-1}|\hat{\psi}_1(2^{-2j}\xi_1)|^2|\hat{\psi}_2(2^j\frac{\xi_2}{\xi_1}-l)|^2 = 1 \quad (4.7)$$

for $(\xi_1,\xi_2) \in D_0$, where D_0 is the horizontal cone: $D_0 = \{(\xi_1,\xi_2) \in \hat{R}^2 : |\xi_1| \geq \frac{1}{8}, |\frac{\xi_2}{\xi_1}| \leq 1\}$. This means that the functions $\hat{\psi}^{(0)}(\xi A_0^{-j}B_0^{-l})$ form a tiling of D_0, which is demonstrated in Fig. 4.1b (the tiling of D_0 is illustrated in solid line while the tiling of D_1 in dashed line). This property, together with the fact that $\hat{\psi}^{(0)}$ is supported in $\left[-\frac{1}{2},\frac{1}{2}\right]^2$, denotes that the collection defined by

$$\left\{\psi_{j,l,k}^{(0)}(x) = 2^{\frac{3j}{2}}\psi^{(0)}(B_0^l A_0^j x - k) : j \geq 0, -2^j \leq l \leq 2^j - 1, k \in Z^2\right\} \quad (4.8)$$

is a Parseval frame for $L^2(D_0)^{\vee} = \{f \in L^2(R^2) : \operatorname{supp} \hat{f} \subset D_0\}$.

Similarly, a Parseval frame can be constructed for $L^2(D_1)^{\vee}$, where D_1 is the vertical cone: $D_1 = \left\{(\xi_1,\xi_2) \in \hat{R}^2 : |\xi_2| \geq \frac{1}{8}, |\frac{\xi_1}{\xi_2}| \leq 1\right\}$. Let

$$A_1 = \begin{pmatrix} 2 & 0 \\ 0 & 4 \end{pmatrix}, \qquad B_1 = \begin{pmatrix} 1 & 0 \\ 1 & 1 \end{pmatrix} \quad (4.9)$$

and $\psi^{(1)}$ be given by

$$\hat{\psi}^{(1)}(\xi) = \hat{\psi}^{(1)}(\xi_1,\xi_2) = \hat{\psi}_1(\xi_2)\hat{\psi}_2\left(\frac{\xi_1}{\xi_2}\right) \quad (4.10)$$

where $\hat{\psi}_1, \hat{\psi}_2$ are defined as above. Then the collection defined by

$$\left\{\psi_{j,l,k}^{(1)}(x) = 2^{\frac{3j}{2}}\psi^{(1)}(B_1^l A_1^j x - k) : j \geq 0, -2^j \leq l \leq 2^j - 1, k \in Z^2\right\} \quad (4.11)$$

is a Parseval frame for $L^2(D_1)^{\vee}$.

Finally, let $\hat{\varphi} \in C_0^{\infty}(R^2)$ be chosen to satisfy

$$G(\xi) = |\hat{\varphi}(\xi)|^2 + \sum_{j\geq 0}\sum_{l=-2^j}^{2^j-1}|\hat{\psi}^{(0)}(\xi A_0^{-j}B_0^{-l})|^2\chi_{D0}(\xi)$$

$$(4.12)$$

$$+ \sum_{j\geq 0}\sum_{l=-2^j}^{2^j-1}|\hat{\psi}^{(1)}(\xi A_0^{-j}B_0^{-l})|^2\chi_{D1}(\xi) = 1, \text{ for } \xi \in \hat{R}^2$$

where χ_D is the indicator function of the set D. This denotes that the support of $\hat{\varphi}$ is contained in $\left[-\frac{1}{8},\frac{1}{8}\right]^2$, with $|\hat{\varphi}(\xi)| = 1$ for $\xi \in \left[-\frac{1}{16},\frac{1}{16}\right]^2$, and the set $\{\varphi_k(x) : k \in Z^2\}$, defined by $\varphi_k(x) = \varphi(x - k)$, is a Parseval frame. Thus, we have the following result:

Theorem 1 Let $\varphi_k(x) = \varphi(x - k)$ and $\psi_{j,l,k}^{(d)}(x) = 2^{\frac{3j}{2}}\psi^{(d)}(B_d^l A_d^j x - k)$, where φ, ψ are given as above, then the collection of shearlets:

$$\{\varphi_k(x) : k \in Z^2\} \cup \left\{\psi_{j,l,k}^{(d)}(x) : j \geq 0, -2^j + 1 \leq l \leq 2^j - 2, k \in Z^2, d = 0, 1\right\}$$

$$\cup \left\{\tilde{\psi}_{j,l,k}^{(d)}(x) : j \geq 0, l = -2^j, 2^j - 1, k \in Z^2, d = 0, 1\right\}$$

$$(4.13)$$

is a Parseval frame for $L^2(R^2)$, where $\hat{\tilde{\psi}}_{j,l,k}^{(d)}(\xi) = \hat{\psi}_{j,l,k}^{(d)}(\xi)\chi_{D_d}(\xi)$.

For $d = 0, 1$, the shearlet transform is mapping $f \in L^2(R^2)$ into the elements $\left\langle f, \psi_{j,l,k}^{(d)}\right\rangle$, where $j \geq 0, -2^j \leq l \leq 2^j - 1, k \in Z^2$.

4.2.3   *The Procedure of the Proposed 3D Surface Separation Method*

4.2.3.1   **Overview of the Proposed Method**

This section provides an overview with respect to the separation of engineering surfaces. The developed method consists of two modules—decomposition of the original surface with NSST and reconstruction of the three surface components (form, waviness, roughness) with inverse NSST. In Module 1, an L-level decomposition is conducted on the original surface to obtain coefficients of shearlet sub-bands in different directions and scales, and these shearlet coefficients in turn contain the texture information of the original surface at different scales with different directions. In Module 2, surface components at different levels are reconstructed by implementing an inverse NSST on the shearlet coefficients and combined into three surface components based on the cutoff wavelengths. The procedure of Module 1 is described in detail in Sect. 4.2.3.2, while the procedure of Module 2 is elaborated in Sect. 4.2.3.3, it can be seen in Fig. 4.2.

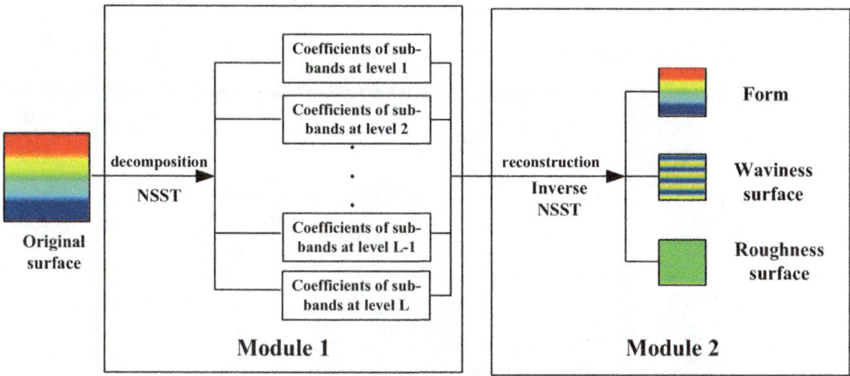

Fig. 4.2 The architecture of the proposed 3D surface separation method. Reprinted from Ref. [33], copyright 2014, with permission from Elsevier

4.2.3.2   The Procedure of Surface Decomposition with NSST

The discrete shift-invariant shearlet transform can be implemented by two steps: multi-scale separation and directional localization. Alien from the standard shearlet transform, the multi-scale separation of shift-invariant shearlet transform is implemented by non-subsampled pyramid filter scheme that is efficient to add shift-invariance to the standard shearlet transform. This also explains why the shift-invariant shearlet transform is called non-subsampled shearlet transform (NSST). The procedure of NSST described above can be summarized as follows:

(1) Initialization: $j = L$ (L is the decomposition level); f is the input image;
(2) By applying non-subsampled Laplacian Pyramid filter (maximally flat filters are adopted as pyramid filter), the original image f is decomposed into a low-pass filtered image f_a^j (image of approximations at level j) and a high-pass filtered image f_d^j (image of details at level j);
(3) $\hat{f}_d^j$ is calculated on a pseudo-polar grid, and a matrix Pf_d^j is generated;
(4) The matrix Pf_d^j is processed by a band-pass filtering;
(5) The Cartesian sampled values are re-assembled directly, and then the inverse two-dimensional fast Fourier transform (FFT) is calculated. Then, the coefficients of shearlets at level j are obtained.
(6) $j = j - 1$, repeat step (2)–(6) until $j = 1$.

In the above procedure, the function of step (2) is to decompose the two-dimensional signals into components of different scales, and the role of step (3)–(5) is to obtain the sub-bands of different orientations. To illustrate this

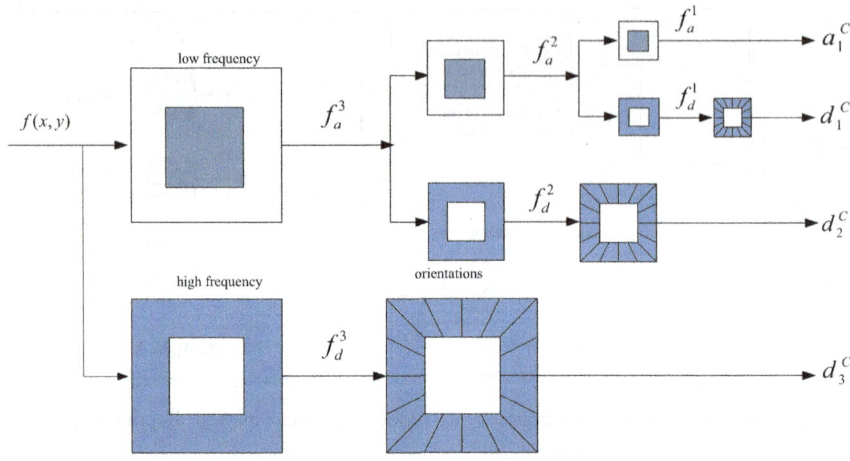

Fig. 4.3 A three-level decomposition using NSST. Reprinted from Ref. [33], copyright 2014, with permission from Elsevier

procedure, a three-level decomposition using NSST is presented in Fig. 4.3. Here, a_1^C is the low-frequency shearlet coefficient, while $d_j^C (j = 1, 2, \ldots, L)$ are the high-frequency shearlet coefficients. Since directional filtering is implemented on detail image at each level, the expression of d_j^C is obtained as follows:

$$d_j^C = \left(d_j^1, d_j^2, \ldots, d_j^{n_j} \right), j = 1, 2, \ldots, L \qquad (4.14)$$

where n_j is the number of directions at level j. As j increases, the texture information contained in the high-frequency shearlet coefficients at level j becomes finer and finer.

Figure 4.4 illustrates a two-level decomposition of the "Zoneplate" image using non-subsample shearlet transform. The direction vector is $\vec{n} = (n_1, n_2) = (4, 6)$. The first-level decomposition generates four-directional sub-bands, and the second-level decomposition generates six-directional sub-bands. It should be noted that the directions of sub-bands are not fixed at each level, and a variety of directions can be obtained through shearlet transform.

4.2.3.3   The Procedure of Surface Reconstruction with INSST

Engineering surfaces are regarded as having fine texture called roughness, superimposed on more general curvature called waviness, and long-range deviations called form. In this way, an engineering surface $S(x, y)$ can be separated as follows:

$$S(x, y) = R(x, y) + W(x, y) + F(x, y) \qquad (4.15)$$

Fig. 4.4 **a** The "Zoneplate" image **b** coefficients of four directions at the first level **c** coefficients of six directions at the second level. Reprinted from Ref. [33], copyright 2014, with permission from Elsevier

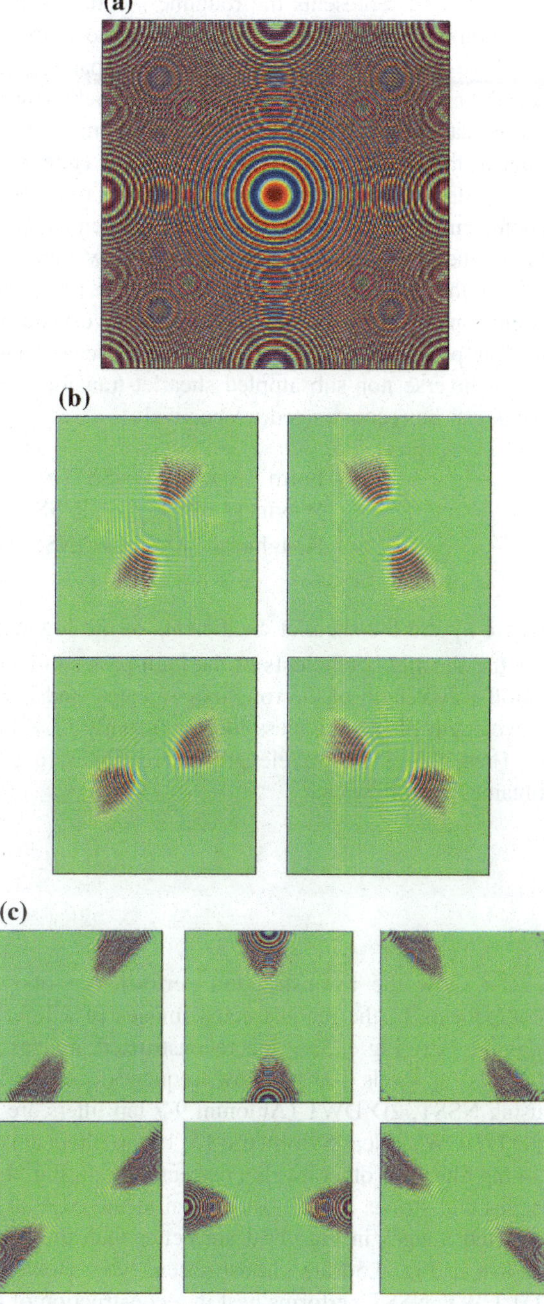

where $R(x,y)$ represents the roughness components, $W(x,y)$ represents the waviness components, and $F(x,y)$ represents the form components.

These surface components can be separated in the multi-scale domain by shearlets. The high-frequency shearlet coefficients at the finest levels (highest levels) can be deemed as the outputs of a high-pass filter band $(1/\lambda_0 \sim 1/\lambda_{rc})$ and refer to the roughness components $R(x,y)$. Here λ_0 denotes the sampling interval, and λ_{rc} denotes the cutoff wavelength of the roughness. The high-frequency shearlet coefficients at the medium levels can be deemed as the outputs of the sub-low-pass filter band $(1/\lambda_{rc} \sim 1/\lambda_{wc})$ and refer to the waviness components $W(x,y)$. Here, λ_{wc} refer to the cutoff wavelength of the waviness. Likewise, the low-frequency shearlet coefficients at the coarsest level (lowest level) can be considered as the outputs of the low-pass filter band $(1/\lambda_{wc} \sim 1/\lambda_l)$. Here, λ_l denotes the sampling length. Based on an inverse non-subsampled shearlet transform (INSST), these surface components can be reconstructed, respectively,

$$\begin{cases} \text{Form} : F(x,y) = \text{INSST}(a_1^C) \\ \text{Waviness} : W(x,y) = \text{INSST}(d_{j_w}^C, \ldots, d_{j_r}^C) \\ \text{Roughness} : R(x,y) = \text{INSST}(d_{j_r}^C, \ldots, d_L^C) \end{cases} \tag{4.16}$$

where a_1^C are the shearlet coefficients of the low-pass sub-bands, $d_j^C (j = 1, \ldots, L)$ are the shearlet coefficients of the high-pass sub-bands, j_r is the level in which the cutoff wavelength of the roughness locates, and j_w is the level in which the cutoff wavelength of the waviness locates (usually $j_w = 1$).

Then, the cutoff wavelengths of roughness and waviness by shearlets can be obtained as follows:

$$\lambda_{rc} = \lambda_0 \times 2^{\frac{3}{2}(L-j_r+1)} \tag{4.17}$$

$$\lambda_{wc} = \lambda_0 \times 2^{\frac{3}{2}(L-j_w+1)} \tag{4.18}$$

To show the reconstruction performance of shearlets, Fig. 4.5 presents the components of the reconstructed images of a light circular disk on a dark background (see Fig. 4.5a), for reconstructed images of high-frequency sub-bands (details) at levels 1–3 and low-frequency sub-bands (approximations) at level 1, using NSST, 2D DWT (Antonini 9-7 tap filters are used as mother wavelets), and 2D DT-CWT (Near-Symmetric 13, 19 tap filters are used for level 1 and Q-Shift 14, 14 tap filters for other levels), respectively. In Fig. 4.5c, it can be seen that irregular edges and stripes are almost normal to the edge of the disk in places. The reconstructed images in Fig. 4.5d are better than those of Fig. 4.5c, since the artifacts shown in Fig. 4.5d are almost absent here. However, in contrast with DWT and DT-CWT, NSST performs best in reconstruction of the disk images, due to the fact that the reconstructed images are smooth, continuous, and clear. The images in Fig. 4.5b demonstrate good shift-invariance because all parts of the disk edge are

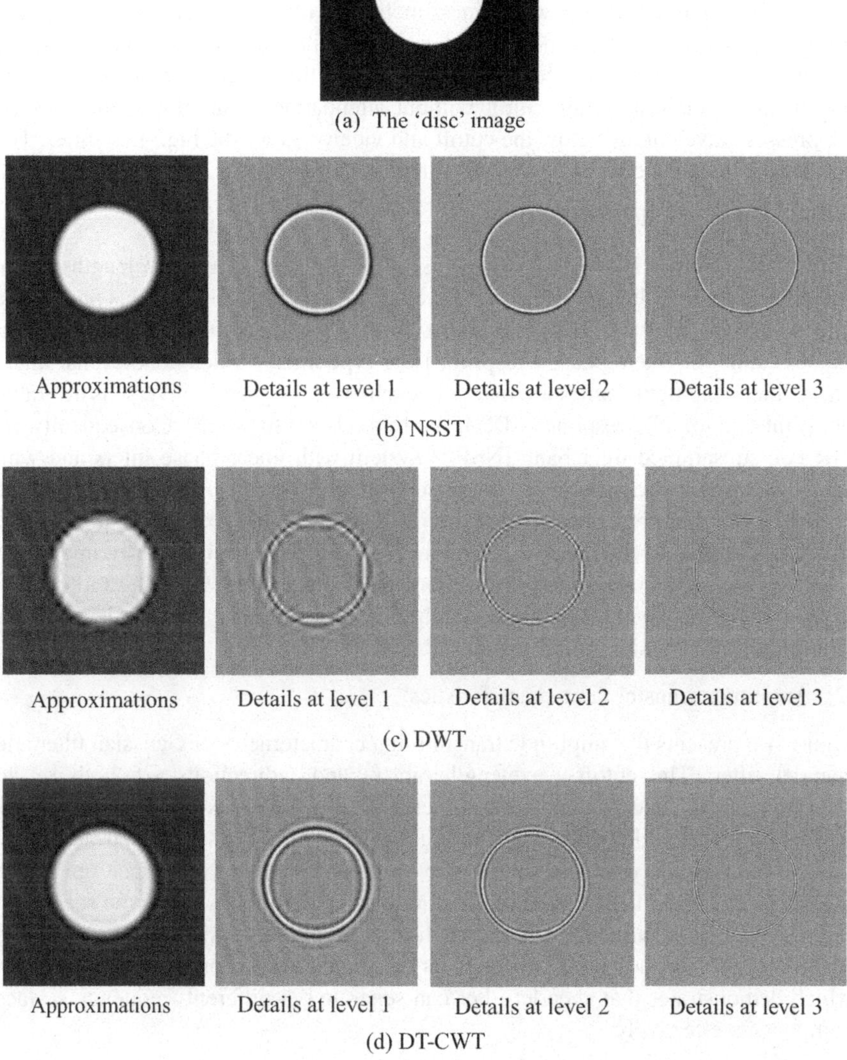

(a) The 'disc' image

Approximations   Details at level 1   Details at level 2   Details at level 3

(b) NSST

Approximations   Details at level 1   Details at level 2   Details at level 3

(c) DWT

Approximations   Details at level 1   Details at level 2   Details at level 3

(d) DT-CWT

Fig. 4.5 a The "disk" image **b** reconstructed images at three levels using NSST **c** reconstructed images at three levels using 2D DWT **d** reconstructed images at three levels using 2D DT-CWT. Reprinted from Ref. [33], copyright 2014, with permission from Elsevier

treated equivalently. Moreover, these images also show good rotational invariance, since each reconstructed image is using coefficients from all the different directional sub-bands at a given level.

4.2.3.4   Transmission Characteristics of Shearlet Filter

To separate different surface components properly, a wavelet filtering method needs to satisfy three properties: (1) finite impulse response (FIR) guarantees that the supporting length of filter's weighting function is finite; (2) zero or linear phase ensures that there will be no distortion in the filtered profiles; and (3) shape amplitude transmission ensures that a low-pass filter transmits almost all the wavelengths above a certain cutoff without attenuating its amplitude and heavily suppresses wavelengths below the cutoff and vice versa for the high-pass filter. The transmission characteristics of shearlet filter will be discussed in the following part.

(1) Phase transmission characteristics

A filter that produces different phase offsets for different input wavelengths is not desirable since the filtered output will have distortions. To overcome this problem, a filter needs to have zero or linear phase. Since FIR filters corresponding to tight frames cannot be linear phase except for Haar-type filters, which denotes that linear phase filters and tight frames are mutually exclusive, the pseudo-inverse is desirable but is infinite impulse response (IIR) if the frame is not tight [23]. Consequently, an FIR non-subsampled filter bank (NSFB) system with linear phase filters and with synthesis filters corresponding to the pseudo-inverse is not possible. However, the pseudo-inverse can be approximated with FIR filters. The maximally flat filters are used as 2D pyramid filter in the procedure of NSST. The design of maximally flat filters is discussed in detail in [24]. Maximally flat filters are zero-phase filters that under a frame which is very close to a tight one, which means that shearlet filter has zero-phase transmission characteristics.

(2) Amplitude transmission characteristics

Figure 4.6 presents the amplitude transmission characteristics of Gaussian filter and shearlet filter. The cutoff wavelengths in x- and y-directions are both set to 0.64 mm, and λx and λy denote the wavelength in x- and y-axes, respectively. It is well known that a desirable 2D filtering method should have a steep amplitude transmission curve near the cutoff wavelength. Likewise, a steeper amplitude transmission profile is preferred in the filtering of 3D surfaces. It can be seen from Fig. 4.6 that both the low-pass and high-pass amplitude transmission profiles of shearlet filter near the cutoff wavelengths are steeper than those of Gaussian filter, which demonstrates that shearlet filter can separate out different frequency surface components effectively.

4.2.4   Numerical Simulation

To verify the performance of NSST, a simulated surface (as shown in Fig. 4.7a) is generated and filtered using the proposed method. The size of the simulated surface

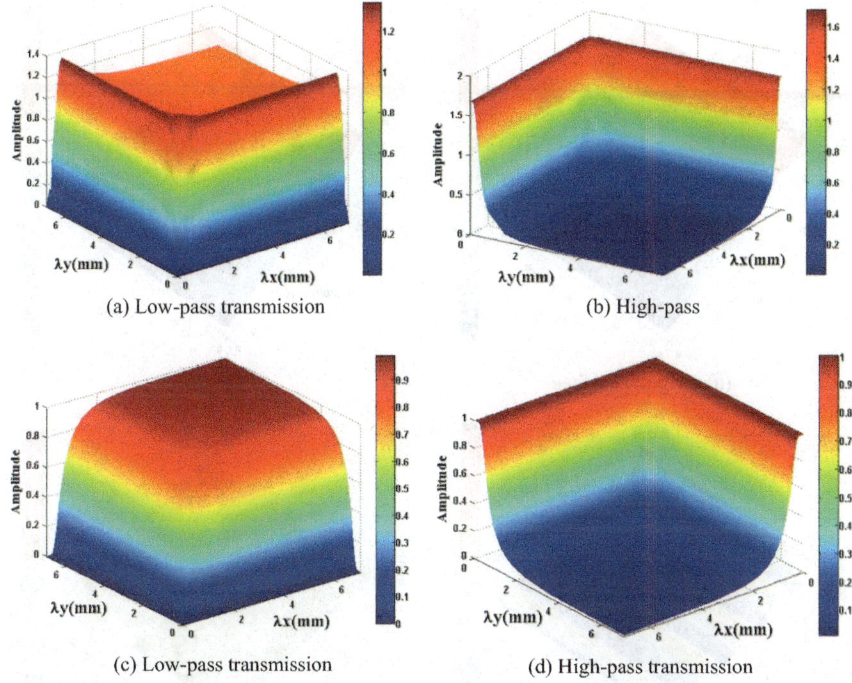

(a) Low-pass transmission   (b) High-pass

(c) Low-pass transmission   (d) High-pass transmission

Fig. 4.6 Transmission characteristics of shearlet filter and Gaussian filter. Reprinted from Ref. [33], copyright 2014, with permission from Elsevier

is 40 mm × 40 mm, the sampling interval is 0.1 mm, and the numerical expression of the surface is as follows:

$$z(x, y) = 0.002x + 0.002 \sin(0.2\pi x) + 0.001 \text{ normrnd}(0, 0.1), 0 \leq x \leq 40 \text{ mm},$$
$$0 \leq y \leq 40 \text{ mm}$$

$$(4.19)$$

where $\text{normrnd}(\mu, \delta)$ denotes random number subjects to the normal distribution with mean parameter μ and standard deviation parameter δ.

The first item on the right side of Eq. (4.19) can be deemed as the form components (an inclined surface), the second item is a sinusoidal function that can be deemed as the waviness components, and the third item are the random noises added to the surface, which can be deemed as the roughness components. Then the NSST is applied to decompose the simulated surface into high-frequency details and low-resolution approximations of five levels. If the two highest levels of details are combined as roughness $(j_r = 4)$ and the three lowest levels of details are combined as the waviness $(j_w = 1)$, Fig. 4.7c, d is obtained. In this way, the cutoff wavelength of the roughness λ_{rc} is $0.1 \times \left(2^{\frac{3}{2}}\right)^2 = 0.8$ mm, and the cutoff

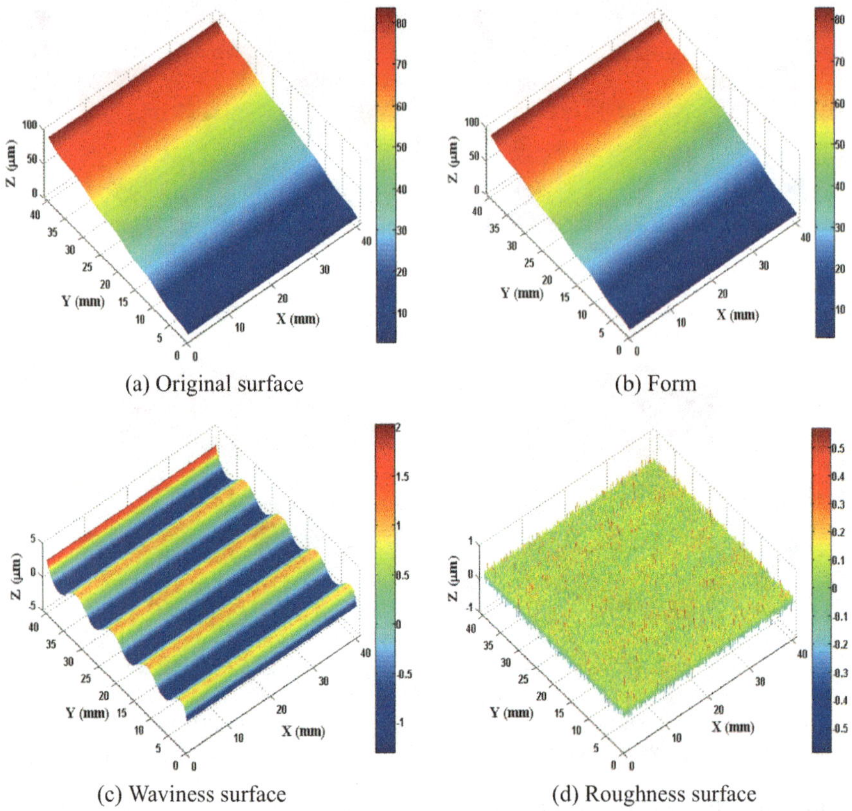

(a) Original surface   (b) Form

(c) Waviness surface   (d) Roughness surface

Fig. 4.7 Surface texture decomposition with shearlet filter. Reprinted from Ref. [33], copyright 2014, with permission from Elsevier

wavelength of the waviness λ_{wc} is $0.1 \times \left(2^{\frac{3}{2}}\right)^5 = 18.1$ mm. The lowest level of the approximation is considered as the form shown in Fig. 4.7b. It can be found from Fig. 4.7 that the proposed method is feasible since different surface components are separated properly by shearlets.

4.2.5   Case Studies

4.2.5.1   Case Study I

Engineering surfaces, which are presented by point cloud data collected by a device called Coherix ShaPix [25], are used to demonstrate the effectiveness of the proposed method for surface texture analysis in practical applications. ShaPix is a laser holographic interferometer which uses a tunable-wavelength laser to gather up to 1 million data points about the part in its micron level. This high-definition metrology (HDM) provides a good platform to develop new surface analysis methods involving surface texture separation and surface feature characterization. In the last few years, some efforts have been made to use it to analyze engineering surfaces [26, 27]. A large number of high-resolution data of engineering surfaces can be obtained by this device, which is served as the input of filtering. All the HDM data of engineering surfaces in case studies are measured by ShaPix, and the engineering surface is the top surface of an engine cylinder block (as shown in Fig. 4.8) processed by a major domestic car manufacturer. The material of the engine cylinder block is cast iron FC250. This top surface is produced by milling. The milling process is carried out on an EX-CELL-O machining center using a milling cutter with a diameter of 200 mm. The milling speed is 816.4 m/min, the depth of milling is 0.5 mm, and feed rate is 3360 mm/min. The 3D height map of this surface is shown in Fig. 4.9. The units of X-axis, Y-axis, and height are mm.

A small piece of surface is randomly selected from the top surface, as shown in Fig. 4.10a, the size of the selected surface is 6.4 mm $\times$ 6.4 mm, and the sampling interval is 0.01 mm. Then, an eight-level decomposition is implemented on the surface data using NSST.

If the reconstruction is applied by combining the four highest levels (levels 5–8) of details as roughness ($j_r = 5$), the four lowest levels of details (levels 1–4) as waviness ($j_w = 1$), and approximations at level 1 as form, the results can be obtained as shown in Fig. 4.10.

For the purpose of comparing the proposed method with the current standard filtering technique, the International Organization for Standardization (ISO) Gaussian filter is implemented. The weighting function of the 3D Gaussian filter is a simple 3D extension of the 2D Gaussian filter:

Fig. 4.8 The top surface of an engine cylinder block. Reprinted from Ref. [33], copyright 2014, with permission from Elsevier

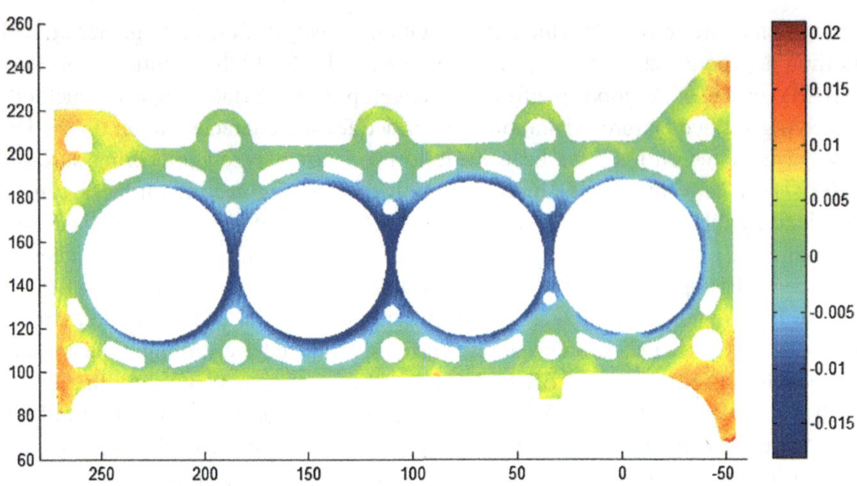

Fig. 4.9 The 3D height map of the top surface. Reprinted from Ref. [33], copyright 2014, with permission from Elsevier

$$S(x, y) = \frac{1}{\alpha^2 \lambda_{xc} \lambda_{yc}} \exp\left[-\left[\pi\left(\frac{x}{\alpha\lambda_{xc}}\right)^2 + \pi\left(\frac{y}{\alpha\lambda_{yc}}\right)^2\right]\right] \qquad (4.20)$$

where λ_{xc} and λ_{yc} are the cutoff wavelengths in the x- and y-directions, and $\alpha = \sqrt{\ln 2/\pi} = 0.4697$. The Gaussian filter is implemented using 2D convolution, which is mathematically convoluting the 2D Gaussian weighting function to the 3D height map. To be consistent with shearlet filter, the cutoff wavelengths in x- and y-directions are both set to $\lambda_{xc} = \lambda_{yc} = 0.01 \times \left(2^{\frac{3}{2}}\right)^4 = 0.64$ mm (This value is also close to 0.8 mm, which is a recommended value from the ASME B46.1 Standard [28]). Then the weighting function can be determined mathematically, which is shown in Fig. 4.11. However, it should be noticed that Gaussian filter is haunted by edge distortion, which arouses excessive distortion near the boundary of the measuring area. And this phenomenon is inevitable due to the fact that the convolution in Gaussian filter is performed on finite-size image. In this study, the first and last cutoff wavelengths are discarded after filtering in order to eliminate the edge effects and the actual size of the evaluation area is 5.12 mm × 5.12 mm. The surface data is separated into two components using Gaussian filter, one is the mean surface and the other is the roughness surface. They are shown in Fig. 4.12. It is noted that the mean surface actually is the combination of two surface components including waviness and form. Figure 4.13 presents a comparison of mean surfaces obtained by Gaussian filter and shearlet filter without discarding any boundary region (the size of evaluation area is 6.4 mm × 6.4 mm). Figure 4.13a shows that there are serious distortions on the boundary of the mean surface obtained by Gaussian filter,

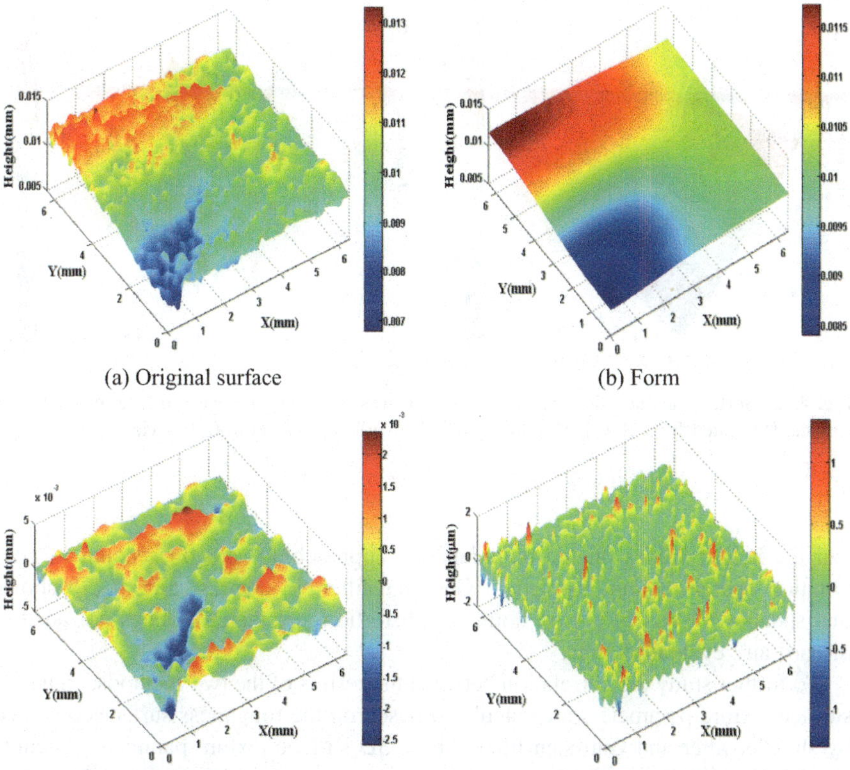

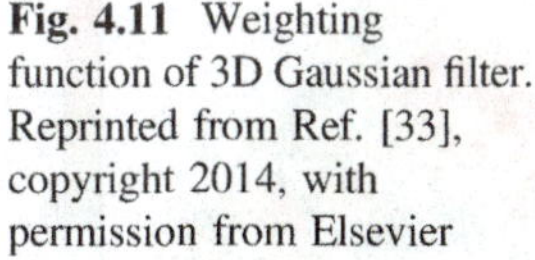

(a) Original surface   (b) Form

(c) Waviness surface   (d) Roughness surface

Fig. 4.10 Surface texture decomposition with shearlet filter. Reprinted from Ref. [33], copyright 2014, with permission from Elsevier

Fig. 4.11 Weighting function of 3D Gaussian filter. Reprinted from Ref. [33], copyright 2014, with permission from Elsevier

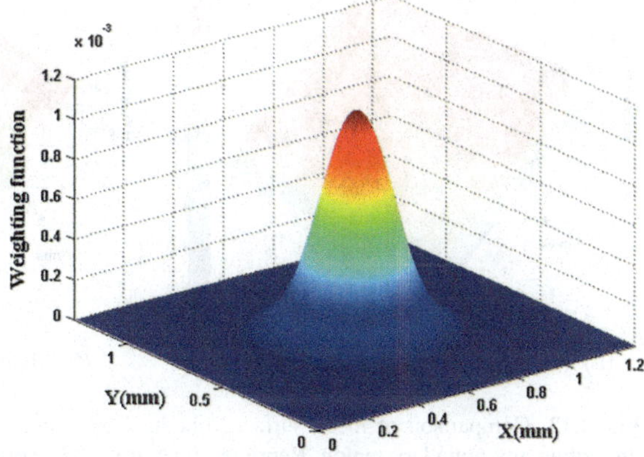

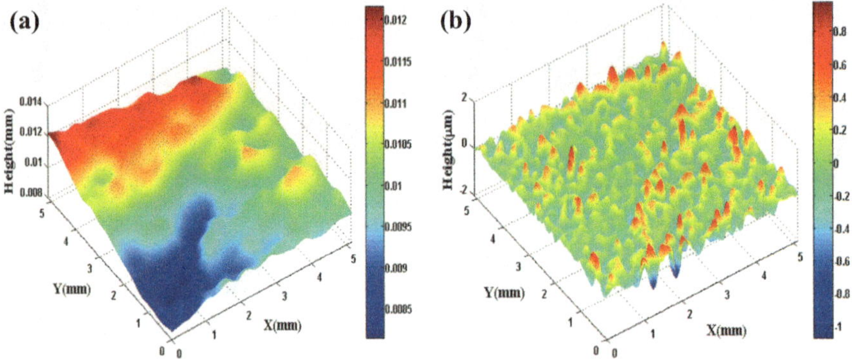

Fig. 4.12 Surface texture decomposition with Gaussian filter: **a** mean surface **b** roughness surface. Reprinted from Ref. [33], copyright 2014, with permission from Elsevier

and the height values on the boundary are approaching zero on account of convolution operation implemented in Gaussian filter. In contrast, Fig. 4.13b shows that shearlet filter is free from boundary distortion and therefore does not need to discard any evaluation area.

To further study the correlation between the results of the two methods, some 3D surface texture parameters are calculated based on the roughness surfaces obtained by shearlet filter and Gaussian filter. These 3D surface texture parameters include amplitude parameters (S_a, S_q, and S_t), shape parameters (S_{ku}, S_{sk}), and spacing parameters (S_{al}, S_{tr}), which are able to compare the two filtering methods in a comprehensive way.

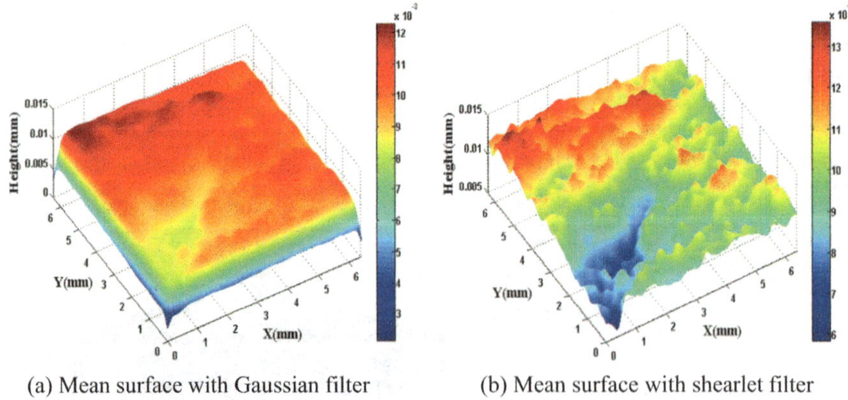

(a) Mean surface with Gaussian filter   (b) Mean surface with shearlet filter

Fig. 4.13 Comparison of mean surfaces obtained by Gaussian filter and shearlet filter without discarding any boundary region. Reprinted from Ref. [33], copyright 2014, with permission from Elsevier

(1) Amplitude parameters

The average roughness parameter S_a is defined as

$$S_a = \frac{1}{A} \iint_A |z(x,y)|\,\mathrm{d}x\mathrm{d}y \tag{4.21}$$

where $z(x,y)$ is the height value of the surface and A is the definition area.

The root-mean-square roughness parameter S_q is defined as

$$S_q = \sqrt{\frac{1}{A} \iint_A z^2(x,y)\,\mathrm{d}x\mathrm{d}y} \tag{4.22}$$

The maximum height of texture surface S_t is defined as

$$S_t = |\max(z(x,y)| + |\min(z(x,y))| \tag{4.23}$$

(2) Shape parameters

The kurtosis of surface height distribution parameter S_{ku} is defined as

$$S_{ku} = \frac{1}{sq^4} \left[\frac{1}{A} \iint_A z^4(x,y)\,\mathrm{d}x\mathrm{d}y \right] \tag{4.24}$$

The skewness of surface height distribution parameter S_{sk} is defined as

$$S_{sk} = \frac{1}{sq^3} \left[\frac{1}{A} \iint_A z^3(x,y)\,\mathrm{d}x\mathrm{d}y \right] \tag{4.25}$$

(3) Spacing parameters

The spacing parameter S_{al} is the horizontal distance of the autocorrelation function that has the fasted decay in any direction to a specified threshold value η, with $0<\eta<1$ ($\eta = 0.1$ in this study).

$$S_{al} = \underset{\tau_x,\tau_y \in R}{\mathrm{Min}} \left(\sqrt{\tau_x^2 + \tau_y^2} \right) \text{ where } R = \{ (\tau_x, \tau_y) : \mathrm{ACF}(\tau_x, \tau_y) \leq \eta \} \tag{4.26}$$

The autocorrelation function (ACF) of a 3D surface is defined as a convolution of the surface with itself, shifted by (τ_x, τ_y):

$$\mathrm{ACF}(\tau_x, \tau_y) = \frac{\iint_A z(x,y)z(x-\tau_x, y-\tau_y)\,\mathrm{d}x\mathrm{d}y}{\iint_A z(x,y)z(x,y)\,\mathrm{d}x\mathrm{d}y} \tag{4.27}$$

The surface texture aspect ratio parameter S_{tr} is defined as

$$S_{tr} = \frac{\underset{\tau_x, \tau_y \in R}{\text{Min}}\left(\sqrt{\tau_x^2 + \tau_y^2}\right)}{\underset{\tau_x, \tau_y \in R}{\text{Max}}\left(\sqrt{\tau_x^2 + \tau_y^2}\right)} \quad \text{where } R = \left\{(\tau_x, \tau_y) : \text{ACF}(\tau_x, \tau_y) \leq \eta\right\} \tag{4.28}$$

Three top surfaces of engine blocks (A, B, C) are selected and eight locations (locations 1–8, as shown in Fig. 4.14) from each top surface are selected for comparing with the same size (6.4 mm × 6.4 mm). To be convenient for comparison, the actual evaluation area with shearlet filter is set to the size of that with Gaussian filter (5.12 mm × 5.12 mm). Figure 4.15 shows the separated roughness surface using two methods on top surface A, respectively. It is observed that the separated roughnesses between shearlet filter and Gaussian filter are matched quite well in most corresponding locations.

Tables 4.2 and 4.3 list all the surface texture parameters using these two methods and the differences between them. The differences in the surface texture parameters are due to the different transmission properties between shearlet filter and Gaussian filter. Nevertheless, it can be found from the last three columns in Table 4.2 that the differences are less than 5% in S_a values, less than 10% in S_q values, and less than 20% in S_t values. It also can be found from the last four columns in Table 4.3 that the average differences in the shape and space parameters between the two methods are less than 25% in general. Since the variation of 10–20% or more in roughness parameters is very common as reported in [26] when using different filter banks, hence, the comparison demonstrates that shearlet filter-based parameters are correlated well with existing standards. However, it is worth noting that part of the evaluation area has to be discarded due to "edge effect"

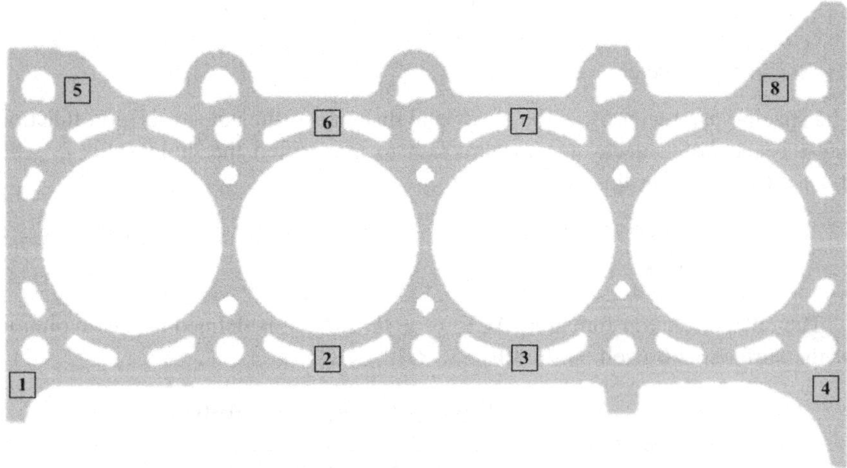

Fig. 4.14 Eight locations selected for comparison. Reprinted from Ref. [33], copyright 2014, with permission from Elsevier

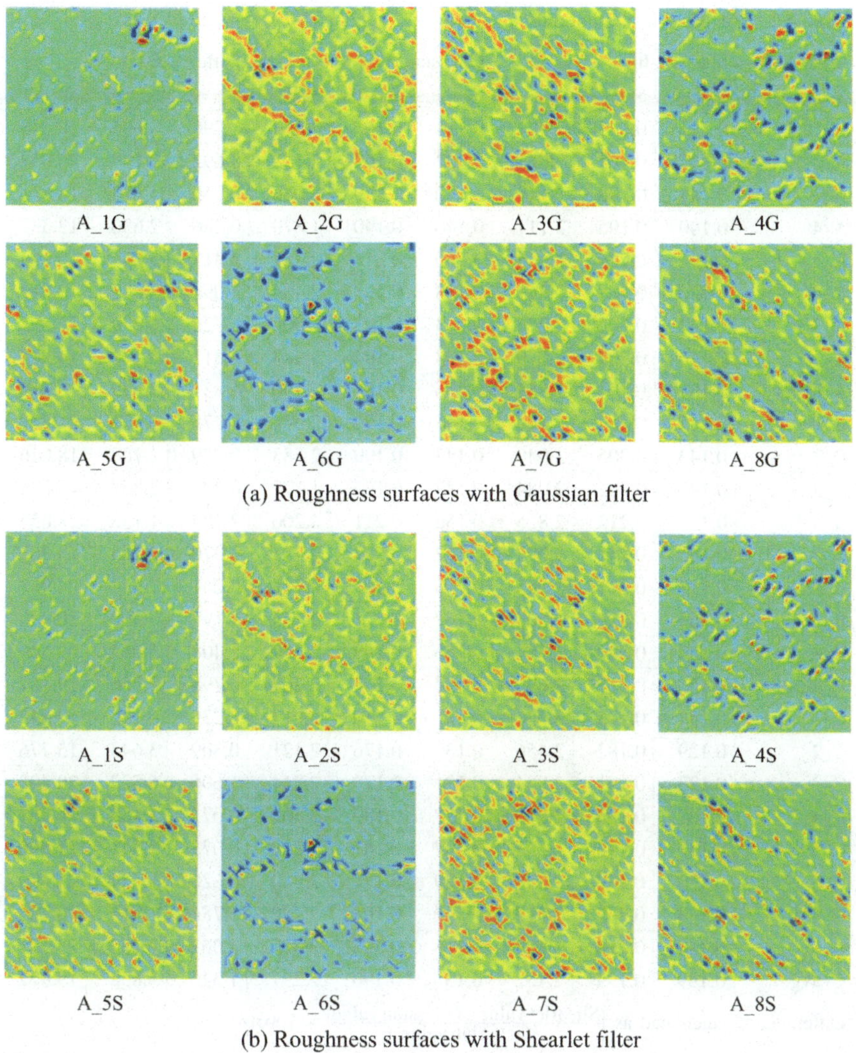

(a) Roughness surfaces with Gaussian filter

(b) Roughness surfaces with Shearlet filter

Fig. 4.15 The separated roughness surfaces using two methods, respectively. Reprinted from Ref. [33], copyright 2014, with permission from Elsevier

when using Gaussian filter, and now shearlet filter overcomes this problem without deviating from the standard values too much.

Since engine block top surface is only one of the many types of machined surfaces, other two different types of engineering surfaces are studied in this section to further validate the correlation between the two filtering methods.

Table 4.2 Comparison of amplitude parameters

Unit (μm)	Shearlet filter			Gaussian filter			Difference (%)[a]		
	S_a	S_q	S_t	S_a	S_q	S_t	S_a	S_q	S_t
A_1	0.117	0.171	2.981	0.116	0.165	2.694	1.298	3.695	10.648
A_2	0.117	0.168	2.661	0.122	0.167	2.128	3.692	0.179	25.038
A_3	0.141	0.198	2.449	0.140	0.189	2.066	0.930	4.873	18.518
A_4	0.140	0.195	2.111	0.140	0.190	1.879	0.286	2.630	12.337
A_5	0.164	0.226	2.828	0.160	0.214	2.457	2.313	5.844	15.104
A_6	0.126	0.188	2.861	0.128	0.183	2.404	1.098	3.063	19.014
A_7	0.147	0.205	2.236	0.143	0.193	1.905	2.722	6.276	17.354
A_8	0.146	0.204	2.279	0.143	0.194	1.983	2.311	5.271	14.912
A_avg.	0.137	0.194	2.551	0.136	0.187	2.190	1.831	3.979	16.616
B_1	0.145	0.207	2.486	0.145	0.199	2.049	0.275	3.962	21.359
B_2	0.143	0.205	3.049	0.143	0.199	2.583	0.279	2.708	18.026
B_3	0.146	0.202	2.084	0.147	0.196	1.777	0.409	2.857	17.234
B_4	0.152	0.213	2.825	0.156	0.211	2.206	2.184	1.138	28.055
B_5	0.162	0.230	3.097	0.162	0.222	2.731	0.370	3.785	13.402
B_6	0.153	0.211	2.965	0.153	0.204	2.372	0.261	3.234	24.968
B_7	0.165	0.229	2.509	0.167	0.221	2.145	1.197	3.300	16.956
B_8	0.159	0.220	2.577	0.156	0.209	2.270	1.408	5.065	13.500
B_avg.	0.153	0.215	2.699	0.154	0.208	2.267	0.798	3.256	19.187
C_1	0.151	0.223	3.431	0.147	0.210	2.919	2.239	6.140	17.542
C_2	0.129	0.182	2.456	0.130	0.176	2.121	0.309	3.643	15.776
C_3	0.133	0.191	2.884	0.131	0.179	2.244	1.606	6.522	28.498
C_4	0.137	0.195	2.716	0.137	0.189	2.362	0.073	3.183	14.968
C_5	0.137	0.196	2.695	0.137	0.187	2.216	0.073	4.541	21.600
C_6	0.130	0.188	2.357	0.129	0.178	1.940	0.620	5.624	21.490
C_7	0.143	0.202	2.578	0.146	0.198	2.220	1.784	1.816	16.127
C_8	0.154	0.218	2.899	0.148	0.203	2.350	4.062	7.185	23.371
C_avg.	0.139	0.199	2.752	0.138	0.190	2.297	1.346	4.832	19.922

[a]Difference is calculated as $\left| \dfrac{\text{Shearlet value} - \text{Gaussian value}}{\text{Gaussian value}} \right| \times 100\%$

4.2.5.2 Case Study II

The second engineering surface is the surface of a cylinder head which is made of aluminum. This surface is produced by milling. The height map of this surface is shown in Fig. 4.16, and the units of X-axis, Y-axis, and height are mm. Eight locations are selected (locations 1–8, as shown in Fig. 4.17) from this surface with the same size, and then eight small surfaces are obtained.

One of these selected surfaces (Location 1) is shown in Fig. 4.18. The size of evaluation area is 5.12 mm × 5.12 mm, and the sampling interval is 0.01 mm. Then, a six-level decomposition is implemented on this surface data using shearlet

Table 4.3 Comparison of shape and space parameters

	Shearlet filter				Gaussian filter				Difference (%)			
	S_{ku}	S_{sk}	S_{al}	S_{tr}	S_{ku}	S_{sk}	S_{al}	S_{tr}	S_{ku}	S_{sk}	S_{al}	S_{tr}
A_1	9.428	0.337	8.485	0.849	9.115	0.413	9.900	0.868	3.438	18.279	14.286	2.269
A_2	6.680	−0.165	8.544	0.854	5.025	−0.158	10.440	0.855	32.944	4.560	18.163	0.105
A_3	5.436	−0.269	7.616	0.595	4.525	−0.237	8.944	0.629	20.143	13.652	14.853	5.498
A_4	5.225	−0.054	8.944	0.970	4.538	−0.052	10.198	0.901	15.132	3.654	12.294	7.621
A_5	5.445	−0.183	7.616	0.633	4.581	−0.157	8.944	0.698	18.867	16.805	14.853	9.436
A_6	7.542	0.208	8.544	0.906	6.216	0.267	10.198	0.943	21.329	21.905	16.219	3.935
A_7	5.313	−0.064	7.616	0.668	4.375	−0.057	9.220	0.755	21.423	12.238	17.395	11.572
A_8	5.083	−0.103	7.071	0.552	4.441	−0.087	8.602	0.608	14.473	17.506	17.800	9.222
A_avg.	6.269	−0.037	8.055	0.753	5.352	−0.009	9.556	0.782	18.469	13.575	15.733	6.207
B_1	5.838	0.016	8.246	0.833	4.726	0.024	9.849	0.818	23.524	33.333	16.273	1.846
B_2	7.268	−0.270	8.544	0.906	6.204	−0.248	10.050	0.881	17.147	9.087	14.984	2.757
B_3	5.068	−0.155	7.616	0.595	4.251	−0.132	9.220	0.649	19.228	16.767	17.395	8.324
B_4	5.543	−0.037	8.944	0.948	4.415	−0.032	10.440	0.934	25.535	15.938	14.329	1.531
B_5	6.147	−0.157	7.616	0.595	5.036	−0.123	9.220	0.678	22.055	27.317	17.395	12.260
B_6	5.402	−0.057	8.485	0.899	4.264	−0.041	10.198	0.943	26.680	38.725	16.794	4.603
B_7	5.088	0.051	7.810	0.685	4.075	0.048	9.849	0.807	24.853	7.158	20.700	15.097
B_8	5.443	−0.119	7.071	0.552	4.538	−0.097	8.944	0.658	19.926	21.891	20.943	16.028
B_avg.	5.724	−0.091	8.042	0.752	4.689	−0.075	9.721	0.796	22.368	21.277	17.352	7.806
C_1	8.965	0.223	8.246	0.833	7.336	0.184	9.849	0.864	22.195	20.967	16.273	3.566

(continued)

Table 4.3 (continued)

	Shearlet filter				Gaussian filter				Difference (%)			
	S_{ku}	S_{sk}	S_{al}	S_{tr}	S_{ku}	S_{sk}	S_{al}	S_{tr}	S_{ku}	S_{sk}	S_{al}	S_{tr}
C_2	6.491	−0.085	8.246	0.874	5.075	−0.062	10.000	0.877	27.894	36.232	17.538	0.342
C_3	7.063	−0.071	7.071	0.526	5.212	−0.077	8.485	0.597	35.514	7.712	16.666	11.960
C_4	6.509	−0.141	8.544	0.927	5.371	−0.105	10.050	0.888	21.195	34.670	14.984	4.323
C_5	6.266	0.122	7.616	0.595	5.071	0.128	9.220	0.685	23.564	5.066	17.395	13.220
C_6	6.800	−0.094	8.062	0.855	5.145	−0.148	9.849	0.911	32.170	36.995	18.140	6.139
C_7	5.740	0.028	7.810	0.685	4.655	0.024	9.849	0.807	23.313	14.583	20.700	15.097
C_8	5.900	0.075	7.071	0.500	5.030	0.092	8.485	0.571	17.284	18.033	16.666	12.404
C_avg.	6.717	0.007	7.833	0.724	5.362	0.005	9.473	0.775	25.391	21.782	17.295	8.381

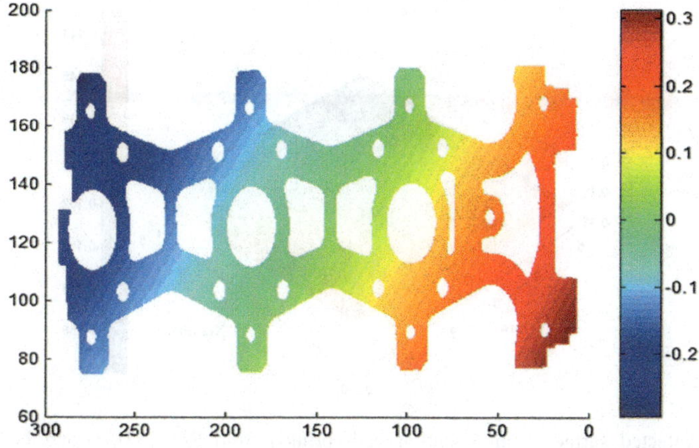

Fig. 4.16 The height map of the cylinder head. Reprinted from Ref. [33], copyright 2014, with permission from Elsevier

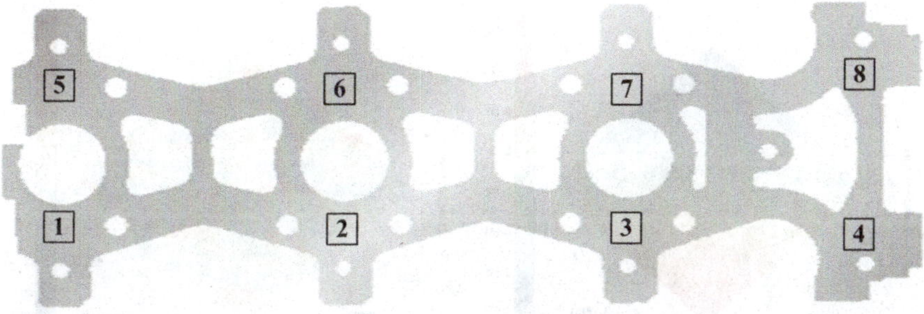

Fig. 4.17 Eight locations selected from this surface. Reprinted from Ref. [33], copyright 2014, with permission from Elsevier

filter. If the reconstruction is applied by combining the four highest levels (levels 3–6) of details as roughness ($j_r = 3$), the two lowest levels of details (levels 1–2) as waviness ($j_w = 1$), and approximations at level 1 as form, the results can be obtained as shown in Fig. 4.19. Likewise, the Gaussian filter is also implemented on this surface $\left(\lambda_{xc} = \lambda_{yc} = 0.01 \times \left(2^{\frac{3}{2}} \right)^4 = 0.64 \text{ mm} \right)$ and the results are shown in Fig. 4.20. Then, the surface texture parameters presented in Case Study I are also calculated using the roughness data obtained by the two methods, respectively. Tables 4.4 and 4.5 list all the surface texture parameters using these two methods and the differences between them. It can be found in Table 4.4 that the average differences are about 7% in S_a values, 5% in S_q values, and 12% in S_t values. Table 4.5 shows that the average differences in the shape and space parameters between the two methods are less than 20% in general.

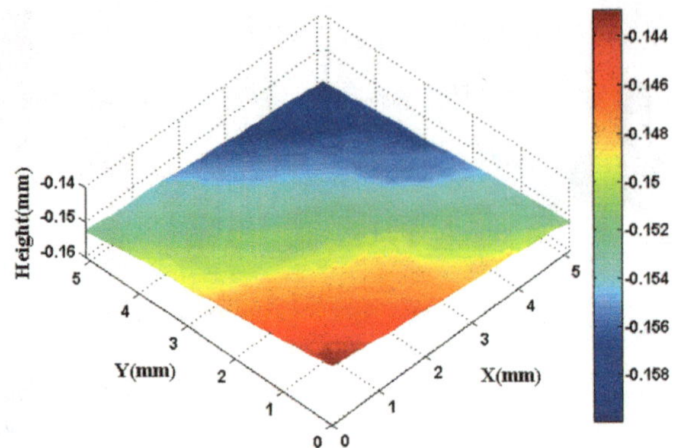

Fig. 4.18 Selected surface from Location 1. Reprinted from Ref. [33], copyright 2014, with permission from Elsevier

(a) Form

(b) Wavinesssurface

(c) Roughness surface

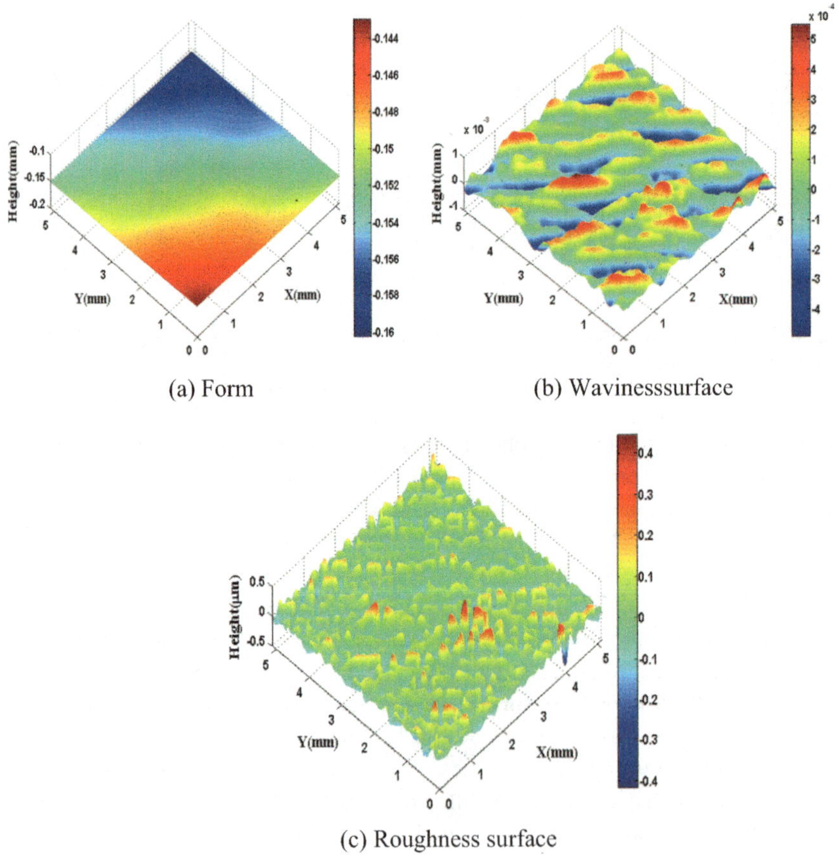

Fig. 4.19 Surface texture decomposition with shearlet filter. Reprinted from Ref. [33], copyright 2014, with permission from Elsevier

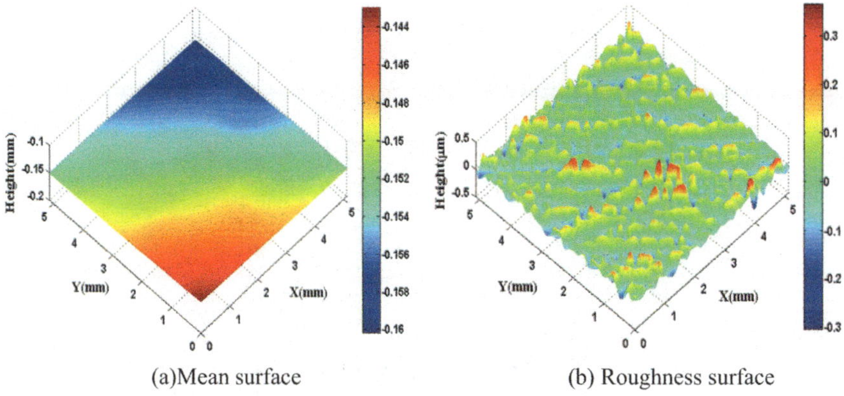

(a)Mean surface   (b) Roughness surface

Fig. 4.20 Surface texture decomposition with Gaussian filter. Reprinted from Ref. [33], copyright 2014, with permission from Elsevier

Table 4.4 Comparison of amplitude parameters

Unit (µm)	Shearlet filter			Gaussian filter			Difference (%)		
	S_a	S_q	S_t	S_a	S_q	S_t	S_a	S_q	S_t
L_1	0.051	0.068	0.860	0.053	0.069	0.668	4.307	1.594	28.704
L_2	0.051	0.066	0.557	0.054	0.068	0.547	4.664	2.504	1.847
L_3	0.052	0.069	0.782	0.056	0.073	0.743	7.348	4.945	5.289
L_4	0.057	0.073	0.687	0.059	0.075	0.598	3.735	2.394	14.890
L_5	0.050	0.067	0.776	0.054	0.070	0.612	7.778	4.585	26.892
L_6	0.059	0.081	1.005	0.065	0.086	0.860	8.756	5.794	16.913
L_7	0.056	0.074	0.814	0.061	0.080	0.811	8.264	7.035	0.382
L_8	0.062	0.082	0.796	0.068	0.088	0.774	8.944	6.818	2.776
Average	0.055	0.073	0.785	0.059	0.076	0.702	6.725	4.459	12.212

4.2.5.3   Case Study III

The third engineering surface is the surface of a pump valve plate which is also produced by milling, and the height map of this surface is shown in Fig. 4.21. The experimental process of this surface is completely the same as that of the second surface except that there are six locations selected from this surface (see Fig. 4.22).

Likewise, Table 4.6 presents the comparison of computed amplitude surface parameters using shearlet filter and Gaussian filter, respectively. The last three columns of it show that the average differences in amplitude parameters are 8–12%. Table 4.7 presents the comparison of computed shape surface parameters and space surface parameters, and it can be found that the differences in the shape and space parameters between the two methods are 15–25%. Taking the two cases in Sects. 4.2.5.2 and 4.2.5.3 into consideration, a conclusion can be drawn that the

Table 4.5 Comparison of shape and space parameters

	Shearlet filter				Gaussian filter				Difference (%)			
	S_{ku}	S_{sk}	S_{al}	S_{tr}	S_{ku}	S_{sk}	S_{al}	S_{tr}	S_{ku}	S_{sk}	S_{al}	S_{tr}
L_1	4.723	−0.029	8.485	0.849	3.790	−0.037	10.296	0.646	24.810	22.581	17.583	31.286
L_2	3.649	−0.054	7.810	0.395	3.196	−0.079	9.434	0.296	14.166	31.258	17.212	33.097
L_3	4.632	−0.084	8.062	0.475	3.983	−0.097	9.900	0.426	16.293	13.669	18.559	11.657
L_4	3.811	−0.032	8.485	0.667	3.328	−0.040	9.900	0.564	14.504	19.799	14.286	18.188
L_5	4.418	−0.022	8.544	0.804	3.630	−0.017	10.630	0.607	21.700	26.437	19.624	32.487
L_6	5.331	−0.180	7.810	0.316	4.265	−0.157	9.900	0.279	24.995	14.522	21.105	13.042
L_7	4.445	−0.198	8.485	0.429	4.188	−0.219	10.000	0.388	6.146	9.332	15.147	10.435
L_8	4.098	0.018	8.602	0.714	3.653	0.014	10.630	0.622	12.193	23.239	19.076	14.855
Average	4.389	−0.073	8.286	0.581	3.754	−0.079	10.086	0.479	16.851	20.105	17.824	20.631

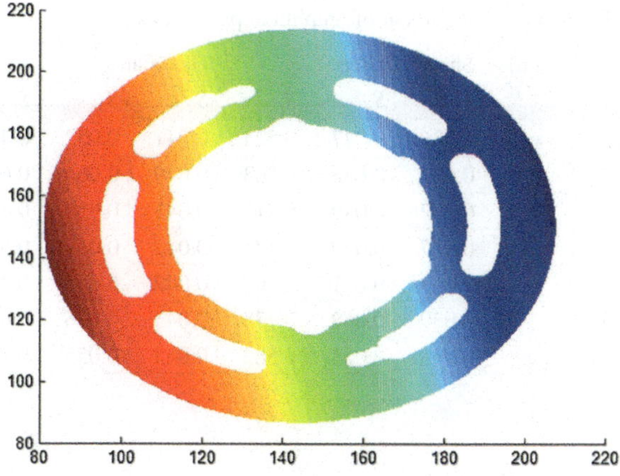

Fig. 4.21 The height map of the pump valve plate. Reprinted from Ref. [33], copyright 2014, with permission from Elsevier

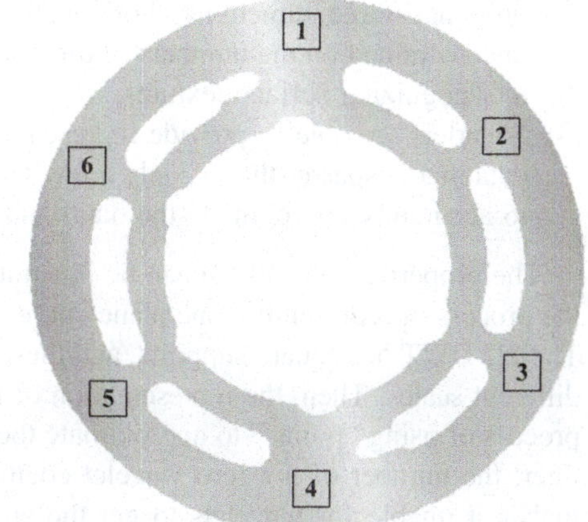

Fig. 4.22 Six locations selected from this surface. Reprinted from Ref. [33], copyright 2014, with permission from Elsevier

differences in amplitude parameters obtained by the two methods are around 10%, while the differences in the shape and space parameters are around 20%. This conclusion is approximate to that of Case Study I, which demonstrates that the differences between the surface parameters through Gaussian filter and shearlet filter are within a reasonable range.

4.2.5.4   Discussions of Properties

Compared with the traditional wavelets, shearlets have the following properties which make it more appropriate in the filtering of 3D engineering surfaces.

(1) Shearlets not only maintain the multi-resolution and time–frequency localization characteristics of wavelets but also perform well in capturing the geometric regularity along singularities of surfaces due to their anisotropic supports.

Table 4.6 Comparison of amplitude parameters

Unit (μm)	Shearlet filter			Gaussian filter			Difference (%)		
	S_a	S_q	S_t	S_a	S_q	S_t	S_a	S_q	S_t
L_1	0.036	0.047	0.524	0.041	0.052	0.437	12.099	9.770	19.876
L_2	0.043	0.058	0.663	0.049	0.064	0.628	12.500	9.798	5.654
L_3	0.037	0.049	0.512	0.041	0.053	0.468	10.412	7.590	9.470
L_4	0.037	0.049	0.525	0.041	0.053	0.539	11.165	8.835	2.579
L_5	0.038	0.051	0.499	0.042	0.055	0.482	9.953	8.015	3.571
L_6	0.039	0.052	0.778	0.043	0.056	0.633	9.602	6.115	22.926
Average	0.038	0.051	0.583	0.043	0.055	0.531	10.955	8.354	10.679

(2) Shearlets exhibit highly directional sensitivity. Compared with 2D DWT, which only has three directions (horizontal, vertical, and diagonal), shearlets are unconstrained on the numbers of directions, which make it directional sensitive in recognizing surface textures.

(3) Shearlets are able to provide representations of larger and higher dimensional data that is sparse (that is, only a few terms of the representations are sufficient to accurately approximate the data) and computationally efficient.

The properties (1) and (3) can be illustrated vividly through Fig. 4.23. It shows the process of representing one plane curve by wavelets and shearlets, respectively. The 2D DWT has square supports, and these square supports have different sizes in different scales. Then, the representation of the plane curve is actually equal to the process of using "points" to approximate the plane curve. When the scale becomes finer, the number of nonzero wavelet coefficients will grow exponentially, which makes it unable for wavelets to get the sparse representation of the function. In contrast, shearlets have elongated supports that can approximate the plane curve with the least amount of coefficients. Since shearlets have anisotropic supports, they can capture geometric smoothness of the curves efficiently, whereas wavelets are only good at capturing point discontinuities. Additionally, NSST not only maintains the merits of shearlets but also possesses the property of fully shift-invariance. Lacking of shift-invariance will lead to the distortion of reconstructed images at different scales after filtering, which is faced by DWT (see Fig. 4.5c). However, NSST is free from this problem and therefore it is suitable to handle complex engineering surfaces.

The role of property (2) is embodied in the separation of engineering surfaces with directional features like lines and curves. To demonstrate this, a worn printed circuit board (PCB) is used for experiment and part of its surface is selected as input of the filtering methods, which is shown in Fig. 4.24. It can be seen that there are many lines and curves on this PCB surface. Both 2D DWT and shearlet filters are used to separate this surface, and the results are shown in Fig. 4.25. It can be found that the mean surface (low-pass components) obtained by shearlet filter is closer to

Table 4.7 Comparison of shape and space parameters

	Shearlet filter				Gaussian filter				Difference (%)			
	S_{ku}	S_{sk}	S_{al}	S_{tr}	S_{ku}	S_{sk}	S_{al}	S_{tr}	S_{ku}	S_{sk}	S_{al}	S_{tr}
L_1	4.333	0.032	8.944	0.640	3.491	0.041	10.770	0.504	24.128	22.794	16.954	27.114
L_2	5.095	0.039	8.544	0.540	4.238	0.031	10.198	0.408	20.211	26.885	16.219	32.483
L_3	4.289	−0.209	8.246	0.592	3.511	−0.207	10.198	0.504	22.156	1.162	19.139	17.554
L_4	4.337	0.012	8.246	0.478	3.840	0.015	10.050	0.377	12.951	19.595	17.947	26.845
L_5	4.464	−0.033	8.246	0.566	3.844	−0.036	10.198	0.471	16.107	7.778	19.139	20.199
L_6	5.751	−0.154	8.246	0.633	4.235	−0.117	10.050	0.558	35.793	31.990	17.947	13.453
Average	4.711	−0.052	8.412	0.575	3.860	−0.046	10.244	0.470	21.891	18.367	17.891	22.941

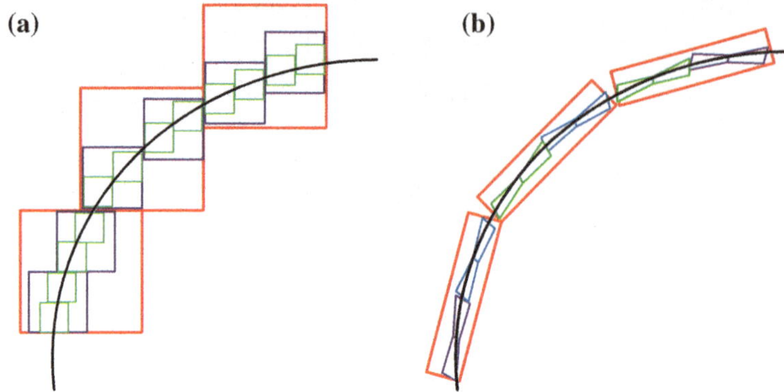

Fig. 4.23 The representation of a plane curve using **a** 2D DWT and **b** shearlets. Reprinted from Ref. [33], copyright 2014, with permission from Elsevier

the form of the raw PCB surface than that obtained by 2D DWT. Besides, the roughness surface (high-pass components) obtained by shearlet filter contains more detail information of texture compared with the one obtained by 2D DWT. Using shearlet filter, all the detail features are extracted precisely and the PCB surface form is preserved without any distortion.

4.2.6 Conclusions

A shearlet-based separation method of 3D engineering surface using high-definition metrology is developed. A 3D engineering surface is decomposed into different sub-bands of coefficients with NSST, and these coefficient sub-bands include high-frequency details and low-frequency approximations. The surface components

Fig. 4.24 Raw PCB surface. Reprinted from Ref. [33], copyright 2014, with permission from Elsevier

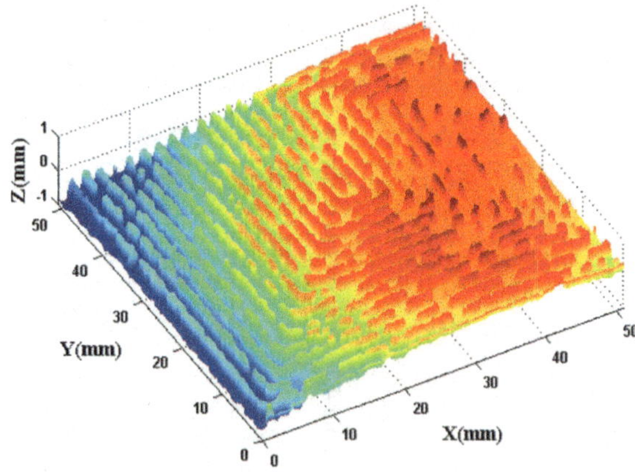

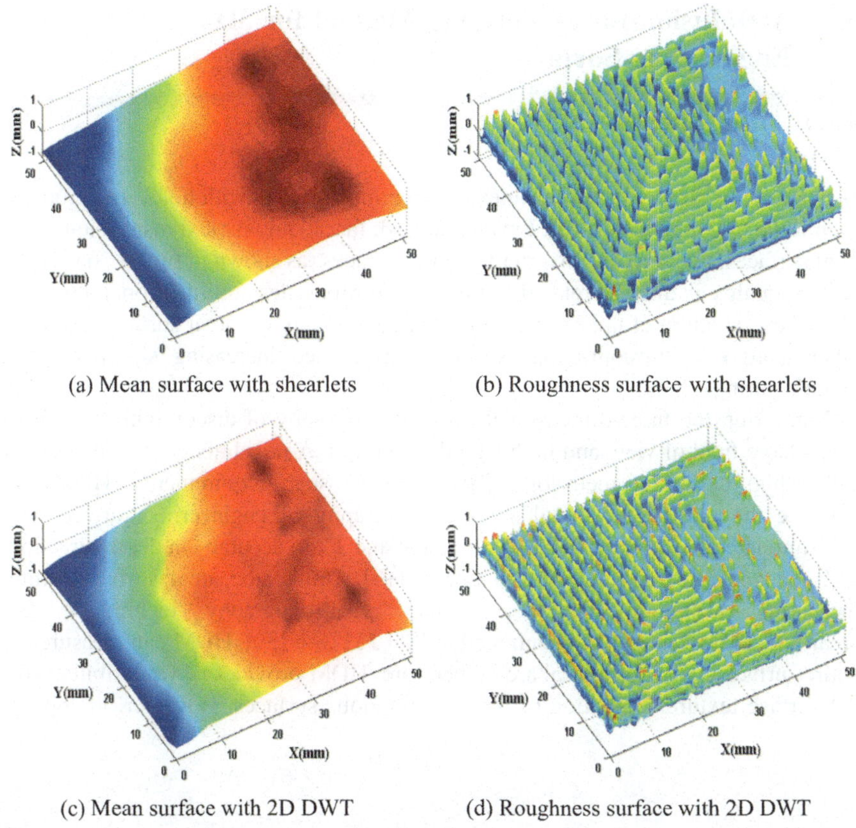

(a) Mean surface with shearlets   (b) Roughness surface with shearlets

(c) Mean surface with 2D DWT   (d) Roughness surface with 2D DWT

Fig. 4.25 Comparison of shearlet and 2D DWT for the separation of the PCB surface. Reprinted from Ref. [33], copyright 2014, with permission from Elsevier

are reconstructed by INSST and surface roughness; waviness and form are obtained by combing different frequency components based on the cutoff wavelengths. The performance of shearlet filter is validated by simulated surface data and real-world 3D surface data using high-definition metrology. Also, several 3D surface texture parameters are computed, respectively, based on the roughness surfaces and compared with those of Gaussian filter. The comparison results have illustrated the effectiveness of shearlet filter. Additionally, considering its ability of multi-resolution, good time–frequency localization, highly directional sensitivity, and shift-invariant property, shearlets are expected to be superior to the current ISO standardized Gaussian filter and wavelet filters like DWT. Shearlets can be a powerful tool in 3D surface texture analysis in the future, when much finer analysis of engineering surfaces becomes necessary.

4.3 A Diffusion-Based Filtering Method for 3D Engineering Surface

4.3.1 Introduction

Discontinuous surfaces are machined engineering surfaced with holes and grooves, such as the engine block faces, the engine desk faces, automatic transmission valve joint surfaces, and torque converter housing surfaces (as shown in Fig. 4.26a, d). As surface quality is directly related to mechanical properties and functional attributes [29], the evaluation of three-dimensional surface texture (or areal surface texture in other words) for discontinuous surfaces has gained increasing significance in manufacturing.

Measuring the three-dimensional surface topography of discontinuous surfaces needs large field of view and high lateral resolution. A novel noncontact metrology called high-definition metrology (HDM) meets such requirements. HDM can measure the three-dimensional surface topography of engineering surface with 0.15 mm lateral resolution in lateral direction and 1 µm accuracy in depth direction, and can generate millions of data points [13]. Unlike the areal surface texture measuring techniques such as those areal-topography techniques and areal-integrating techniques mentioned in ISO 25178-6 [30], HDM can measure the entire surface instead of a local area. Therefore, HDM provides new possibilities for 3D surface texture evaluation of the discontinuous surfaces. However, in spite of

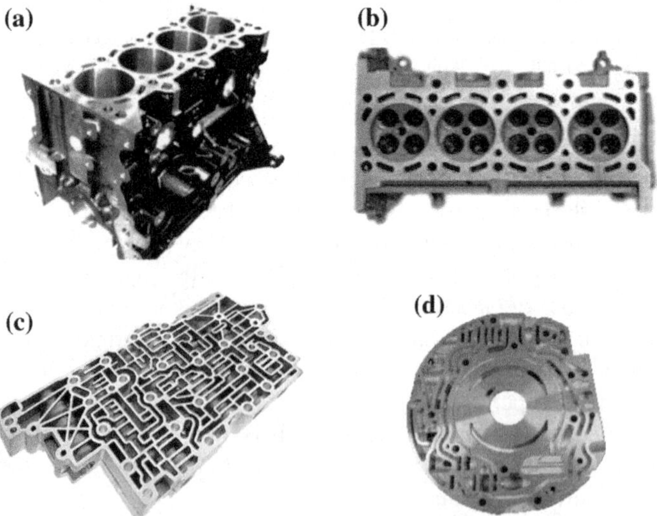

Fig. 4.26 Examples of discontinuous engineering surfaces. Reprinted from Ref. [44], copyright 2015, with permission from Springer Nature

the development of measurement metrology, there is limited research that addresses the evaluation of discontinuous surfaces.

Filtering techniques are well established to separate the measured surface into various scale-limited surfaces for further analysis. Then, surface features including form, waviness, and roughness can be calculated from the large-, medium-, and small-scale surface. Nevertheless, traditional filters such as Gaussian filters are only appropriate for continuous surfaces that do not have nonempty data within the evaluation area. As for discontinuous surfaces, Gaussian filters cause "boundary distortion" or "end effects" problems due to surface discontinuity caused by empty data at the hole areas. Although adding extra data points such as zero-padding or extrapolation is a possible way for signal processing, it may be inaccurate for surface texture evaluation.

Specifically, take, for example, the filtering of HDM measured automobile engine block face (Fig. 4.27). With the roughness and waviness specifications $Rz = 12.5$, $R_{max} = 15$, and $Wt = 10$, the cutoff wavelengths of roughness are 0.008 mm and 2.5 mm, and the cutoff wavelengths of waviness are 2.5 mm and 25 mm, respectively [31]. Therefore, the HDM data contains the 3D surface form and 3D surface waviness information (HDM cannot measure roughness due to the lateral resolution). To separate waviness from surface form using Gaussian filtering, the minimum traversing length should include half a waviness cutoff on each end of the evaluation length according to ASME B46.1 [15] to reduce the end effects, which means the traverse should be at least twice the long-cutoff wavelength of waviness, i.e., 50 mm. However, for engine block faces with cylinder holes, bolt holes, and cooling holes, there is no 50 mm × 50 mm continuous evaluation area (as shown in red boxes in Fig. 4.27). Therefore, the surface texture at the boundaries of the holes will be distorted. Consequently, it is difficult to evaluate 3D surface texture of discontinuous surfaces.

To overcome the problem of edge distortions, as well as outlier-induced distortions, new filters such as robust Gaussian filters [32], robust spline filters [5], wavelet filters [18], and shearlet filters [33] have been proposed. The progress on

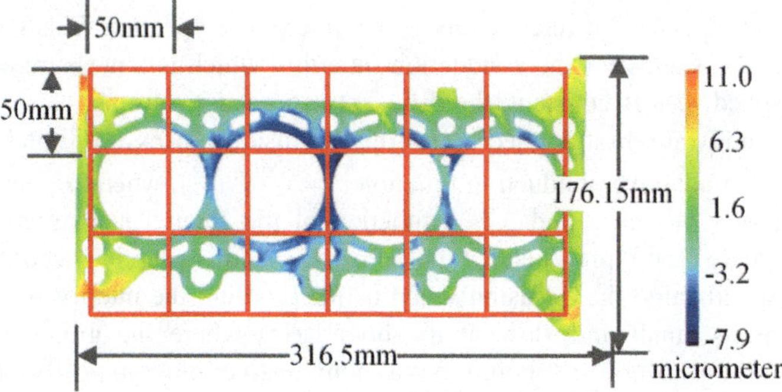

Fig. 4.27 Engine block face measured by HDM. Reprinted from Ref. [44], copyright 2015, with permission from Springer Nature

filtration in surface metrology is illustrated in a thorough review by Jiang and Whitehouse [34]. Among these new filters, the anisotropic diffusion filter is worth exploring due to its edge-preserving abilities. Anisotropic diffusion filters, which are formulated in terms of heat diffusion process, have been successfully developed for image edge detection in which the image intensity can only diffuse inside of the objects but not across the boundaries. Therefore, the image objects are smoothed without blurring their boundaries.

Recently, anisotropic diffusion filters have been adopted for the filtering of structured engineering surfaces that have deterministic structure with high aspect ratio geometric features. Zeng et al. employed the diffusion filter for MEMs' surfaces [35]. Jiang et al. adopted the diffusion filter for freeform engineering surfaces [36]. The features of MEMs' surface and freeform surfaces are well preserved with high accuracy. However, for discontinuous surfaces, the boundaries of the objects are known in advance. Therefore, we propose a modified anisotropic diffusion filter by adding edge detectors to stop any diffusion across the boundaries.

4.3.2   Diffusion Filtering

In this section, a modified anisotropic diffusion filter with edge detectors is presented for the filtering of discontinuous engineering surfaces.

4.3.2.1   Anisotropic Diffusion Filtering

The anisotropic diffusion filter comes from the heat diffusion equation which describes the changes of heat within a specified region with respect to time and space. The general model of the diffusion equation is

$$\frac{\partial u}{\partial t} = \mathrm{div}(c\nabla u) \tag{4.29}$$

where u is the quantity of interest, div is the divergence operator, ∇ is the gradient operator, and c stands for the conduction function, which is a measurement of the diffusion speed. If c is constant, the diffusion process is isotropic.

Perona and Malik first applied anisotropic diffusion process to digital image by introducing an adaptive conduction function $c = g(|\nabla u_t|^2)$, where u_t represents the image intensity at time t and g is a function of the Euclidean magnitude of the gradient of u_t [37]. It represents that the diffusion speed is adaptive according to the gradient. Specifically, the diffusion speed is quick within the interior region where the gradient is small and slow at the boundaries where the gradient is large. Therefore, the function g should be a nonnegative monotonically decreasing function of the gradient. Two typical formulations for conduction function g are

$$g_1 = \exp(-|\nabla u|^2 / K^2) \tag{4.30}$$

$$g_2 = 1 / (1 + |\nabla u|^2 / K^2) \tag{4.31}$$

where parameter K controls the direction of the diffusion process. The regions with gradient smaller than K are smoothed; the regions with gradient larger than K are enhanced. K can either be a value depending on the gradient or set by hand. Weickert J. proposed a different conduction function in which the diffusion speed is much more rapid in the interior region [38].

$$g_3 = \begin{cases} 1 - \exp\left(-3.315 / (|\nabla u| / K)^8\right), & |\nabla u| \neq 0 \\ 1, & |\nabla u| = 0 \end{cases} \tag{4.32}$$

Figure 4.28 illustrates the three conduction functions. It can be seen that g_3 is much larger than g_1 and g_2 for $|\nabla u|/K < 1$. As it is calculated at each iteration, the conduction coefficient depends on both space and time variables.

The diffusion filter needs to be discretized for practical use. Let $I(i,j)$ be the surface height at location (i,j) at time zero, and $I_{i,j}^{t+1}$ be the diffused surface height at location (i,j) and time t. The discrete scheme of the anisotropic diffusion filter is as follows:

$$\begin{aligned} I_{i,j}^{t+1} &= I_{i,j}^t + \lambda [c_N \nabla_N I + c_S \nabla_S I + c_E \nabla_E I + c_W \nabla_W I]_{i,j}^t \\ I_{i,j}^0 &= I(i,j) \end{aligned} \tag{4.33}$$

where $0 \leq \lambda \leq 1/4$ for the scheme to be stable, and the subscripts N, S, E, and W are the abbreviation of North, South, East, and West. The symbol ∇ here indicates the difference between each pixel and its nearest neighbors in four directions:

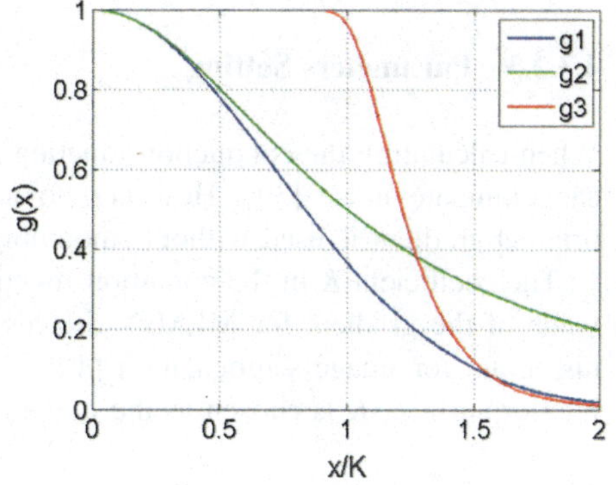

Fig. 4.28 Three conduction functions. Reprinted from Ref. [44], copyright 2015, with permission from Springer Nature

$$\nabla_N I_{i,j}^t = I_{i-1,j}^t - I_{i,j}^t \quad \nabla_S I_{i,j}^t = I_{i+1,j}^t - I_{i,j}^t$$
$$\nabla_E I_{i,j}^t = I_{i,j+1}^t - I_{i,j}^t \quad \nabla_W I_{i,j}^t = I_{i,j-1}^t - I_{i,j}^t \tag{4.34}$$

The conduction coefficients are updated at each iteration:

$$c_{N_{i,j}}^t = g\left(\left|\nabla_N I_{i,j}^t\right|\right) \quad c_{S_{i,j}}^t = g\left(\left|\nabla_S I_{i,j}^t\right|\right)$$
$$c_{S_{i,j}}^t = g\left(\left|\nabla_S I_{i,j}^t\right|\right) \quad c_{S_{i,j}}^t = g\left(\left|\nabla_S I_{i,j}^t\right|\right) \tag{4.35}$$

4.3.2.2   Anisotropic Diffusion Filter with Edge Detectors

One major difference between the filtering of discontinuous surfaces and structured engineering surfaces is that the surface boundary of the former is already known in advance. Therefore, we can completely prevent the diffusion across the boundaries by setting the conduction coefficient equal to zero at the boundary. Specifically, if two adjacent points belong to the surface, diffusion can take place between them. Otherwise, diffusion between two empty points or between one empty point and one boundary point is prohibited. This is achieved by edge detectors W_N, W_S, W_E, W_w. First, let $w_{i,j}$ be the window function of the discontinuous surface. $w_{i,j} = 1$ if $I_{i,j}$ belong to the surface; otherwise, $w_{i,j} = 0$. Therefore, the edge detectors are defined as follows:

$$W_{N_{i,j}} = w_{i-1,j} \cdot w_{i,j} \quad W_{S_{i,j}} = w_{i+1,j} \cdot w_{i,j}$$
$$W_{E_{i,j}} = w_{i,j+1} \cdot w_{i,j} \quad W_{W_{i,j}} = w_{i,j-1} \cdot w_{i,j} \tag{4.36}$$

Hence, the anisotropic diffusion filter with edge detectors is

$$I_{i,j}^{t+1} = I_{i,j}^t + \lambda[W_N c_N \nabla_N I + W_S c_S \nabla_S I + W_E c_E \nabla_E I + W_W c_W \nabla_W I]_{i,j}^t \tag{4.37}$$

The diffused surface is deemed as the areal form, and the difference between the original surface and diffused surface is the areal surface texture.

4.3.2.3   Parameters Setting

When calculating the conduction function g, smoothing the gradient with a kernel can eliminate noise [39]. However, to avoid any possible edge distortion, the original gradient is used without smoothing.

The coefficient K in the condition function can be chosen as the mean absolute value of the gradient for MEMs' surfaces [35] or 70% percentile of the gradient histogram for image segmentation [40]. However, to allow smoothing within the interior surface, K is chosen as the maximum gradient within the surface:

$$K = \max(|\mathbf{W}_N \nabla_N I|, |\mathbf{W}_S \nabla_S I|, |\mathbf{W}_E \nabla_E I|, |\mathbf{W}_W \nabla_W I|) \tag{4.38}$$

To be consistent with Gaussian filters, the time t should be assigned carefully. The two-dimensional areal Gaussian filter is

$$S(x,y) = \frac{1}{\alpha^2 \lambda^2} \exp\left[-\frac{\pi(x^2 + y^2)}{\alpha^2 \lambda^2} \right] \tag{4.39}$$

where $\alpha = \sqrt{\ln 2/\pi}$, and λ is the cutoff wavelength in x- and y-directions. The relation between the standard error of the Gaussian filter σ and cutoff wavelength λ is

$$\sigma = \frac{\alpha}{\sqrt{2\pi}}\lambda = \sqrt{\frac{\ln 2}{2}}\frac{\lambda}{\pi} \tag{4.40}$$

Let σ_l denote the standard deviation in pixels, and l denote the sampling distance between two adjacent pixels. Therefore, σ_l is given by

$$\sigma_l = \frac{\sigma}{l} \tag{4.41}$$

A linkage exists between Gaussian filter and diffusion filter. In the case of the Gaussian scale space, the convolution of an image with a Gaussian kernel of standard deviation σ_l is equivalent to diffuse the image for some time t with the following correspondence [40]:

$$t = \sigma_l^2/2 \tag{4.42}$$

Note that Eq. (4.42) does not indicate a theoretical relation between the Gaussian filter of standard deviation σ_l and the diffusion filter of time t. Nevertheless, it provides a sound basis for practical use. Combining the Eqs. (4.41) and (4.43) gives the relation between diffusion time t and Gaussian cutoff wavelength λ:

$$t = \frac{\ln 2}{4\pi^2}\left(\frac{\lambda}{l}\right)^2 = 0.0176\left(\frac{\lambda}{l}\right)^2 \tag{4.43}$$

4.3.3   Simulation

A discontinuous areal surface is simulated to evaluate the performance of the proposed filter: $z(x,y) = -0.5\cos(2\pi x/6.4) + 0.05x + 0.05y$. The cosine term is regarded as surface waviness, and last two terms are regarded as the surface form. The wavelength of surface waviness is. The area of the simulated surface is

51.2 mm $\times$ 51.2 mm. The sampling spacing is $l = 0.2$ mm in x- and y-directions. The simulated surface has 256×256 data points. To simulate the discontinuous surface, a square area in the center is removed (as shown in Fig. 4.29).

Gaussian filter and the proposed anisotropic diffusion filter are applied to the simulated surface, respectively. The cutoff wavelength for Gaussian filter is 25 mm. For the diffusion filter, conduction function g_3 is used and the corresponding diffusion time is $t = 0.0176(25/0.2)^2 = 275$. The filtered results are displayed in Fig. 4.30.

The Gaussian filtered surface form and waviness show serious end effects near the inner and outer boundaries. In practice use, half of the cutoff wavelength is discarded at the boundaries. However, the simulated discontinuous surface cannot afford the boundary removal due to the small evaluation area. On the other hand, the areal surface form and waviness are separated satisfactorily with no boundary distortion by diffusion filtering.

The root-mean-square error of the difference between the diffusion filtered waviness and nominal cosine surface is calculated. Figure 4.31 shows the plot of root-mean-square error versus diffusion time t. It can be seen that the root-mean-square error reaches its minimum around the time 275, which implies that the setting of diffusion time is reasonable.

4.3.4 Experiment

An experiment on the face milling process of engine blocks was conducted to demonstrate the ability of the proposed filter for practical surfaces. The depth of cut was 0.5 mm, the cutting speed was 1300 rpm, and feed rate was 3360 mm/min. The face milling cutter had a diameter of 200 mm with 15 cutting inserts intercalated by three wiper inserts. The finished engine block faces were measured by HDM. Raw HDM data was converted into grid image with resampling interval $l = 0.2$ mm using the HDM data preprocessing method proposed by Wang et al. [27].

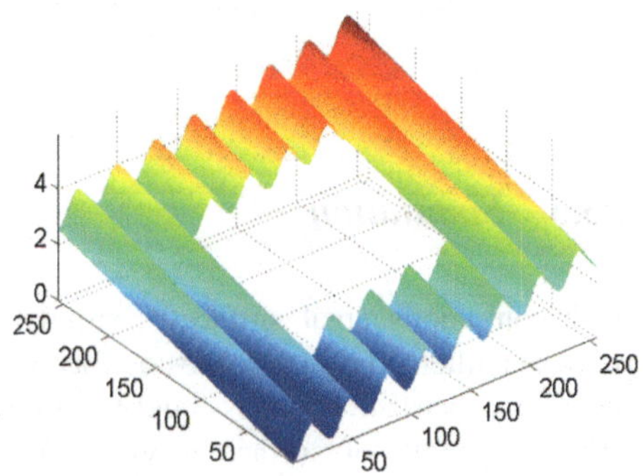

Fig. 4.29 Simulated discontinuous surface. Reprinted from Ref. [44], copyright 2015, with permission from Springer Nature

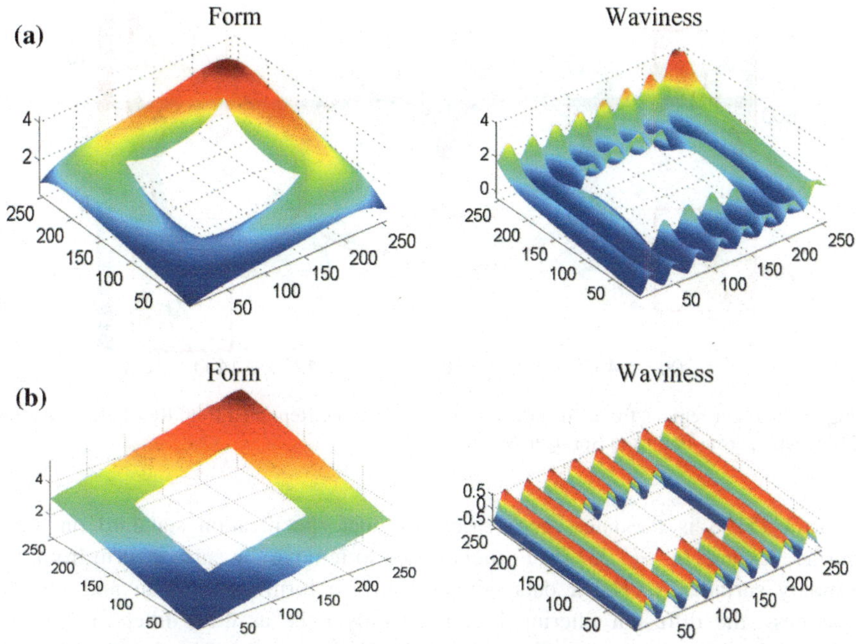

Fig. 4.30 Filtered surface form and waviness: **a** Gaussian filter, **b** diffusion filter. Reprinted from Ref. [44], copyright 2015, with permission from Springer Nature

Fig. 4.31 Plot of the root-mean-square error versus diffusion time. Reprinted from Ref. [44], copyright 2015, with permission from Springer Nature

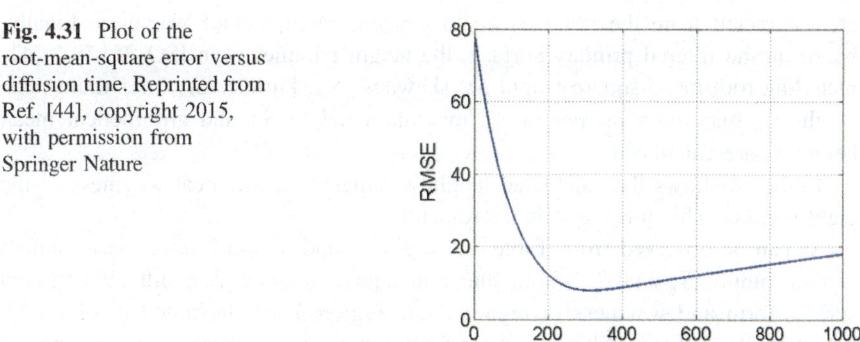

Figure 4.32 displays the HDM converted image of the engine block face. Eight regions are selected to evaluate the areal surface texture. From Fig. 4.32, we can see that the four corner areas of the engine block face have large surface variation, so square regions with an area of 50 mm × 50 mm at the four corners are selected as region 1 to region 4. Moreover, the surface texture around the cylinder bore has a vital influence on the sealing performance, so region 5 to region 8 are circular regions around the cylinder bore with a radius of 80 mm.

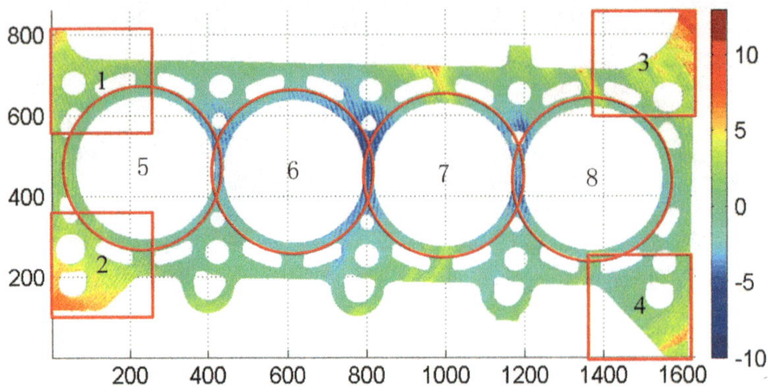

Fig. 4.32 Eight regions for areal surface texture evaluation. Reprinted from Ref. [44], copyright 2015, with permission from Springer Nature

ISO 25178 is the latest geometrical product specification standard on areal surface texture. In ISO 25178, F-filter is used to remove the surface form from the primary surface. As HDM data contains surface form information and surface waviness, the diffusion filtering is applied only once as the F-filter to separate surface waviness and surface form. As the waviness cutoff is 25 mm, the diffusion time is $t = 275$. Conduction function g_3 is used and K is set as the maximum gradient within each region. To begin with, the proposed diffusion filter is performed on the eight regions to acquire the surface form. Then the filtered surfaces are subtracted from the raw surface to generate areal surface waviness. Finally, based on the filtered primary surface, the height parameters in ISO 25178-2 [41] including root-mean-square height S_q, skewness S_{sk}, kurtosis S_{ku}, maximum peak height S_p, maximum pit height S_v, maximum height S_z, and arithmetical mean height S_a are calculated.

Table 4.8 shows the calculated height parameters of the areal waviness of the eight regions. The dimension is micrometer.

As can be observed from Table 4.8, region 2 and region 3 have the minimum and maximum S_q and S_a among the four square regions. The diffusion filtered surface form and waviness of region 2 and region 3 are illustrated in Fig. 4.33. Similarly, the diffusion filtered surface form and waviness of region 6 and region 8 are presented in Fig. 4.34. It can be seen that the areal waviness is separated accurately from the surface form at these regions.

Region 8 has the largest S_q and S_a value of the four circular regions, which may imply that the sealing performance of the cylinder bore at this region could be worse than the other three. In light of this, evaluation of local areal surface texture at critical areas and global areal texture over the entire surface are both suggested to access the susceptibility to leakage of combustion gases, coolant, or lubricant.

Table 4.8 Calculated areal surface textures

	S_q	S_{sk}	S_{ku}	S_p	S_v	S_z	S_a
Region 1	0.74	−0.01	3.04	3.26	4.29	7.55	0.59
Region 2	0.63	−0.10	3.22	2.69	3.88	6.57	0.50
Region 3	0.82	−0.01	3.22	3.27	4.02	7.29	0.65
Region 4	0.74	−0.01	3.04	3.26	4.29	7.55	0.59
Region 5	0.78	0.05	2.92	3.19	3.40	6.59	0.63
Region 6	0.77	0.03	2.98	2.93	3.29	6.22	0.62
Region 7	0.81	0.07	2.97	4.14	4.24	8.38	0.65
Region 8	0.90	0.09	3.03	3.79	3.82	7.61	0.72

4.3.5   Conclusions

In this section, a modified anisotropic diffusion filter is introduced to filter discontinuous engineering surfaces into various scale-limited surfaces. The edge detector is adopted to ensure that the diffusion process will only take place inside the surface but not across the boundaries. The diffusion time is specified considering its relation with the cutoff wavelength of Gaussian filters. Simulated and experimental discontinuous surfaces demonstrate that the proposed anisotropic diffusion filter can separate areal surface waviness from areal surface form accurately. Thus, three-dimensional surface texture parameters instead of the two-dimensional surface texture can be calculated based on the filtered surface, which helps fully evaluate the surface quality and functional performance.

4.4   A Fast and Adaptive Bidimensional Empirical Mode Decomposition Based Filtering Method for 3D Engineering Surface

4.4.1   Introduction

The surface texture is an important index to evaluate the quality of workpieces [42, 43] and is generally described from the small to the large scale: roughness, waviness, and form. It is well known that different components of the surface texture have different influences on the functional performance of workpieces. To be specific, roughness is a good indicator of the surface irregularities and thus can be applied to detect errors in the material removal process, and also it has great influence on the workpiece functionality such as wear and friction. Waviness, which may occur from machine or work deflections, chatter, residual stress, vibrations, or heat treatment, has influence on tightness of workpieces. Form may directly affect the assembling performance of workpieces. Therefore, the motivation for separating these components derives from the fact that they have different

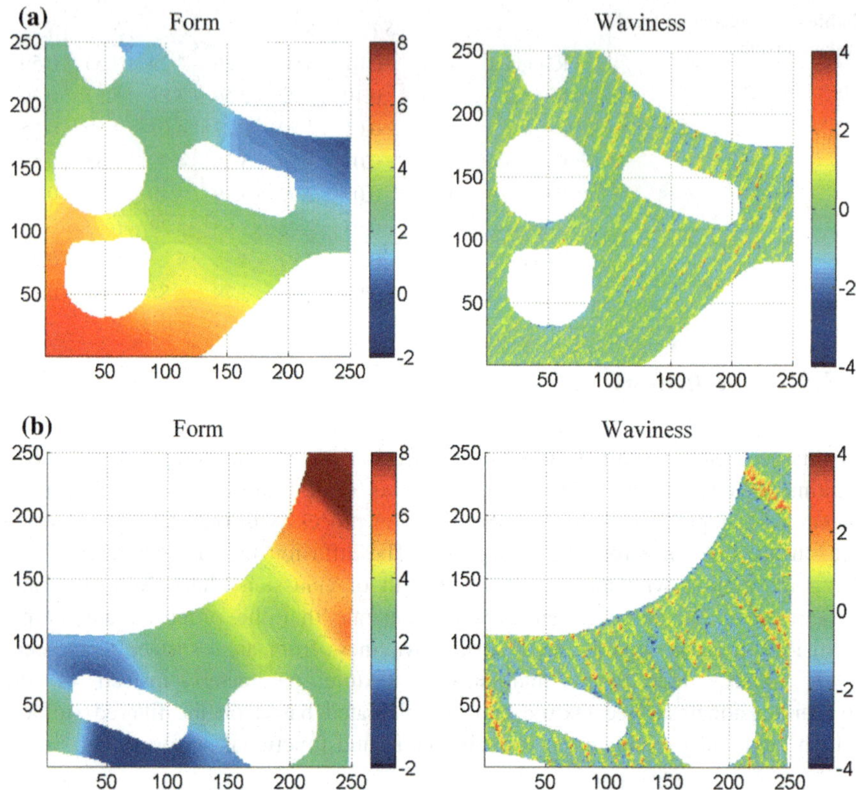

Fig. 4.33 Filtered square regions: **a** region two, **b** region three. Reprinted from Ref. [44], copyright 2015, with permission from Springer Nature

origins and influences on workpieces functionalities in different ways. It is very important to separate the surface texture into different components before surface topography analysis.

Digital filtering is an essential step to realize the separation process. Filtering of workpiece surfaces has been a hot research topic on account of its importance for surface texture analysis. The traditional filtering approaches such as 2CR filter and Gaussian filter have been first studied, and the Gaussian filter is one of the most widely used standard filtering approaches. However, it is well recognized that it is not robust against outliers. To overcome the shortcoming, some modified approaches such as robust regression Gaussian filter [3], spline filter [4], robust spline filter [5], and morphological filter [6] have been developed. Recent advances in filtering approaches are reviewed in [1, 9].

Several researchers develop wavelet-based filtering approaches and apply them to analyze workpiece surfaces. Different from the previous filtering approaches, wavelet-based filters can provide multi-scale analysis since they can divide a surface profile into different frequency components and investigate each component

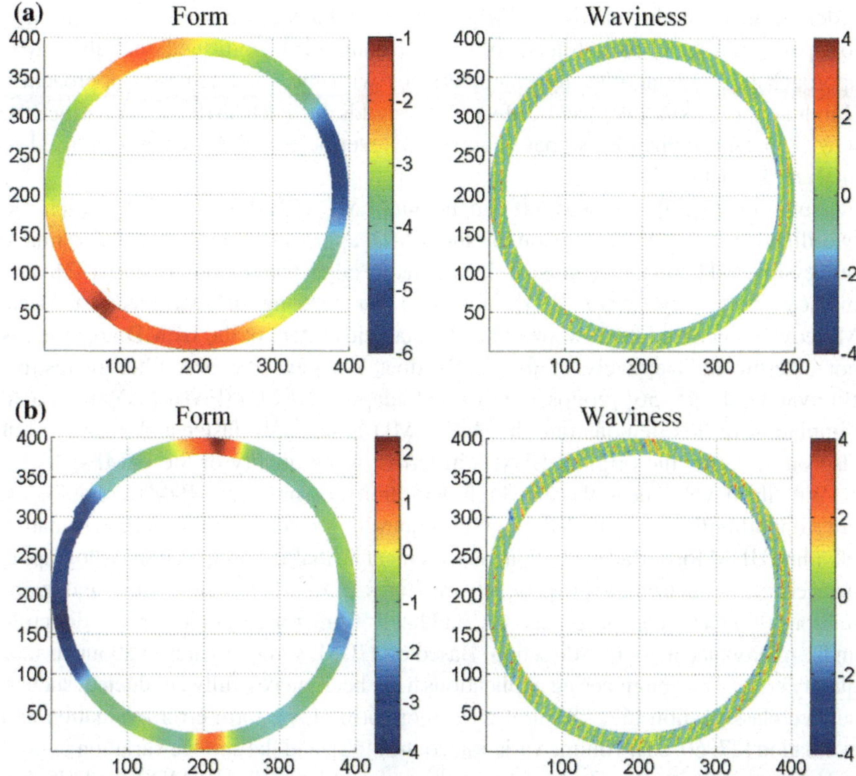

Fig. 4.34 Filtered circular regions: **a** region six, **b** region eight. Reprinted from Ref. [44], copyright 2015, with permission from Springer Nature

with a resolution matched to its scale. Fu et al. [10] adopted the wavelet transform to surface topography analysis and compared different wavelet bases. Orthogonal wavelet bases and biorthogonal wavelet bases were recommended due to their transmission characteristics of the corresponding filters. Jiang et al. [11] proposed a lifting wavelet representation for characterization of surface topography. Josso et al. [12] proposed a frequency normalized wavelet transform for surface roughness analysis and characterization. Wang et al. [44] proposed a modified anisotropic diffusion filter to separate workpiece surfaces into various scale-limited surfaces. Most recently, Du, Liu and Huang [33] presented a shearlet-based filtering approach. The workpiece surface was decomposed into different sub-bands of coefficients with non-subsampled shearlet transform (NSST). Then the surface components at different levels were reconstructed based on inverse NSST.

Recently, Huang et al. [45] and Du et al. [46] introduced and improved the empirical mode decomposition (EMD) approach to analyze one-dimensional non-stationary and nonlinear signals based on instantaneous frequency. Flandrin et al. [47] proposed the concept of filter banks based on EMD and the corresponding order intrinsic mode functions (IMFs) were combined to achieve the high-pass, low-pass, and band-pass filters. Wu and Huang [48] confirmed that the EMD approach had similar filtering characteristics with the wavelet-based approaches. Boudraa and Cexus [49] used different thresholds for each IMF to reconstruct the new filter and realize the signal denoising. Nevertheless, EMD cannot be used to analyze 3D data.

Nunes et al. [50] proposed a bidimensional EMD (BEMD) approach, which is a two-dimensional (2D) extension of the EMD approach, mainly used for image processing [51], image denoising [52], image edge pattern processing [53], and medical image registration [54], not used for filtering of workpiece surfaces. Moreover, since the window size of order statistics filters in the BEMD approach is not determined adaptively, it frequently does not have the best filtering results. Bhuiyan et al. [55, 56] proposed a fast and adaptive BEMD (FABEMD) approach. Simulation results demonstrate that FABEMD is not only faster and adaptive but also outperforms the original BEMD in terms of the quality of the BIMFs.

With the development of online high-definition measurement (HDM) technologies, great opportunities are provided for online controlling surface quality. A representative of online HDM for surface variation measurement is ShaPix based on laser holographic interferometry metrology [13], which measures 3D surface height map and gains millions of data points within seconds, and has 150-micron resolution in x–y-direction and 1-micron accuracy in z-direction. Based on HDM, some researches about surface quality control and engineering applications have been successfully conducted, such as surface classification [57, 58], tool wear monitoring [59], form error evaluation and estimation [27, 60, 61], volume variation control [62], and flat surface variation control [63]. However, to the best knowledge of the authors, there is no BEMD-based filtering approach for workpiece surfaces using HDM. The high-density point cloud data of HDM is large. About one million measurement points are collected from a cylinder head by HDM system. So, HDM needs a fast and adaptive analysis. Therefore, this section presents a novel fast and adaptive bidimensional empirical mode decomposition (FABEMD) approach for filtering of workpiece surfaces using HDM.

4.4.2 Brief Introduction to BEMD

The BEMD approach decomposes a signal into its bidimensional IMFs (BIMFs) and a residue based on the local spatial scales. Let the original signal be denoted as $I(x, y)$, a BIMF as $F(x, y)$, and the residue as $R(x, y)$. The original bidimensional signal $I(x, y)$ can be decomposed by BEMD

$$I(x, y) = \sum F_i(x, y) + R(x, y) \tag{4.44}$$

where $F_i(x, y)$ is the ith BIMF obtained from its source signal $S_i(x, y)$, and $S_i(x, y) = S_{i-1}(x, y) - R_{i-1}(x, y)$.

It requires one iteration or more to obtain $F_i(x,y)$, and the intermediate state of a BIMF in jth iteration can be denoted as $F_{Tj}(x,y)$. The decomposition steps of the BEMD approach are summarized as follows:

Step 1: Set $i = 1$ and $S_i(x,y) = I(x,y)$.
Step 2: Set $j = 1$ and $ST_j(x,y) = S_i(x,y)$. ST_j represents the input signal of the jth decomposition.
Step 3: Obtain the local maxima map of $F_{Tj}(x,y)$, denoted as $P_j(x,y)$.
Step 4: Interpolate the maxima points in $P_i(x,y)$ and generate the upper envelope, denoted as $U_{Ej}(x,y)$.
Step 5: Obtain the local minima map of $F_{Tj}(x,y)$, denoted as $Q_j(x,y)$.
Step 6: Interpolate the minima points in $Q_i(x,y)$ and generate the lower envelope, denoted as $L_{Ej}(x,y)$.
Step 7: Calculate the mean envelope $M_{Ej}(x,y) = (U_{Ej}(x,y) + L_{Ej}(x,y))/2$.
Step 8: Calculate the details of the signal in the decomposition process, $F_{Tj+1}(x,y) = F_{Tj}(x,y) - M_{Ej}(x,y)$.
Step 9: Check whether $F_{Tj+1}(x,y)$ follows the BIMF properties by finding the standard deviation (SD), denoted as D [Eq. (4.45)], between $F_{Tj+1}(x,y)$ and $F_{Tj}(x,y)$, and compare it with the desired threshold.

$$D = \sum_{x=1}^{M} \sum_{j=1}^{N} \frac{\left| F_{Tj+1}(x,y) - F_{Tj}(x,y) \right|^2}{\left| F_{Tj}(x,y) \right|^2} \tag{4.45}$$

where (x, y) is the coordinate, M is the total number of rows, and N is the total number of columns of the 2D data. The value of D is usually chosen to be 0.5 to ensure that the mean value of BIMF is close to 0.
Step 10: If $F_{Tj+1}(x,y)$ meets the criteria according to step 9, then $F_j(x,y) = F_{Tj+1}(x,y)$, set $S_{i+1}(x,y) = S_i(x,y)$ and $i = i+1$, and go to step 11. Otherwise, set $j = j+1$, go to step 3, and continue up to step 10.
Step 11: Determine whether $S_i(x,y)$ has less than three extrema points, and if so, the residual $R(x,y) = S_i(x,y)$, and the decomposition is complete. Otherwise, go to step 2 and continue up to step 11.

In the process of extracting BIMFs, the number of extreme points in $S_{i+1}(x,y)$ should be less than the number of extreme points in $S_i(x,y)$. Let the BIMFs and the residual of a signal together be named as bidimensional empirical mode components (BEMCs). All the BEMCs compose the original 2D signal as follows:

$$\sum F(x,y) = \sum_{i=1}^{K+1} F_{i+1}(x,y) = I(x,y) \tag{4.46}$$

where $F_i(x,y)$ is the ith BEMC, and K is the total number of BEMCs except the residual.

4.4.3 The Proposed Method

4.4.3.1 Overview of the Proposed Method

This section proposes an FABEMD approach for filtering of workpiece surfaces using HDM. Compared with the original BEMD approach, three aspects are improved: (1) The proposed approach does not need to calculate the minimum Euclidean distance between adjacent extreme points and does not need to calculate adjacent maxima and minima distance array. (2) The proposed approach uses the adaptive window algorithm to optimally select the window size of the order statistics filters. (3) The envelope surface is drawn by the extremum filters and the average filters, so the computation time is greatly saved.

The framework of the proposed approach is shown in Fig. 4.35, and the procedure involves the following steps:

Step 1: Read the 3D engineering surface data collected by HDM.
Step 2: Use the neighboring window algorithm to find local maxima and local minima of the original surface, and generate the extrema spectrum.

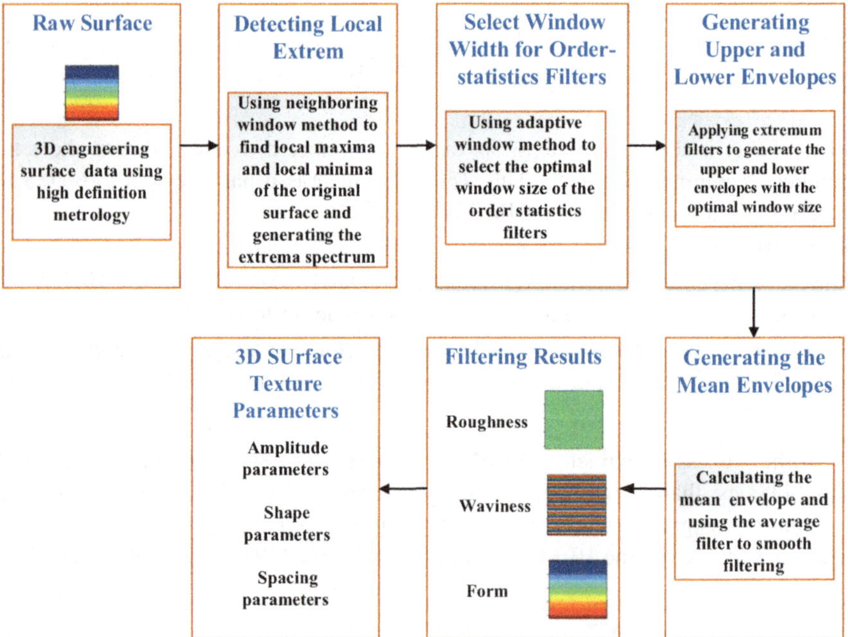

Fig. 4.35 The framework of the proposed approach

Step 3: Use the developed adaptive window algorithm to select the optimal window size of the order statistics filters.
Step 4: Apply extremum filters to generate the upper and lower envelopes with the optimal window size.
Step 5: Calculate the mean envelope and use the average filter to smooth filtering.
Step 6: Decompose into roughness, waviness, and form.
Step 7: Calculate the 3D surface parameters of the decomposed components, amplitude parameters, shape parameters, and spacing parameters for surface texture analysis.

4.4.3.2   Detecting Local Extrema

In the detection of local extrema, the local maxima and minima points from the given data need to be found. The 2D array of local maxima (minima) points is called a maxima (minima) map. This section uses the neighboring window algorithm to find local maxima points P_{ij} and local minima points Q_{ij}. In this algorithm, a data point is considered as a local maximum (minimum), if its value is strictly higher (lower) than all of its neighbors. Let $A(x, y)$ be an $M \times N$ 2D matrix represented by

$$
A(x, y) = \begin{bmatrix} a_{11} & a_{12} & \cdots & a_{1N} \\ a_{21} & a_{22} & \cdots & a_{2N} \\ \vdots & \vdots & \cdots & \vdots \\ a_{M1} & a_{M2} & \cdots & a_{MN} \end{bmatrix}
\tag{4.47}
$$

where a_{mn} is the element of $A(x, y)$ located in the mth row and nth column.

Let the window size for local extrema determination be $w_{ex} \times w_{ex}$. Then, the extrema points are calculated by

$$
\begin{aligned}
a_{mn} &\triangleq \text{Local Maximum, if } a_{mn} > a_{kl} \\
&\quad\;\; \text{Local Minimum, if } a_{mn} < a_{kl}
\end{aligned}
\tag{4.48}
$$

where

$$
\begin{aligned}
k &= m - \frac{W_{ex} - 1}{2} : m + \frac{W_{ex} - 1}{2}, \quad (k \neq m) \\
l &= n - \frac{W_{ex} - 1}{2} : n + \frac{W_{ex} - 1}{2}, \quad (l \neq n)
\end{aligned}
\tag{4.49}
$$

Generally, a given 2D data only needs a 3×3 window to get an optimum extrema map. A higher window size may result in a lower number of local extrema points for a given data matrix. In order to find extrema points at the boundary or corner, the neighboring points within the window that are beyond the boundary are neglected. For illustration purposes, consider an 8×8 matrix given in Fig. 4.36a. The maxima map given in Fig. 4.36b and minima map given in Fig. 4.36c are obtained through applying a 3×3 neighboring window for every point in the matrix.

4.4.3.3   Adaptive Window Algorithm to Select Window Width for Order Statistics Filters

After obtaining the extrema map, the next step is to draw the upper and lower envelopes. The core of the traditional BEMD approach is the order statistics filter, in which the maximum value filter (MAX Filter) is used to draw the upper envelope, and the minimum value filter (MIN Filter) is used to draw the lower envelope. Order statistics filters are spatial filters whose responses are determined based on ordering (ranking) the elements contained within the data area encompassed by the filters. The response of the filters at any point is determined by the ranking result. For the envelope estimation approach, the most important part is the order statistics filter, and for the order statistics filter, the crucial part is to select an appropriate window size for the filter.

There are four types of the window sizes $(d_1 \leq d_2 \leq d_3 \leq d_4)$ for an order statistics filter determined by the extrema spectrum [Eq. (4.50)].

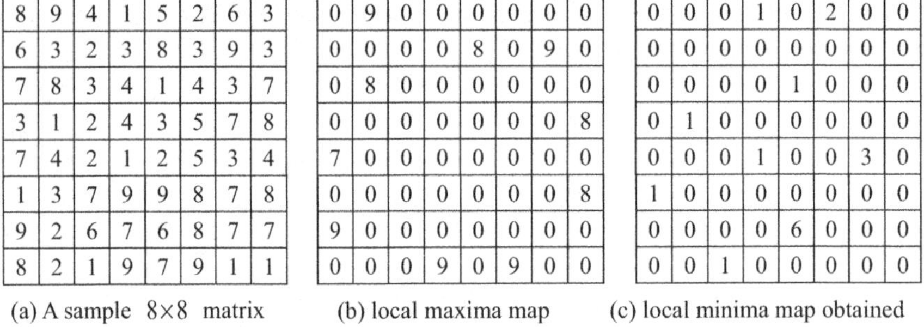

8	9	4	1	5	2	6	3
6	3	2	3	8	3	9	3
7	8	3	4	1	4	3	7
3	1	2	4	3	5	7	8
7	4	2	1	2	5	3	4
1	3	7	9	9	8	7	8
9	2	6	7	6	8	7	7
8	2	1	9	7	9	1	1

(a) A sample 8×8 matrix

0	9	0	0	0	0	0	0
0	0	0	0	8	0	9	0
0	8	0	0	0	0	0	0
0	0	0	0	0	0	0	8
7	0	0	0	0	0	0	0
0	0	0	0	0	0	0	8
9	0	0	0	0	0	0	0
0	0	0	9	0	9	0	0

(b) local maxima map obtained from (a)

0	0	0	1	0	2	0	0
0	0	0	0	0	0	0	0
0	0	0	0	1	0	0	0
0	1	0	0	0	0	0	0
0	0	0	1	0	0	3	0
1	0	0	0	0	0	0	0
0	0	0	0	6	0	0	0
0	0	1	0	0	0	0	0

(c) local minima map obtained from (a)

Fig. 4.36 A sample 8×8 data matrix using the neighboring window algorithm

$$w_{en} = d_1 = \text{minimum}\{\text{minimum}\{d_{adj-max}\},\ \text{minimum}\{d_{adj-min}\}\}$$
$$w_{en} = d_2 = \text{maximum}\{\text{minimum}\{d_{adj-max}\},\ \text{minimum}\{d_{adj-min}\}\}$$
$$w_{en} = d_3 = \text{minimum}\{\text{maximum}\{d_{adj-max}\},\ \text{maximum}\{d_{adj-min}\}\}$$
$$w_{en} = d_4 = \text{maximum}\{\text{maximum}\{d_{adj-max}\},\ \text{maximum}\{d_{adj-min}\}\} \tag{4.50}$$

where w_{en} is the window size of an order statistics filter, $d_{adj-max}$ is the adjacent maxima distance array, and $d_{adj-min}$ is the adjacent minima distance array.

There is always an optimal window size of an order statistics filter in the interval $[d_1, d_4]$ for a workpiece surface, which makes the FABEMD approach have the best filtering results. It is usually difficult to adaptively choose an optimal window size for a workpiece surface. Therefore, an adaptive window algorithm is developed to automatically select the optimal window size of order statistics filters. The flowchart of the proposed adaptive window algorithm is shown in Fig. 4.37, and the specific implementation process is described as follows.

Step 1: Read the workpiece surfaces data, obtain roughness using Gaussian filter, and calculate the root-mean-square roughness parameter S_{qg} as the reference value.

$$S_{qg} = \sqrt{\frac{1}{A} \iint_A z^2(x, y)\mathrm{d}x\mathrm{d}y} \tag{4.51}$$

Step 2: Calculate the window size d of order statistics filters and the original window size d as shown in Eq. (4.52).

$$d = \frac{1}{2} \times \sqrt{\frac{M \times N}{n_{\text{extrem}}}} \tag{4.52}$$

where $M \times N$ is the image size, and n_{extrem} is the sum of the numbers of localized maxima and minima. A value of d corresponds to the distance between extrema in case of its uniform distribution.

Step 3: According to the original window size d, obtain workpiece surface roughness using FABEMD filter and calculate the root-mean-square roughness parameter S_{qf}.

$$S_{qf} = \sqrt{\frac{1}{A} \iint_A z^2(x, y)\mathrm{d}x\mathrm{d}y} \tag{4.53}$$

Step 4: Calculate the deviation ratio δ [Eq. (4.54)]. When $\delta < 0.1$, go to step 6, otherwise, go to step 5.

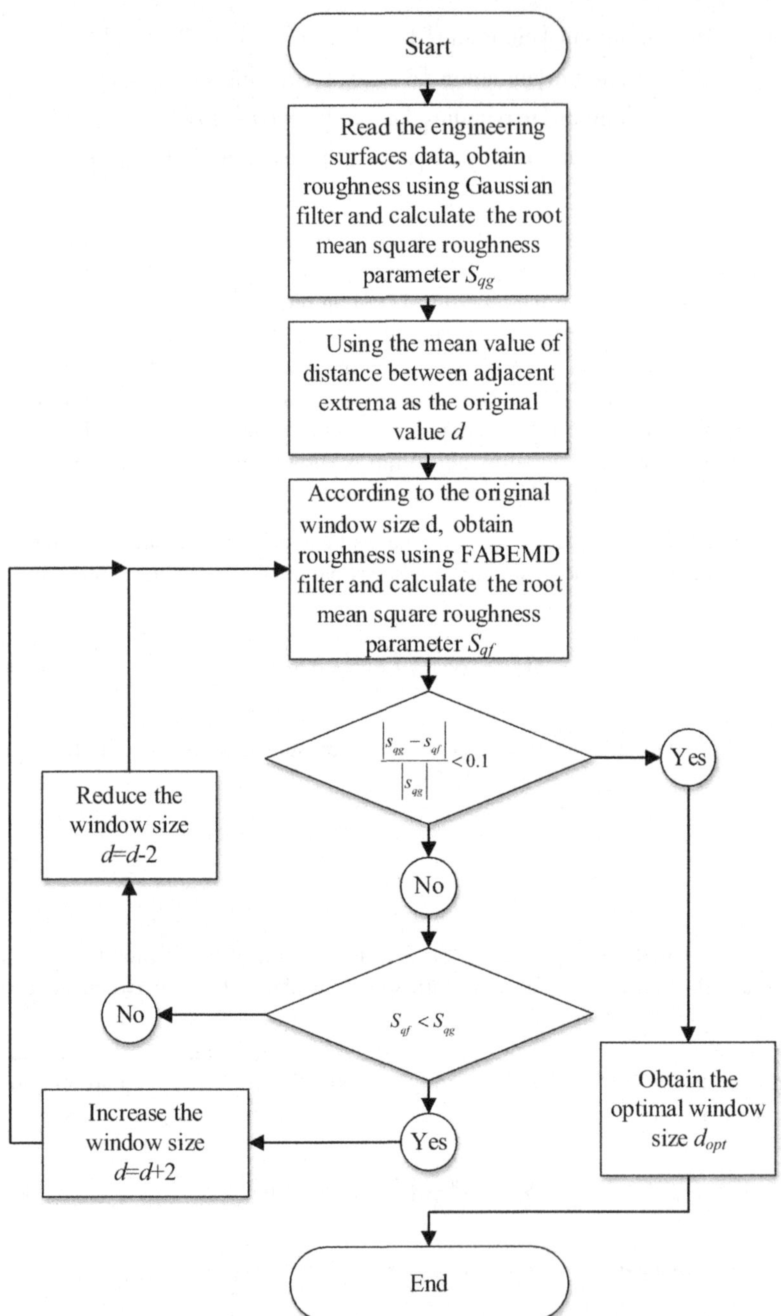

Fig. 4.37 Flowchart of the adaptive window algorithm

$$\delta = \frac{\left|S_{qg} - S_{qf}\right|}{\left|S_{qg}\right|} \times 100\% \tag{4.54}$$

Step 5: If $S_{qf} < S_{qg}$, then increase the window size of order statistics filters, set $d = d + 2$, and return to step 3 until $\delta < 0.1$. If $S_{qf} > S_{qg}$, reduce the window size of order statistics filters, set $d = d - 2$, and return to step 3 until $\delta < 0.1$.

Step 6: If $\delta < 0.1$, stop searching, obtain the optimal window size d_{opt}, and the program ends.

4.4.3.4   Generating Envelopes

After determining the optimal window size of order statistics filter for envelope information, the maximum and minimum filters are applied to obtain the upper and lower envelopes, UE_{ij} and LE_{ij}, respectively,

$$UE_{ij}(x, y) = \underset{(s,t)\in Z_{xy}}{\mathrm{MAX}} \left\{ S_{ij}(s, t) \right\} \tag{4.55}$$

$$LE_{ij}(x, y) = \underset{(s,t)\in Z_{xy}}{\mathrm{MIN}} \left\{ S_{ij}(s, t) \right\} \tag{4.56}$$

In Eq. (4.55), the value of the upper envelope UE_{ij} at any point (x, y) is the maximum value of the elements in S_{ij} in the region defined by Z_{xy}. Z_{xy} is the square region of size $W_{en} \times W_{en} \left(W_{en} = d_{\mathrm{opt}} \right)$ centered at any point (x, y) of S_{ij}. Similarly, the value of the lower envelope LE_{ij} at any point (x, y) is the minimum value of the elements in S_{ij} in the region defined by Z_{xy} in Eq. (4.56). It should be noted that the maximum filter and minimum filter produce new 2D matrices for upper and lower envelope surfaces from the given 2D data matrix, and they do not alter the actual 2D data. Since smooth continuous surfaces for upper and lower envelopes are preferable, averaging smoothing operations are carried out on both UE_{ij} and LE_{ij} by applying the same window size for corresponding order statistics filters. The averaging smoothing operations are expressed as

$$UE_{ij}(x, y) = \frac{1}{W_{sm} \times W_{sm}} \sum_{(s,t)\in Z_{xy}} UE_{ij}(s, t) \tag{4.57}$$

$$LE_{ij}(x, y) = \frac{1}{W_{sm} \times W_{sm}} \sum_{(s,t)\in Z_{xy}} LE_{ij}(s, t) \tag{4.58}$$

where Z_{xy} is the square region of size $W_{sm} \times W_{sm}$ centered at any point (x, y) of UE_{ij} or LE_{ij}, and W_{sm} is the window width of the averaging smoothing filter and $W_{sm} = W_{en} = d_{\mathrm{opt}}$.

9	9	9	8	8	9	9	9
9	9	9	8	8	9	9	9
8	8	8	8	8	9	9	9
8	8	8	4	5	7	8	8
7	7	9	9	9	9	8	8
9	9	9	9	9	9	8	8
9	9	9	9	9	9	9	8
9	9	9	9	9	9	9	7

(a) upper envelope matrix using maximum filter before smoothing

3	2	1	1	1	2	2	3
3	2	1	1	1	1	2	3
1	1	1	1	1	1	3	3
1	1	1	1	1	1	3	3
1	1	1	1	1	2	3	3
1	1	1	1	1	2	3	3
1	1	1	1	6	1	1	1
2	1	1	1	6	1	1	1

(b) lower envelope matrix using minimum before smoothing

Fig. 4.38 The upper and lower envelope matrixes using maximum filter and minimum filter before smoothing

The operation in Eq. (4.57) is the arithmetic mean filtering. From the smoothed envelopes UE_{ij} and LE_{ij}, the mean or average envelope ME_{ij} is calculated as

$$ME_{ij} = (UE_{ij} + LE_{ij})/2 \tag{4.59}$$

A 3×3 window for maximum and minimum filters is applied to the data matrix of Fig. 4.36a, and the results in the upper and lower envelope matrices are shown in Fig. 4.38a, b, respectively. Window width W_{en} obtained by Type-3 is 3. The averaging smoothing operations are applied to Fig. 4.38a, b, and the results in the smoothed upper and lower envelope matrices are shown in Fig. 4.39a, b, respectively. The mean envelope matrix produced by averaging the matrices in Fig. 4.39a, b is shown in Fig. 4.39c according to Eq. (4.58).

In the BEMD approach, SD criterion [Eq. (4.45)] is employed as the most important stopping criterion, and the maximum number of allowable iterations is used as an auxiliary stopping criterion to prevent the occurrence of over-sifting.

The correlation coefficient is used to evaluate the similarity between two vectors.

$$r = \frac{\sum_m \sum_n (A_{mn} - \overline{A})(B_{mn} - \overline{B})}{\sqrt{\left(\sum_m \sum_n (A_{mn} - \overline{A})^2 \right)\left(\sum_m \sum_n (B_{mn} - \overline{B})^2 \right)}} \tag{4.60}$$

where r is utilized to evaluate the correlation coefficient of the matrixes or vectors with the same size. $\overline{A}(\overline{B})$ is the mean value of $A(B)$. The numbers of rows and columns of A and B are M and N, respectively. The larger the correlation coefficient is, the bigger the similarity is. Conversely, the smaller the correlation coefficient is, the smaller the similarity is.

9.0	9.0	8.7	8.3	8.3	8.7	9.0	9.0
8.7	8.7	8.4	8.2	8.3	8.7	9.0	9.0
8.3	8.3	7.8	7.3	7.3	8.0	8.6	8.7
7.7	7.9	7.7	7.6	7.6	8.0	8.3	8.3
8.0	8.2	8.0	7.9	7.8	8.0	8.1	8.0
8.3	8.6	8.8	9.0	9.0	8.8	8.4	8.2
9.0	9.0	9.0	9.0	9.0	8.9	8.4	8.2
9.0	9.0	9.0	9.0	9.0	9.0	8.5	8.3

(a) upper envelope matrix after smoothing

2.5	2.0	1.3	1.0	1.2	1.5	2.2	2.5
2.0	1.7	1.2	1.0	1.1	1.6	2.2	2.7
1.5	1.3	1.1	1.0	1.0	1.6	2.2	2.8
1.0	1.0	1.0	1.0	1.1	1.8	2.4	3.0
1.0	1.0	1.0	1.0	1.2	1.9	2.6	3.0
1.0	1.0	1.0	1.6	1.8	2.2	2.1	2.3
1.2	1.1	1.0	2.1	2.2	2.4	1.6	1.7
1.3	1.2	1.0	2.7	2.7	2.7	1.0	1.0

(b) lower envelope matrix after smoothing

5.8	5.5	5.0	4.7	4.8	5.1	5.6	5.8
5.3	5.2	4.8	4.6	4.7	5.1	5.6	5.8
4.9	4.8	4.4	4.2	4.2	4.8	5.4	5.8
4.3	4.4	4.3	4.3	4.3	4.9	5.4	5.7
4.5	4.6	4.5	4.4	4.5	4.9	5.3	5.5
4.7	4.8	4.9	5.3	5.4	5.5	5.3	5.3
5.1	5.1	5.0	5.6	5.6	5.7	5.0	4.9
5.1	5.1	5.0	5.8	5.8	5.8	4.8	4.6

(c) mean envelope matrix obtained by averaging the data in (a) and (b)

Fig. 4.39 The envelope matrix after smoothing

4.4.4 Simulation Experiment

To verify the performance of the proposed approach, a simulated 3D surface is randomly generated and filtered using the proposed FABEMD approach. Then, the filtering result is compared with the original BEMD approach. The size of the simulated surface is 42 mm × 42 mm, the sampling interval is 0.1mm, and the numerical expression of the surface is described as follows:

$$z(x, y) = 0.8 \times x + 0.8 \times \sin(0.4 \times \pi \times y) + 0.5 \times \text{normrnd}(0, 0.1)$$
$$0 \leq x \leq 42 \text{ mm}, 0 \leq y \leq 42 \text{ mm} \tag{4.61}$$

where $\text{normrnd}(\mu, \delta)$ denotes random number following the normal distribution with mean parameter $\mu = 0$ and standard deviation parameter $\delta = 0.1$. $0.5 \times \text{normrnd}(0, 0.1)$ (a random noise) is the roughness component, $0.8 \times \sin(0.4 \times \pi \times y)$ (a sinusoidal function) is the waviness component, and $0.8 \times x$ (an inclined surface) is the form component. The simulated surface and its components are shown in Fig. 4.40.

Figure 4.41 shows the filtering results using the FABEMD approach to the simulated surface. From Fig. 4.41, it can be seen that Figs. 4.40b–d and 4.41b–d,

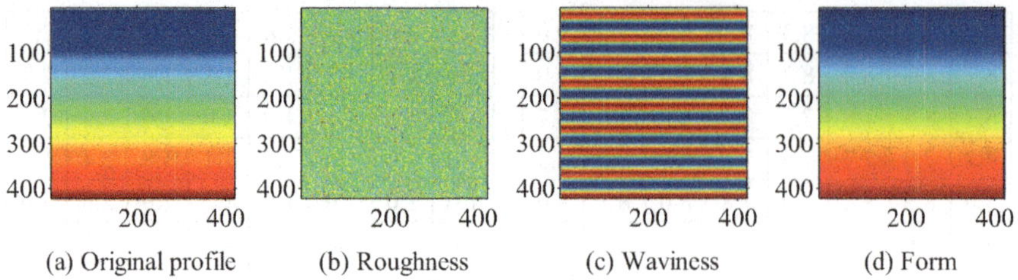

(a) Original profile   (b) Roughness   (c) Waviness   (d) Form

Fig. 4.40 Simulated surface and its components

have obvious similarity. The correlation coefficients of Figs. 4.40b–d and 4.41b–d are 0.9423, 0.9701, and 0.9997, respectively. Therefore, the proposed FABEMD approach can properly separate the simulated surface into different surface components and can effectively eliminate mode mixing.

The simulated surface is decomposed by the original BEMD approach, and the decomposition results are shown in Fig. 4.42. From Fig. 4.42, it can be seen that two sets of BIMF components (BIMF1 and BIMF2, BIMF3 and BIMF4) have a similar scale and mutual influence. According to Eq. (4.60), the correlation coefficients of BIMF1 and BIMF2, and BIMF3 and BIMF4 are 0.9349 and 0.7596, respectively. The three simulated components cannot be separated into different BIMFs, and the mode mixing problem occurs.

4.4.5 Case Studies

In order to further validate the effectiveness of the proposed approach, the point cloud data of three different workpiece surfaces collected by HDM is used for filtering analysis. Gaussian filter (ISO 11562:1996, ASME B46.1-2002), the FABEMD-based filter, and the BEMD-based filter are applied, respectively, and then the filtering performances of these three filters are analyzed.

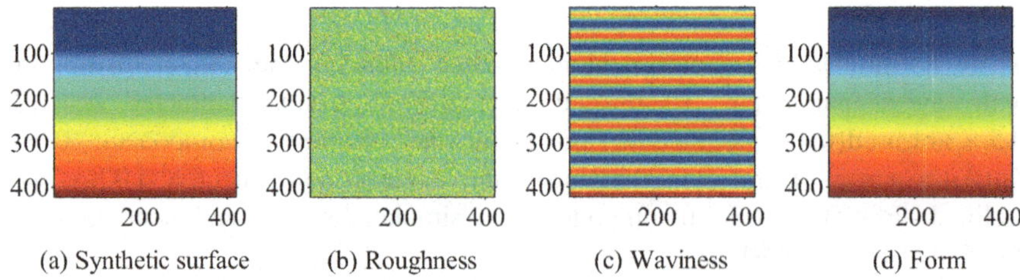

(a) Synthetic surface   (b) Roughness   (c) Waviness   (d) Form

Fig. 4.41 Filtering result of simulated surface using the proposed FABEMD approach

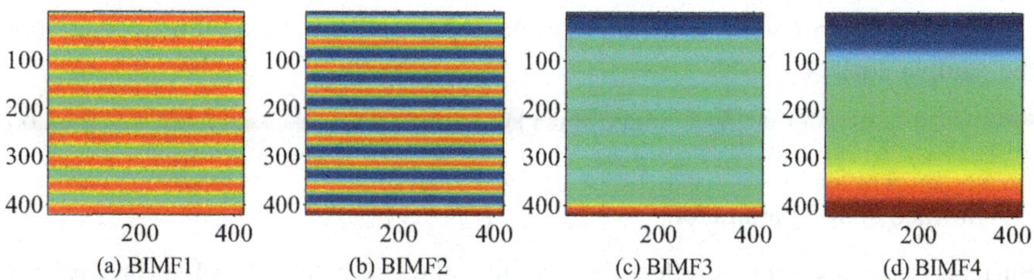

Fig. 4.42 Filtering result of simulated surface using the original BEMD approach

The 3D Gaussian filter is a 3D extension of 2D Gaussian filter, which is realized by the convolution of 3D surface data and the 2D weight function. The weighting function of Gaussian filter for surface texture analysis is

$$S(x, y) = \frac{1}{\alpha^2 \lambda_{xc} \lambda_{yc}} \exp\left[-\left[\pi\left(\frac{x}{\alpha \lambda_{xc}}\right)^2 + \pi\left(\frac{y}{\alpha \lambda_{yc}}\right)^2\right]\right] \tag{4.62}$$

where λ_{xc} and λ_{yc} are the cutoff wavelengths in the x- and y-directions, and $\alpha = \sqrt{\ln 2 / \pi} = 0.4697$. Set $\lambda_{xc} = \lambda_{xy} = 0.64$ mm.

In order to evaluate the characteristics of 3D surface topography, some representative 3D surface parameters are selected, including surface amplitude parameters S_a, S_q, and S_t, surface shape parameters S_{ku}, and surface spacing parameters S_{al} and S_{tr}.

(1) Surface amplitude parameters

Surface amplitude parameter is the description of surface height deviations. The parameters S_a, S_q, and S_t are selected.

The average roughness parameter S_a reflects the deviation of arithmetic mean distribution of the mean surface data

$$S_a = \frac{1}{A} \iint\limits_A |z(x, y)| \mathrm{d}x\mathrm{d}y \tag{4.63}$$

where $Z(x, y)$ is the height value of the point located in line x and column y and A is the evaluation area.

The root-mean-square roughness S_q is the standard deviation of the sample to reflect the surface roughness

$$S_q = \sqrt{\frac{1}{A} \iint\limits_A z^2(x, y) \mathrm{d}x\mathrm{d}y} \tag{4.64}$$

The maximum height of texture surface represents the maximum fluctuation of the sample surface

$$S_t = |\max(z(x, y))| + |\min(z(x, y))| \qquad (4.65)$$

(2) Surface shape parameters

Surface shape parameters reflect the height shape characteristics of surface texture. Surface kurtosis S_{ku} reflects the surface height distribution

$$S_{ku} = \frac{1}{sq^4} \left[\frac{1}{A} \iint_A z^4(x, y)\mathrm{d}x\mathrm{d}y \right] \qquad (4.66)$$

(3) Surface spacing parameters

Surface spacing parameters reflect the spatial distribution of a sample surface. The fastest autocorrelation decay length S_{al} represents the composition of surface components, and $S_{al} \in (0, 1)$. The smaller the value of S_{al} is, the bigger the probability of the surface with low-frequency components is. The larger the value of S_{al} is, the bigger the probability of the surface with high-frequency components is. S_{al} is the horizontal distance of the autocorrelation function (ACF) that has the fastest decay in any direction to a specified threshold value η, and $0 < \eta < 1$ ($\eta = 0.1$ in this section).

$$S_{al} = \min_{\tau_x, \tau_y \in R} \left(\sqrt{\tau_x^2 + \tau_y^2} \right) \quad \text{where} \quad R = \{ (\tau_x, \tau_y) : \mathrm{ACF}(\tau_x, \tau_y) \leq \eta \} \qquad (4.67)$$

The ACF of a 3D surface is defined as a convolution of the surface with itself

$$\mathrm{ACF}(\tau_x, \tau_y) = \frac{\iint_A z(x, y)z(x - \tau_x, y - \tau_y)\mathrm{d}x\mathrm{d}y}{\iint_A z(x, y)z(x, y)\mathrm{d}x\mathrm{d}y} \qquad (4.68)$$

where τ_x, τ_y are the x-direction and y-direction shifts, respectively.

The surface texture aspect ratio parameter S_{tr} [Eq. (4.69)] reflects the surface texture characteristics of sample surface. The smaller the value of S_{tr} is, the more obvious the surface texture is. The larger the value of S_{tr} is, the less obvious the surface texture is.

$$S_{tr} = \frac{\min\limits_{\tau_x, \tau_y \in R} \left(\sqrt{\tau_x^2 + \tau_y^2} \right)}{\max\limits_{\tau_x, \tau_y \in R} \left(\sqrt{\tau_x^2 + \tau_y^2} \right)} \quad \text{where} \quad R = \{ (\tau_x, \tau_y) : \mathrm{ACF}(\tau_x, \tau_y) \leq \eta \} \qquad (4.69)$$

4.4.5.1   Case Study I

The first workpiece surface is the surface of a pump valve plate, and its height map is shown in Fig. 4.43. A surface sample is arbitrarily selected from the surface, the size of the surface sample is 6.4 mm × 6.4 mm (shown in Fig. 4.44a), and the

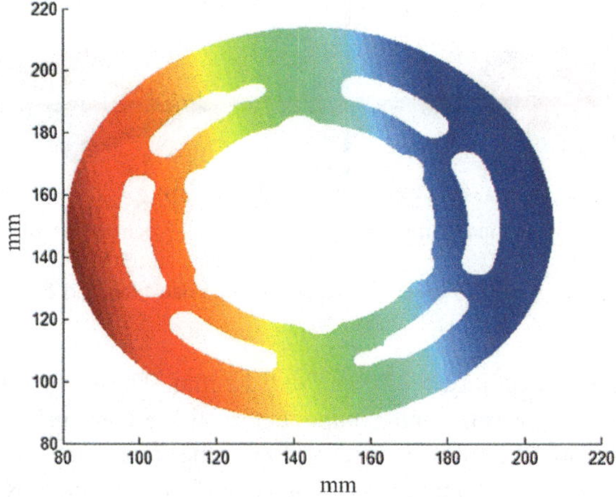

Fig. 4.43 The height map of the valve plate surface

sampling interval is 0.01 mm. The results of the FABEMD-based filter are shown in Fig. 4.44. Figure 4.44b–d represents roughness, waviness, and form, respectively.

Because of the boundary effects caused by the local weighted average in the Gaussian filter, the evaluation area of the Gaussian filter needs to be subtracted the first and last wavelengths from the sum in order to obtain good filtering results, and

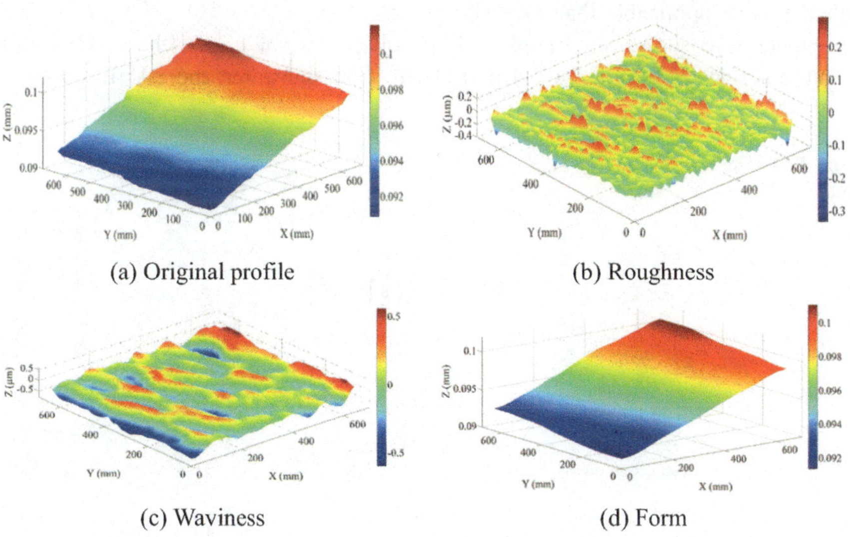

Fig. 4.44 Filtering results of a surface sample using the FABEMD filter

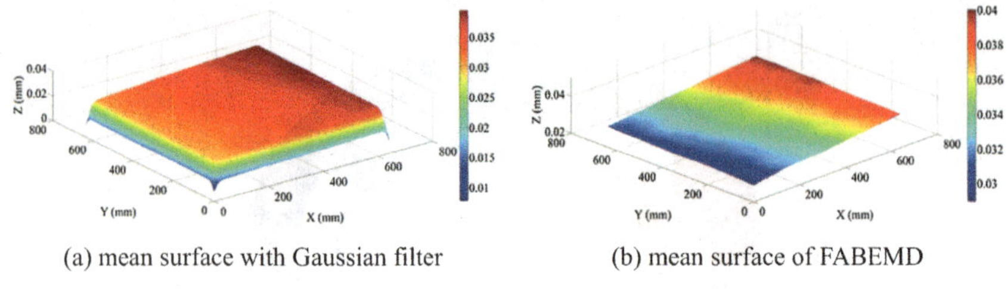

(a) mean surface with Gaussian filter   (b) mean surface of FABEMD

Fig. 4.45 Comparison of mean surfaces obtained by Gaussian filter and FABEMD

then the actual area becomes 5.12 mm × 5.12 mm. The surface sample is separated into two components: one component is the mean surface including the waviness surface and the form surface, and the other one is the roughness surface.

Figure 4.45 presents a comparison of mean surfaces obtained by Gaussian filter and FABEMD-based filter without discarding any boundary region, and the actual size of evaluation area is 6.4 mm × 6.4 mm. Figure 4.45a illustrates that the values on the boundary of the mean surface by Gaussian filter are close to zero, and there are obvious distortions on the boundary. This is the result of the convolution operation. As shown in Fig. 4.45b, there is no distortion in the FABEMD filtering results. Therefore, due to the boundary effect of Gaussian filter, it is necessary to remove the boundary in practical application. It is feasible for the original data to be long enough. However, when the original data length is limited, and then the boundary area is given up, there will be inevitably some influences on the evaluation results. The FABEMD approach does not need to predict the extrema points on the boundary, so the boundary distortions are avoided. The FABEMD-based filter is more applicable than Gaussian filter.

Figure 4.46 shows six samples of the oil pump valve, and the size of each sample surface is 6.4 mm × 6.4 mm. Then, Gaussian filter, the FABEMD-based

Fig. 4.46 The distribution map of six sample surface

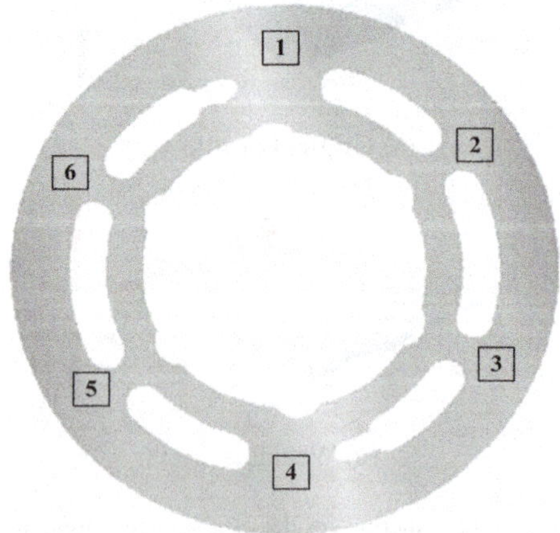

filter, and the BEMD-based filter are applied to the six samples, and the filtering results are compared. To be convenient for the comparison, the actual evaluation area of three filters is set as 5.12 mm × 5.12 mm and the sampling interval is 0.01 mm.

Table 4.9 shows the surface amplitude parameters obtained by Gaussian filter and FABEMD-based filter. It can be found from the last three columns in Table 4.9 that the mean differences are close to 5% in S_a values, close to 4% in S_q values, and close to 10% in S_t values.

Table 4.10 shows the surface amplitude parameters obtained by Gaussian filter and the BEMD-based filter. It can be found from the last three columns in Table 4.10 that the mean differences are close to 200% in S_a values, close to 190% in S_q values, and close to 130% in S_t values. It can be seen that the mean differences in the surface amplitude parameters obtained by Gaussian filter and the FABEMD-based filter are less than 10%. Nevertheless, the mean differences in the surface amplitude parameters obtained by Gaussian filter and the original BEMD-based filter are about 130–200%. Therefore, the FABEMD-based filter can reflect the sample surface amplitude, but the BEMD-based filter cannot reflect the sample surface amplitude.

Table 4.11 shows the surface shape and spacing parameters obtained by Gaussian filter and the FABEMD-based filter. It can be found from the last three columns in Table 4.11 that the mean differences are close to 11% in S_{ku} values, close to 12% in S_{al} values, and close to 11% in S_{tr} values. In general, the mean differences in shape parameters and spacing parameters are within 20%.

Table 4.12 shows the surface shape and spacing parameters obtained by Gaussian filter and the FABEMD-based filter. It can be found from the last three columns in Table 4.12 that the mean differences are close to 18% in S_{ku} values, close to 100% in S_{al} values, and close to 15% in S_{tr} values.

Table 4.13 lists the comparison results of the window width and computational time using the BEMD-based filter and FABEMD-based filter. It can be found that the optimal window width of the order statistics filters searched by the

Table 4.9 The comparison of surface amplitude parameters using the FABEMD-based filter and Gaussian filter

Unit (μm)	Gaussian filter			FABEMD-based filter			Difference		
	S_a	S_q	S_t	S_a	S_q	S_t	S_a (%)	S_q (%)	S_t (%)
L_1	0.0405	0.0522	0.4367	0.0387	0.0507	0.4866	4.460	2.779	11.437
L_2	0.0488	0.0643	0.6279	0.0468	0.0619	0.6239	4.090	3.651	0.632
L_3	0.0413	0.0527	0.4678	0.0364	0.0478	0.5888	11.965	9.315	25.848
L_4	0.0412	0.0532	0.5390	0.0397	0.0522	0.5570	3.600	1.955	3.340
L_5	0.0422	0.0549	0.4816	0.0415	0.0542	0.5611	1.594	1.224	16.506
L_6	0.0427	0.0556	0.6329	0.0411	0.0548	0.6145	3.809	1.371	2.900
Mean	0.0428	0.0555	0.5310	0.0407	0.0536	0.5720	4.920	3.382	10.111

Difference is calculated by $\left| \dfrac{\text{NFABEMD value} - \text{Gaussian value}}{\text{Gaussian value}} \right| \times 100\%$

Table 4.10 The comparison of surface amplitude parameters using original BEMD-based filter and Gaussian filter

Unit (µm)	Gaussian filter			Original BEMD filter			Difference		
	S_a	S_q	S_t	S_a	S_q	S_t	S_a (%)	S_q (%)	S_t (%)
L_1	0.0405	0.0522	0.4367	0.1276	0.1632	1.1715	215.208	212.657	168.277
L_2	0.0488	0.0643	0.6279	0.1453	0.1811	1.1899	197.975	181.751	89.513
L_3	0.0413	0.0527	0.4678	0.1235	0.1558	1.1604	199.075	195.436	148.022
L_4	0.0412	0.0532	0.5390	0.1304	0.1653	1.2846	216.821	210.692	138.344
L_5	0.0422	0.0549	0.4816	0.1215	0.1535	1.0845	187.952	179.418	125.184
L_6	0.0427	0.0556	0.6329	0.1250	0.1573	1.2452	192.632	182.914	96.762
Mean	0.0428	0.0555	0.5310	0.1289	0.1627	1.1894	201.610	193.811	127.684

Table 4.11 The comparison of surface shape and surface space parameters using the FABEMD-based filter and Gaussian filter

	Gaussian filter			FABEMD-based filter			Difference		
	S_{ku}	S_{al}	S_{tr}	S_{ku}	S_{al}	S_{tr}	S_{ku} (%)	S_{al} (%)	S_{tr} (%)
L_1	3.4910	10.7703	0.5038	3.9946	9.4868	0.4543	14.424	11.917	9.821
L_2	4.2383	10.1980	0.4079	4.3547	9.4340	0.3594	2.746	7.492	11.893
L_3	3.5110	10.1980	0.5036	4.4378	8.5440	0.4220	26.399	16.219	16.219
L_4	3.8398	10.0499	0.3766	4.3204	8.9443	0.3720	12.516	11.001	1.222
L_5	3.8443	10.1980	0.4122	3.9735	8.9443	0.3874	3.360	12.294	6.019
L_6	4.2349	10.0499	0.5575	4.4616	8.9443	0.4339	5.353	11.001	22.173
Mean	3.8599	10.2440	0.4603	4.2571	9.0496	0.4048	10.800	11.654	11.224

Table 4.12 The comparison of surface shape and surface space parameters using the BEMD-based filter and Gaussian filter

	Gaussian filter			BEMD-based filter			Difference		
	S_{ku}	S_{al}	S_{tr}	S_{ku}	S_{al}	S_{tr}	S_{ku} (%)	S_{al} (%)	S_{tr} (%)
L_1	3.4910	10.7703	0.5038	3.3864	21.1896	0.4156	2.998	96.741	17.517
L_2	4.2383	10.1980	0.4079	2.8532	18.9737	0.3518	32.680	86.052	13.761
L_3	3.5110	10.1980	0.5036	3.1626	19.6469	0.4154	9.923	92.654	17.522
L_4	3.8398	10.0499	0.3766	3.3659	21.8403	0.3608	12.342	117.319	4.201
L_5	3.8443	10.1980	0.4122	3.0063	19.1050	0.3835	21.799	87.340	6.974
L_6	4.2349	10.0499	0.5575	3.0917	20.2485	0.4129	26.996	101.480	25.935
Mean	3.8599	10.2440	0.4603	3.1443	20.1673	0.3900	17.790	96.931	14.318

FABEMD-based filter is 15 or 17. That is to say, the window width of the order statistics filter is relatively stable in the filtering process, which means that the surface topography of the workpieces is more uniform. However, the window width

Table 4.13 The comparison of window width and computational time using the BEMD-based filter and FABEMD-based filter

Comparative item	Gaussian filter	BEMD-based filter		FABEMD-based filter		Timescale (%)
	Time (s)	Original window size	Time (s)	Optimal window size	Time (s)	
L_1	88	59	412	17	94	22.82
L_2	145	57	407	17	172	42.26
L_3	93	63	407	15	96	23.59
L_4	90	69	546	15	90	16.48
L_5	87	59	326	17	99	30.37
L_6	89	61	437	17	95	21.74
Mean	98.67	–	422.5	–	107.67	26.20

Timescale is calculated by $\left| \dfrac{\text{NFABEMD Time} - \text{BEMD Time}}{\text{BEMD Time}} \right| \times 100\%$

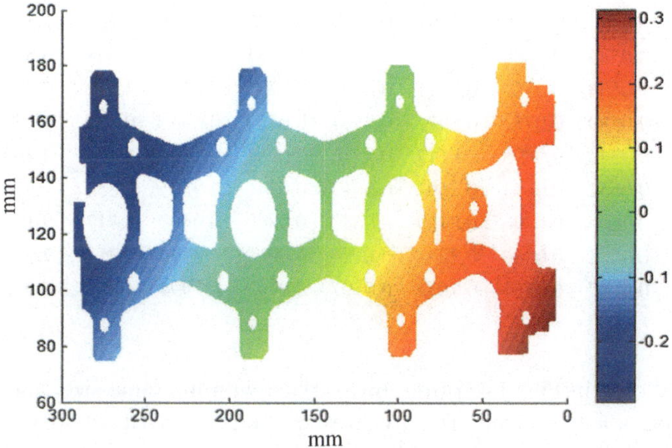

Fig. 4.47 The height map of the engine cylinder head

of the order statistics filters searched by the BEMD-based ranges from 57 to 69 and the filter window width changes greatly, which cannot reflect the surface topography. Moreover, the filtering time of the FABEMD-based filter is less than 30% of the filtering time of the BEMD-based filter, so the computational efficiency is improved.

4.4.5.2   Case Study II

The second surface is the joint surface of an engine cylinder head, which is made of aluminum. The height map of this surface is shown in Fig. 4.47. Eight locations are

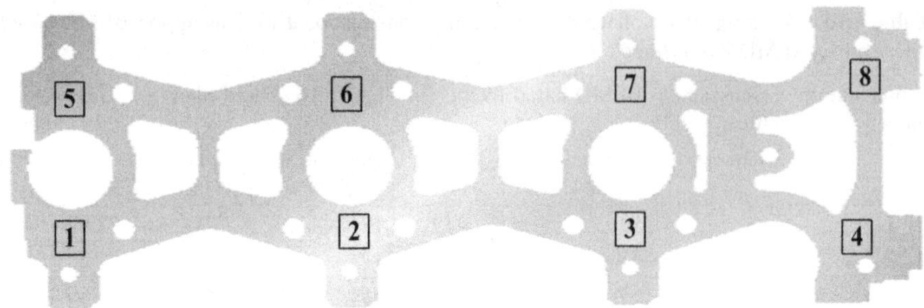

Fig. 4.48 The distribution map of eight sample surfaces

Table 4.14 The comparison of surface amplitude parameters using the FABEMD-based filter and Gaussian filter

Unit (μm)	Gaussian filter			FABEMD-based filter			Difference		
	S_a	S_q	S_t	S_a	S_q	S_t	S_a (%)	S_q (%)	S_t (%)
L_1	0.0534	0.0690	0.6682	0.0539	0.0707	0.7618	0.976	2.507	14.008
L_2	0.0539	0.0685	0.5922	0.0565	0.0731	0.636	4.838	6.595	7.396
L_3	0.0558	0.0728	0.7431	0.0583	0.0785	0.8556	4.482	7.783	15.139
L_4	0.0589	0.0752	0.5977	0.0591	0.0764	0.655	0.266	1.650	9.587
L_5	0.0540	0.0698	0.6117	0.0516	0.0681	0.7514	4.344	2.469	22.838
L_6	0.0651	0.0863	0.8597	0.0617	0.0846	0.9325	5.253	1.976	8.468
L_7	0.0605	0.0796	0.8109	0.0640	0.0869	0.9489	5.717	9.196	17.018
L_8	0.0682	0.0880	0.7744	0.0724	0.0949	0.854	6.253	7.950	10.279
Mean	0.0587	0.0761	0.7072	0.0597	0.0792	0.7994	4.016	5.016	13.092

selected (shown in Fig. 4.48) from this surface with the same size, and then eight small surfaces are obtained. Then, Gaussian filter, the FABEMD-based filter, and BEMD-based filter are applied to the eight sample surfaces, and the surface texture parameters are obtained by the three filters.

Table 4.14 shows the surface amplitude parameters obtained by Gaussian filter and the FABEMD-based filter. It can be found from the last three columns in Table 4.6 that the mean differences are close to 4% in S_a values, close to 5% in S_q values, and close to 13% in S_t values. Table 4.10 shows the surface amplitude parameters obtained by Gaussian filter and the BEMD-based filter. It can be found from the last three columns in Table 4.15 that the mean differences are close to 320% in S_a values, close to 320% in S_q values, and close to 250% in S_t values.

Table 4.16 shows the surface shape and spacing parameters obtained by Gaussian filter and the FABEMD-based filter. It can be found from the last three

Table 4.15 The comparison of surface amplitude parameters using the BEMD-based filter and Gaussian filter

Unit (µm)	Gaussian filter			BEMD-based filter			Difference		
	S_a	S_q	S_t	S_a	S_q	S_t	S_a (%)	S_q (%)	S_t (%)
L_1	0.0534	0.0690	0.6682	0.2002	0.2601	2.0383	274.797	276.852	205.047
L_2	0.0539	0.0685	0.5922	0.2136	0.2861	2.2826	295.948	317.471	285.447
L_3	0.0558	0.0728	0.7431	0.2466	0.3125	2.4807	341.940	329.037	233.842
L_4	0.0589	0.0752	0.5977	0.2733	0.3446	2.3799	363.976	358.349	298.162
L_5	0.0540	0.0698	0.6117	0.2565	0.3266	2.8264	375.294	367.791	362.017
L_6	0.0651	0.0863	0.8597	0.2912	0.3512	2.2874	347.392	307.100	166.080
L_7	0.0605	0.0796	0.8109	0.2234	0.2915	2.2083	269.052	266.361	172.310
L_8	0.0682	0.0880	0.7744	0.2888	0.3607	2.6552	323.665	310.142	242.867
Mean	0.0587	0.0761	0.7072	0.2492	0.3167	2.3948	324.008	316.638	245.721

Table 4.16 The comparison of surface space parameters using FABEMD filter and Gaussian filter

	Gaussian filter			FABEMD-based filter			Difference		
	S_{ku}	S_{al}	S_{tr}	S_{ku}	S_{al}	S_{tr}	S_{ku} (%)	S_{al} (%)	S_{tr} (%)
L_1	3.7896	10.2956	0.3463	4.2491	8.4853	0.2666	12.125	17.584	23.010
L_2	3.2838	9.4340	0.2964	3.8343	8.4853	0.2606	16.763	10.056	12.073
L_3	3.9827	9.8995	0.3255	5.1213	8.4853	0.2666	28.588	14.286	18.093
L_4	3.3280	9.8995	0.3041	3.6743	8.4853	0.2553	10.404	14.286	16.049
L_5	3.6304	10.6301	0.3067	4.1058	8.4853	0.2352	13.094	20.177	23.306
L_6	4.2649	9.8995	0.2791	5.0170	8.4853	0.2306	17.633	14.286	17.380
L_7	4.1878	10.0000	0.2881	5.4650	8.4853	0.2599	30.497	15.147	9.784
L_8	3.6529	10.6301	0.3220	3.8548	9.8489	0.2616	5.528	7.350	18.740
Mean	3.7650	10.0860	0.3085	4.4152	8.6557	0.2546	16.829	14.146	17.305

columns in Table 4.16 that the mean differences are close to 17% in S_{ku} values, close to 15% in S_{al} values, and close to 18% in S_{tr} values. In general, the mean differences in shape parameters and spacing parameters are within 20% [7, 8].

Table 4.17 shows the surface shape and spacing parameters obtained by Gaussian filter and the FABEMD-based filter. It can be found from the last three columns in Table 4.17 that the mean differences are close to 17% in S_{ku} values, close to 175% in S_{al} values, and close to 32% in S_{tr} values.

Table 4.18 shows the comparison results of the window width and computational time using the BEMD-based filter and FABEMD-based filter. It can be seen that the optimal window width of the order statistics filters searched by the FABEMD-based filter is 15 or 17. That is to say, the window width of the order statistics filter is relatively stable in the filtering process, which means that the

Table 4.17 The comparison of surface space parameters using the BEMD-based filter and Gaussian filter

	Gaussian filter			BEMD-based filter			Difference		
	S_{ku}	S_{al}	S_{tr}	S_{ku}	S_{al}	S_{tr}	S_{ku} (%)	S_{al} (%)	S_{tr} (%)
L_1	3.7896	10.2956	0.3463	3.6244	25.0000	0.4595	4.359	142.821	32.699
L_2	3.2838	9.4340	0.2964	4.3492	28.2843	0.4650	32.445	199.813	56.875
L_3	3.9827	9.8995	0.3255	3.6657	27.2029	0.4415	7.960	174.791	35.629
L_4	3.3280	9.8995	0.3041	2.9795	30.4795	0.4771	10.473	207.889	56.896
L_5	3.6304	10.6301	0.3067	3.5005	29.1548	0.3586	3.578	174.265	16.907
L_6	4.2649	9.8995	0.2791	2.3775	29.6985	0.3420	44.254	200.000	22.531
L_7	4.1878	10.0000	0.2881	3.6887	21.6333	0.3529	11.919	116.333	22.517
L_8	3.6529	10.6301	0.3220	3.0149	30.4138	0.3679	17.466	186.109	14.272
Mean	3.7650	10.0860	0.3085	3.4000	27.7334	0.4081	16.557	175.253	32.291

Table 4.18 The comparison of window width and computational time using the BEMD-based filter and FABEMD-based filter

Comparative item	Gaussian filter	BEMD-based filter		FABEMD-based filter		Timescale (%)
	Time (s)	Original window size	Time (s)	Optimal window size	Time (s)	
L_1	115	77	602	15	130	21.59
L_2	79	83	677	15	82	12.11
L_3	129	81	1521	15	134	8.81
L_4	247	99	1546	15	267	17.27
L_5	88	89	864	15	88	10.19
L_6	59	75	587	15	57	9.71
L_7	79	63	400	15	101	25.25
L_8	56	91	840	17	61	7.26
Mean	106.5	–	879.63	–	115	14.02

surface topography of the workpieces is more uniform. The window width of the order statistics filters searched by the BEMD-based filter ranges from 63 to 99 and the filter window width changes greatly, which cannot reflect the surface topography of the workpieces. Moreover, the filtering time of the FABEMD-based filter is less than 15% of the BEMD-based filter's filtering time, so the computational efficiency is improved.

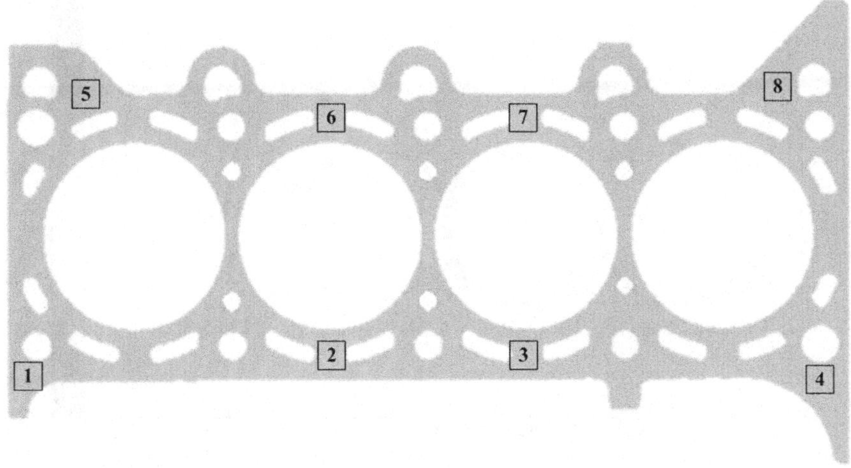

Fig. 4.49 The distribution map of eight sample surfaces

4.4.5.3   Case Study III

The third surface is the top surface of an engine cylinder block, which is made of cast iron FC250. Eight locations (shown in Fig. 4.49) from three top surfaces of engine blocks A, B, and C using HDM (shown in Fig. 4.50) are selected. Then, Gaussian filter, the FABEMD-based filter, and BEMD-based filter are applied to the eight sample surfaces, and the surface texture parameters are obtained by the three filters.

Table 4.19 shows the surface amplitude parameters obtained by Gaussian filter and the FABEMD-based filter. It can be found from the last three columns in Table 4.19 that the mean differences are close to 6% in S_a values, close to 6% in S_q values, and close to 15% in S_t values. Table 4.20 shows the surface amplitude parameters obtained by Gaussian filter and the BEMD-based filter. It can be found from the last three columns in Table 4.20 that the mean differences are close to 80% in S_a values, close to 70% in S_q values, and close to 20% in S_t values.

Table 4.21 shows the surface shape and spacing parameters obtained by Gaussian filter and the FABEMD filter. It can be found from the last three columns in Table 4.21 that the mean differences are close to 15% in S_{ku} values, close to 10% in S_{al} values, and close to 10% in S_{tr} values. Table 4.22 shows the surface shape and spacing parameters obtained by Gaussian filter and the FABEMD-based filter. It can be found from the last three columns in Table 4.22 that the mean differences are close to 40% in S_{ku} values, close to 60% in S_{al} values, and close to 20% in S_{tr} values.

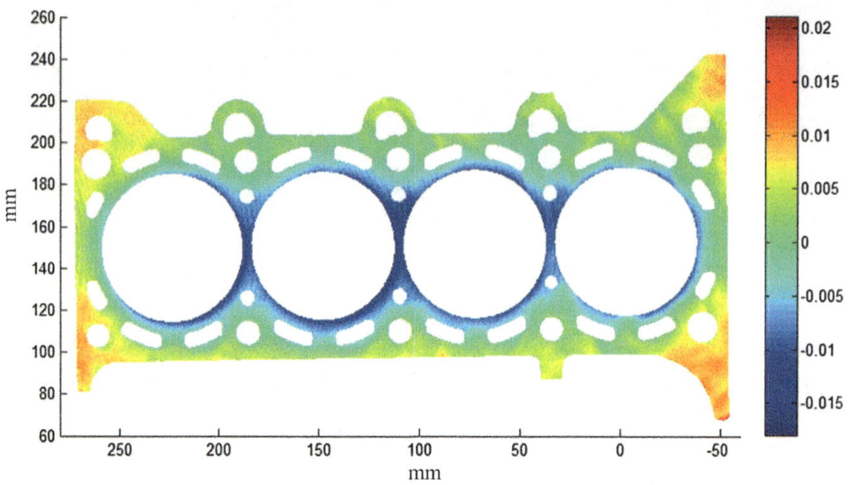

Fig. 4.50 The height map of the top surface of an engine cylinder block

Table 4.23 shows the comparison results of the window width and computational time using the BEMD-based filter and FABEMD-based filter. It can be seen that the optimal window width of the order statistics filters searched by the FABEMD filter is 17 or 19. That is to say, the window width of the order statistics filter is relatively stable in the filtering process, which means that the surface topography of the workpieces is more uniform. However, the window width of the order statistics filters searched by the BEMD-based filter ranges from 41 to 53, and the filter window width changes greatly, which cannot reflect the surface topography of the parts. Moreover, the filtering time of FABEMD filter is less than 50% of the BEMD filter's filtering time, so the computational efficiency is improved.

It can be seen from the above three cases that the differences in the surface texture parameters between Gaussian filter and the BEMD-based filter are relatively large. So the BEMD-based filter cannot be directly applied to workpiece surface separations. Nevertheless, for Gaussian filter and the FABEMD-based filter, their differences in the average roughness parameter S_a and the root-mean-square roughness parameter S_q are close to 6%, and the maximum height of texture surface S_t is less than 20%. Moreover, the surface shape parameters S_{ku} and the surface spacing parameters S_{al} and S_t are also less than 20%. Because the difference of 10–20% is very common as reported in [7, 8], the results of three real-world surface data show that the performance of the FABEMD-based filter is similar to the performance of the standard Gaussian filer, but superior to the BEMD-based filter, moreover, the FABEMD-based filter does not have the boundary distortions.

Table 4.19 The comparison of surface amplitude parameters using the FABEMD-based filter and Gaussian filter

Unit (μm)	Gaussian filter			FABEMD-based filter			Difference		
	S_a	S_q	S_t	S_a	S_q	S_t	S_a (%)	S_q (%)	S_t (%)
A_1	0.1156	0.1651	2.6943	0.1116	0.1554	2.4564	3.459	5.898	8.828
A_2	0.1183	0.1622	2.1284	0.1104	0.1563	1.9070	6.710	3.620	10.401
A_3	0.1398	0.1887	2.0661	0.1315	0.1756	1.7277	5.955	6.957	16.380
A_4	0.1399	0.1901	1.8789	0.1322	0.1819	2.1565	5.529	4.310	14.774
A_5	0.1600	0.2139	2.4569	0.1513	0.2017	2.1235	5.440	5.699	13.569
A_6	0.1119	0.1581	2.1261	0.1096	0.1544	2.4022	2.108	2.296	12.988
A_7	0.1379	0.1848	1.9595	0.1321	0.1746	1.5903	4.233	5.538	18.843
A_8	0.1428	0.1935	1.9830	0.1318	0.1772	1.7560	7.735	8.422	11.447
Mean	0.1333	0.1821	2.1616	0.1263	0.1721	2.0150	5.146	5.343	13.404
B_1	0.1452	0.1994	2.0488	0.1361	0.1858	2.2308	6.262	6.818	8.883
B_2	0.1361	0.1854	2.3379	0.1220	0.1679	2.2269	10.320	9.461	4.748
B_3	0.1468	0.1960	1.7773	0.1392	0.1860	1.6384	5.163	5.115	7.816
B_4	0.1557	0.2109	2.2060	0.1455	0.1970	2.0811	6.557	6.572	5.663
B_5	0.1622	0.2219	2.7310	0.1513	0.2060	2.2255	6.761	7.162	18.509
B_6	0.1447	0.1910	2.0815	0.1415	0.1877	1.7823	2.198	1.749	14.374
B_7	0.1601	0.2105	2.0696	0.1507	0.1985	2.1003	5.897	5.686	1.484
B_8	0.1563	0.2093	2.2703	0.1497	0.1975	1.8803	4.191	5.596	17.178
Mean	0.1509	0.2031	2.1903	0.1420	0.1908	2.0207	5.918	6.020	9.832
C_1	0.1474	0.2101	2.9192	0.1390	0.1925	2.4003	5.738	8.388	17.776
C_2	0.1266	0.1719	2.0016	0.1320	0.1765	1.7630	4.277	2.661	11.924
C_3	0.1308	0.1794	2.2440	0.1267	0.1709	2.0207	3.153	4.749	9.950
C_4	0.1370	0.1885	2.3623	0.1362	0.1902	2.8634	0.615	0.925	21.213
C_5	0.1369	0.1872	2.2162	0.1389	0.1864	1.9199	1.441	0.429	13.370
C_6	0.1211	0.1648	1.7526	0.1192	0.1605	1.6505	1.577	2.596	5.822
C_7	0.1397	0.1883	2.1512	0.1419	0.1898	1.9103	1.522	0.784	11.198
C_8	0.1476	0.2032	2.3495	0.1473	0.2023	2.1238	0.188	0.419	9.604
Mean	0.1359	0.1867	2.2496	0.1351	0.1836	2.0815	2.314	2.619	12.607

4.4.5.4   Comparison with the Shearlet-Based Filter

Take the surface of case study I as an example. The recent developed shearlet-based filter [13] has good filtering performances and is applied to further comparison analysis. The cutoff wavelength of shearlet is 0.64 mm. Table 4.24 shows the surface amplitude parameters obtained by Gaussian filter and the shearlet-based filter. It can be found from the last three columns in Table 4.24 that the mean

Table 4.20 The comparison of surface amplitude parameters using the BEMD-based filter and Gaussian filter

Unit (μm)	Gaussian filter			BEMD-based filter			Difference		
	S_a	S_q	S_t	S_a	S_q	S_t	S_a (%)	S_q (%)	S_t (%)
A_1	0.1156	0.1651	2.6943	0.1833	0.2410	3.0396	58.471	45.918	12.816
A_2	0.1183	0.1622	2.1284	0.2337	0.3000	2.5406	97.484	84.992	19.369
A_3	0.1398	0.1887	2.0661	0.2306	0.2873	2.2990	64.982	52.206	11.273
A_4	0.1399	0.1901	1.8789	0.2288	0.2944	2.3113	63.567	54.861	23.011
A_5	0.1600	0.2139	2.4569	0.2410	0.3034	2.4455	50.683	41.843	0.463
A_6	0.1119	0.1581	2.1261	0.2089	0.2708	2.9939	86.706	71.322	40.821
A_7	0.1379	0.1848	1.9595	0.2240	0.2786	2.0193	62.363	50.697	3.050
A_8	0.1428	0.1935	1.9830	0.2379	0.2992	2.3661	66.599	54.662	19.320
Mean	0.1333	0.1821	2.1616	0.2235	0.2843	2.5019	68.857	57.062	16.265
B_1	0.1452	0.1994	2.0488	0.2356	0.3034	2.6825	62.252	52.123	30.926
B_2	0.1361	0.1854	2.3379	0.2404	0.3073	2.5096	76.705	65.719	7.342
B_3	0.1468	0.1960	1.7773	0.2428	0.3052	2.1047	65.424	55.710	18.425
B_4	0.1557	0.2109	2.2060	0.2516	0.3191	2.2775	61.613	51.315	3.243
B_5	0.1622	0.2219	2.7310	0.2836	0.3584	2.7291	74.820	61.539	0.068
B_6	0.1447	0.1910	2.0815	0.2413	0.3031	2.6128	66.786	58.665	25.527
B_7	0.1601	0.2105	2.0696	0.2443	0.3056	2.4334	52.525	45.148	17.581
B_8	0.1563	0.2093	2.2703	0.2435	0.3045	2.6625	55.805	45.525	17.274
Mean	0.1509	0.2031	2.1903	0.2479	0.3133	2.5015	64.491	54.468	15.048
C_1	0.1474	0.2101	2.9192	0.2371	0.3012	2.7601	60.801	43.347	5.452
C_2	0.1266	0.1719	2.0016	0.2377	0.2972	2.0208	87.728	72.879	0.955
C_3	0.1308	0.1794	2.2440	0.2027	0.2565	2.1661	54.949	42.966	3.471
C_4	0.1370	0.1885	2.3623	0.2533	0.3296	3.5874	84.893	74.879	51.859
C_5	0.1369	0.1872	2.2162	0.2471	0.3083	2.3451	80.539	64.743	5.816
C_6	0.1211	0.1648	1.7526	0.2093	0.2595	1.9339	72.831	57.537	10.347
C_7	0.1397	0.1883	2.1512	0.2631	0.3264	2.1417	88.277	73.282	0.439
C_8	0.1476	0.2032	2.3495	0.2584	0.3273	2.5157	75.072	61.081	7.074
Mean	0.1359	0.1867	2.2496	0.2386	0.3008	2.4338	75.636	61.339	10.677

Table 4.21 The comparison of surface shape and surface space parameters using the FABEMD-based filter and Gaussian filter

	Gaussian filter			FABEMD-based filter			Difference		
	S_{ku}	S_{al}	S_{tr}	S_{ku}	S_{al}	S_{tr}	S_{ku} (%)	S_{al} (%)	S_{tr} (%)
A_1	9.1148	9.8995	0.8682	6.7925	9.8489	0.8705	25.478	0.512	0.263
A_2	5.0939	10.2956	0.8435	5.6636	9.2195	0.7199	11.185	10.452	14.645
A_3	4.5250	8.9443	0.6293	4.0407	8.4853	0.5970	10.703	5.132	5.132
A_4	4.5382	10.1980	0.9014	4.7112	9.8489	0.9105	3.811	3.424	1.014
A_5	4.5811	8.9443	0.6984	4.1437	8.6023	0.6717	9.549	3.823	3.823

(continued)

Table 4.21 (continued)

	Gaussian filter			FABEMD-based filter			Difference		
	S_{ku}	S_{al}	S_{tr}	S_{ku}	S_{al}	S_{tr}	S_{ku} (%)	S_{al} (%)	S_{tr} (%)
A_6	5.6777	9.4340	0.9434	5.7261	8.9443	0.9701	0.853	5.191	2.835
A_7	4.3867	9.2195	0.7656	3.8753	9.2195	0.8086	11.657	0.000	5.612
A_8	4.4407	8.6023	0.6083	4.1878	7.8102	0.5254	5.695	9.208	13.629
Mean	5.2948	9.4422	0.7823	4.8926	8.9974	0.7592	9.867	4.718	5.869
B_1	4.7258	9.8489	0.8179	4.7322	9.8489	0.8638	0.136	0.000	5.612
B_2	5.3655	9.8489	0.9105	5.2404	9.2195	0.8673	2.331	6.390	4.747
B_3	4.2510	9.2195	0.6487	4.0750	8.4853	0.5970	4.140	7.964	7.964
B_4	4.4152	10.4403	0.9338	4.5338	10.0499	0.9454	2.686	3.740	1.243
B_5	5.0360	9.2195	0.6778	4.3764	9.2195	0.7199	13.098	0.000	6.210
B_6	3.9728	10.0499	0.9761	3.8823	9.8489	0.9849	2.278	2.000	0.897
B_7	3.9866	9.4340	0.8274	3.9565	9.4340	0.8274	0.755	0.000	0.000
B_8	4.5384	8.9443	0.6576	3.9084	8.4853	0.6307	13.881	5.132	4.089
Mean	4.5364	9.6257	0.8062	4.3381	9.3239	0.8046	4.913	3.153	3.845
C_1	7.3363	9.8489	0.8638	5.3702	9.8489	0.9105	26.800	0.000	5.409
C_2	5.0942	9.8489	0.8638	4.2327	9.8489	0.8179	16.911	0.000	5.314
C_3	5.2124	8.4853	0.5970	4.4603	8.4853	0.5708	14.430	0.000	4.395
C_4	5.3706	10.0499	0.8883	6.0444	10.0499	0.9454	12.545	0.000	6.430
C_5	5.0714	9.2195	0.6853	4.2062	8.9443	0.6648	17.061	2.986	2.986
C_6	4.7760	9.8489	0.9849	4.2548	9.8489	0.9849	10.913	0.000	0.000
C_7	4.4940	9.4340	0.7834	4.0786	9.8489	0.8638	9.243	4.398	10.256
C_8	5.0301	8.4853	0.5708	4.4833	8.4853	0.5708	10.871	0.000	0.000
Mean	5.2981	9.4026	0.7797	4.6413	9.4200	0.7911	14.847	0.923	4.349

differences are close to 11% in S_a values, close to 9% in S_q values, and close to 11% in S_t values. Table 4.25 shows the surface shape and spacing parameters obtained by Gaussian filter and shearlet-based filter. It can be found from the last three columns in Table 4.17 that the mean differences are close to 21% in S_{ku} values, close to 18% in S_{al} values, and close to 23% in S_{tr} values. The results in Tables 4.9 and 4.24 indicate that the FABEMD-based filter and shearlet-based filter have no distinct difference from each other in the surface amplitude parameters. The results in Tables 4.11 and 4.25 indicate that the FABEMD filter is better than the shearlet-based filter in the surface shape and spacing parameters. Table 4.26 shows the differences in the surface amplitude parameters S_a between Gaussian filter and the shearlet filter, BEMD-based filter, and FABEMD-based filter. It can be found that the differences between Gaussian filter and FABEMD-based filter are close to 4.9% in S_a values, which is the smallest. It can also be found that the time of FABEMD-based filter is shorter than shearlet filter and BEMD-based filter, but slightly longer than Gaussian filter.

Table 4.22 The comparison of surface shape and surface space parameters using BEMD filter and Gaussian filter

	Gaussian filter			BEMD-based filter			Difference		
	S_{ku}	S_{al}	S_{tr}	S_{ku}	S_{al}	S_{tr}	S_{ku} (%)	S_{al} (%)	S_{tr} (%)
A_1	9.1148	9.8995	0.8682	4.6746	13.8924	0.8786	48.715	40.335	1.197
A_2	5.0939	10.2956	0.8435	3.5070	15.0000	0.4779	31.152	45.693	43.335
A_3	4.5250	8.9443	0.6293	2.9224	13.4164	0.6448	35.416	50.000	2.453
A_4	4.5382	10.1980	0.9014	3.3547	14.0357	0.8158	26.079	37.631	9.494
A_5	4.5811	8.9443	0.6984	3.0721	12.5300	0.7640	32.940	40.089	9.383
A_6	5.6777	9.4340	0.9434	4.0292	15.2971	0.7351	29.035	62.148	22.076
A_7	4.3867	9.2195	0.7656	2.8634	14.1421	0.8771	34.725	53.393	14.552
A_8	4.4407	8.6023	0.6083	3.0115	12.2066	0.6028	32.183	41.898	0.894
Mean	5.2948	9.4422	0.7823	3.4294	13.8150	0.7245	33.781	46.398	12.923
B_1	4.7258	9.8489	0.8179	3.3730	13.9284	0.7726	28.627	41.421	5.538
B_2	5.3655	9.8489	0.9105	3.1891	15.5242	0.8082	40.562	57.624	11.243
B_3	4.2510	9.2195	0.6487	2.9632	13.0000	0.6055	30.294	41.005	6.662
B_4	4.4152	10.4403	0.9338	3.1491	13.9284	0.7726	28.676	33.410	17.263
B_5	5.0360	9.2195	0.6778	3.0794	15.0333	0.7622	38.853	63.059	12.449
B_6	3.9728	10.0499	0.9761	3.0651	14.3178	0.6501	22.848	42.468	33.396
B_7	3.9866	9.4340	0.8274	2.9785	13.6015	0.9212	25.288	44.175	11.336
B_8	4.5384	8.9443	0.6576	2.9789	12.2066	0.7334	34.363	36.473	11.531
Mean	4.5364	9.6257	0.8062	3.0970	13.9425	0.7532	31.189	44.954	13.677
C_1	7.3363	9.8489	0.8638	3.4618	14.1421	0.8623	52.813	43.592	0.178
C_2	5.0942	9.8489	0.8638	2.8783	15.2315	0.6325	43.499	54.653	26.782
C_3	5.2124	8.4853	0.5970	3.1265	12.0416	0.6303	40.019	41.912	5.572
C_4	5.3706	10.0499	0.8883	3.8724	15.8114	0.8380	27.896	57.329	5.661
C_5	5.0714	9.2195	0.6853	2.9632	14.5602	0.7828	41.570	57.928	14.225
C_6	4.7760	9.8489	0.9849	2.8417	15.1327	0.7921	40.501	53.650	19.576
C_7	4.4940	9.4340	0.7834	2.7657	16.0312	0.7491	38.456	69.931	4.386
C_8	5.0301	8.4853	0.5708	3.0339	13.4164	0.6202	39.685	58.114	8.653
Mean	5.2981	9.4026	0.7797	3.1179	14.5459	0.7384	40.555	54.638	10.629

Table 4.23 The comparison of window width and computational time using the BEMD-based filter and FABEMD-based filter

Comparative item	Gaussian filter	BEMD-based filter		FABEMD-based filter		Timescale (%)
	Time (s)	Original window size	Time (s)	Optimal window size	Time (s)	
A_1	132	47	292	19	149	51.03
A_2	136	47	278	17	140	50.36
A_3	213	49	622	19	220	35.37

(continued)

Table 4.23 (continued)

Comparative item	Gaussian filter	BEMD-based filter		FABEMD-based filter		Timescale (%)
	Time (s)	Original window size	Time (s)	Optimal window size	Time (s)	
A_4	176	45	268	19	204	76.12
A_5	179	43	249	19	185	74.30
A_6	197	47	484	17	220	45.45
A_7	248	47	293	19	294	100.34
A_8	101	47	361	17	109	30.19
Mean	172.75	–	355.88	–	190.13	57.90
B_1	144	51	460	19	154	33.48
B_2	145	51	393	17	150	38.17
B_3	157	53	479	19	170	35.49
B_4	138	47	312	19	151	48.40
B_5	196	55	385	19	200	51.95
B_6	143	47	330	19	162	49.09
B_7	123	41	260	19	140	53.85
B_8	157	47	285	19	174	61.05
Mean	150.375	–	363	–	162.63	46.43
C_1	233	45	437	19	258	59.04
C_2	212	47	587	19	240	40.89
C_3	199	41	442	19	210	47.51
C_4	317	51	525	19	321	61.14
C_5	123	51	407	19	152	37.35
C_6	137	47	312	19	142	45.51
C_7	159	51	366	19	172	46.99
C_8	163	51	367	19	189	51.50
Mean	192.86	–	430.38	–	210.5	48.74

Table 4.24 The comparison of surface amplitude parameters using shearlet-based filter and Gaussian filter

Unit (μm)	Shearlet filter			Gaussian filter			Difference		
	S_a	S_q	S_t	S_a	S_q	S_t	S_a (%)	S_q (%)	S_t (%)
L_1	0.036	0.047	0.524	0.041	0.052	0.437	12.099	9.770	19.876
L_2	0.043	0.058	0.663	0.049	0.064	0.628	12.500	9.798	5.654
L_3	0.037	0.049	0.512	0.041	0.053	0.468	10.412	7.590	9.470
L_4	0.037	0.049	0.525	0.041	0.053	0.539	11.165	8.835	2.579
L_5	0.038	0.051	0.499	0.042	0.055	0.482	9.953	8.015	3.571
L_6	0.039	0.052	0.778	0.043	0.056	0.633	9.602	6.115	22.926
Mean	0.038	0.051	0.583	0.043	0.055	0.531	10.955	8.354	10.679

Table 4.25 The comparison of surface shape and surface space parameters using shearlet-based filter and Gaussian filter

	Shearlet filter			Gaussian filter			Difference		
	S_{ku}	S_{al}	S_{tr}	S_{ku}	S_{al}	S_{tr}	S_{ku} (%)	S_{al} (%)	S_{tr} (%)
L_1	4.333	8.944	0.640	3.491	10.770	0.504	24.128	16.954	27.114
L_2	5.095	8.544	0.540	4.238	10.198	0.408	20.211	16.219	32.483
L_3	4.289	8.246	0.592	3.511	10.198	0.504	22.156	19.139	17.554
L_4	4.337	8.246	0.478	3.840	10.050	0.377	12.951	17.947	26.845
L_5	4.464	8.246	0.566	3.844	10.198	0.471	16.107	19.139	20.199
L_6	5.751	8.246	0.633	4.235	10.050	0.558	35.793	17.947	13.453
Mean	4.711	8.412	0.575	3.860	10.244	0.470	21.891	17.891	22.941

Table 4.26 The comparison of surface amplitude parameters and time using Gaussian filter, shearlet-based filter, BEMD-based filter, and FABEMD-based filter

Unit (μm)	Gaussian filter		Shearlet filter		BEMD-based filter		FABEMD-based filter	
	S_a	Difference	Time (s) (%)	Time (s)	Difference (%)	Time (s)	Difference (%)	Time (s)
L_1	0.041	88	12.099	115	215.208	412	4.460	94
L_2	0.049	145	12.500	121	197.975	407	4.090	172
L_3	0.041	93	10.412	99	199.075	407	11.965	96
L_4	0.041	90	11.165	97	216.821	546	3.600	90
L_5	0.042	87	9.953	106	187.952	326	1.594	99
L_6	0.043	89	9.602	109	192.632	437	3.809	95
Mean	0.043	98.67	10.955	129.4	201.610	422.5	4.920	107.67

4.4.6 Conclusions

A novel filter based on FABEMD approach for workpiece surfaces using HDM is proposed. The neighboring window algorithm is presented to extract local extrema and draw the extrema spectrum. The adaptive window algorithm is developed to realize the automatic acquisition of the window size of the order statistics filters. The average smoothing filter is presented for smooth filtering and generating of the mean envelope. The performance of FABEMD-based filter is validated by the simulated surface data and real-world 3D surface data using HDM. Some conclusions can be drawn.

(1) The FABEMD-based filter can effectively decompose the simulated surface into three components: roughness, waviness, and form, and has good applicability for workpiece surface filtering.

(2) Compared with the BEMD-based filter, the optimal window width of the order statistics filters searched by the adaptive window algorithm shows its stability

and effectiveness, which indicates that the FABEMD-based filter has higher accuracy. The envelope surface in the FABEMD-based filter is drawn by the extremum filters and the average filters, so it has higher efficiency than the BEMD-based filter has. Moreover, the proposed approach does not need to calculate the minimum Euclidean distance between adjacent extreme points and calculate adjacent maxima distance array and adjacent minima distance array.

(3) Compared with Gaussian filter that has obvious distortion at the surface boundary, the FABEMD-based filter has no boundary distortions, and the filtering results are basically the same as those of Gaussian filter.

(4) Compared with the shearlet-based filter, the results of the FABEMD-based filter on the surface amplitude parameters are similar, and the results on the surface shape and spacing parameters are better than the shearlet-based filter.

References

1. Raja J, Muralikrishnan B, Fu S (2002) Recent advances in separation of roughness, waviness and form. Precis Eng 26(2):222–235
2. ISO 11562 Geometrical product specification (GPS)—surface texture: profile method—metrological characteristics of phase correct filters (1996)
3. Brinkmann S, Bodschwinna H, Lemke HW (2001) Accessing roughness in three-dimensions using Gaussian regression filtering. Int J Mach Tools Manuf 41(13–14):2153–2161
4. Krystek M (1996) Form filtering by splines. Measurement 18(1):9–15
5. Goto T, Miyakura J, Umeda K, Kadowaki S, Yanagi K (2005) A robust spline filter on the basis of L-2-norm. Precis Eng J Int Soc Precis Eng Nanotechnol 29(2):157–161
6. Srinivasan V (1998) Discrete morphological filters for metrology. In: Proceedings of the 6th ISMQC symposium on metrology for quality control in production, pp 623–628
7. Chen QH, Yang SN, Li Z (1999) Surface roughness evaluation by using wavelets analysis. Precis Eng J Am Soc Precis Eng 23(3):209–212
8. Chen X, Raja J, Simanapalli S (1995) Multi-scale analysis of engineering surfaces. Int J Mach Tools Manuf 35(2):231–238
9. Muralikrishnan B, Raja J (2008) Computational surface and roundness metrology. Springer Science & Business Media
10. Fu SY, Muralikrishnan B, Raja J (2003) Engineering surface analysis with different wavelet bases. J Manuf Sci Eng Trans Asme 125(4):844–852
11. Jiang XQ, Blunt L, Stout KJ (2000) Development of a lifting wavelet representation for surface characterization. Proc R Soc Math Phys Eng Sci 456(2001):2283–2313
12. Josso B, Burton DR, Lalor MJ (2002) Frequency normalised wavelet transform for surface roughness analysis and characterisation. Wear 252(5–6):491–500
13. Huang Z, Shih AJ, Ni J (2006) Laser interferometry hologram registration for three-dimensional precision measurements. J Manuf Sci Eng 128(4):1006
14. Blunt L, Jiang X (2003) Advanced techniques for assessment surface topography: development of a basis for 3D surface texture standards "surfstand". Elsevier
15. ASME B46.1 Surface Texture (Surface Roughness, Waviness, and Lay) (2009)
16. Chen MJ, Pang QL, Wang JH, Cheng K (2008) Analysis of 3D microtopography in machined KDP crystal surfaces based on fractal and wavelet methods. Int J Mach Tools Manuf 48(7–8):905–913
17. Liao Y, Stephenson DA, Ni J (2012) Multiple-scale wavelet decomposition, 3D surface feature exaction and applications. J Manuf Sci Eng Trans Asme 134(1)

18. Zeng W, Jiang X, Scott P (2005) Metrological characteristics of dual-tree complex wavelet transform for surface analysis. Meas Sci Technol 16(7):1410–1417
19. Guo K, Labate D (2007) Optimally sparse multidimensional representation using shearlets. Siam J Math Anal 39(1):298–318
20. Easley G, Labate D, Lim WQ (2008) Sparse directional image representations using the discrete shearlet transform. Appl Comput Harmon Anal 25(1):25–46
21. Wang L, Li B, Tian LF (2013) A novel multi-modal medical image fusion method based on shift-invariant shearlet transform. Imaging Sci J 61(7):529–540
22. Danti A, Poornima K (2012) Face recognition using shearlets. In: Proceedings of the 2012 7th IEEE International Conference on Industrial and Information Systems (ICIIS). IEEE, pp 1–6
23. Mallat S (2009) A wavelet tour of signal processing[M]. Academic Press
24. da Cunha AL, Zhou JP, Do MN (2006) The nonsubsampled contourlet transform: theory, design, and applications. IEEE Trans Image Process 15(10):3089–3101
25. http://www.coherix.com
26. Thomas T, Charlton G (1981) Variation of roughness parameters on some typical manufactured surfaces. Precis Eng 3(2):91–96
27. Wang M, Xi L, Du S (2014) 3D surface form error evaluation using high definition metrology. Precis Eng 38(1):230–236
28. ASME B46.1 Surface texture (surface roughness, waviness, and lay) (2002)
29. Kamguem R, Tahan SA, Songmene V (2013) Evaluation of machined part surface roughness using image texture gradient factor. Int J Precis Eng Manuf 14(2):183–190
30. ISO 25178-6. Geometrical product specifications (GPS)—surface texture: areal—Part 6: classification of methods for measuring surface texture (2010)
31. ISO 4288. Geometrical product specifications (GPS)—surface texture: profile method: rules and procedures for the assessment of surface texture (1996)
32. Seewig J (2005) Linear and robust Gaussian regression filters. In: 7th international symposium on measurement technology and intelligent instruments, vol 13, pp 254–257
33. Du S, Liu C, Huang D (2015) A shearlet-based separation method of 3D engineering surface using high definition metrology. Precis Eng 40:55–73
34. Jiang XJ, Whitehouse DJ (2012) Technological shifts in surface metrology. CIRP Ann Manuf Technol 61(2):815–836
35. Zeng W, Jiang X, Scott PJ, Blunt L (2012) Diffusion filtration for the evaluation of MEMs surface. Int J Precis Eng Manuf
36. Jiang X, Cooper P, Scott PJ (2011) Freeform surface filtering using the diffusion equation. Proc R Soc Math Phys Eng Sci 467(2127):841–859
37. Perona P, Malik J (1990) Scale-space and edge detection using anisotropic diffusion. IEEE Trans Pattern Anal Mach Intell 12(7):629–639
38. Weickert J (2001) Efficient image segmentation using partial differential equations and morphology. Pattern Recogn 34(9):1813–1824
39. Alvarez L, Lions P-L, Morel J-M (1992) Image selective smoothing and edge detection by nonlinear diffusion. II. SIAM J Numer Anal 29(3):845–866
40. Alcantarilla PF, Bartoli A, Davison AJ (2012) KAZE features. In: Computer vision—ECCV 2012, Part VI, vol 7577, pp 214–227
41. ISO 25178-2. Geometrical product specifications (GPS)—surface texture: areal—Part 2: terms, definitions and surface texture parameters (2012)
42. Barari A (2013) Inspection of the machined surfaces using manufacturing data. J Manuf Syst 32(1):107–113
43. Ramasamy SK, Raja J (2013) Performance evaluation of multi-scale data fusion methods for surface metrology domain. J Manuf Syst 32(4):514–522
44. Wang M, Shao Y-P, Du S-C, Xi L-F (2015) A diffusion filter for discontinuous surface measured by high definition metrology. Int J Precis Eng Manuf 16(10):2057–2062
45. Huang NE, Shen Z, Long SR, Wu MLC, Shih HH, Zheng QN, Yen NC, Tung CC, Liu HH (1998) The empirical mode decomposition and the Hilbert spectrum for nonlinear and non-stationary time series analysis. Proc R Soc Math Phys Eng Sci 454(1971):903–995

46. Du SC, Liu T, Huang DL, Li GL (2017) An optimal ensemble empirical mode decomposition method for vibration signal decomposition. J Vib Acoust Trans Asme 139(3)

47. Flandrin P, Rilling G, Goncalves P (2004) Empirical mode decomposition as a filter bank. IEEE Signal Process Lett 11(2):112–114

48. Wu ZH, Huang NE (2004) A study of the characteristics of white noise using the empirical mode decomposition method. Proc R Soc Math Phys Eng Sci 460(2046):1597–1611

49. Boudraa AO, Cexus JC (2007) EMD-Based signal filtering. IEEE Trans Instrum Meas 56 (6):2196–2202

50. Nunes JC, Bouaoune Y, Delechelle E, Niang O, Bunel P (2003) Image analysis by bidimensional empirical mode decomposition. Image Vis Comput 21(12):1019–1026

51. Nunes J, Guyot S, Delechelle E (2005) Texture analysis based on local analysis of the bidimensional empirical mode decomposition. Mach Vis Appl 16(3):177–188

52. Zhou Y, Li HG (2011) Adaptive noise reduction method for DSPI fringes based on bi-dimensional ensemble empirical mode decomposition. Opt Express 19(19):18207–18215

53. Trusiak M, Patorski K, Wielgus M (2012) Adaptive enhancement of optical fringe patterns by selective reconstruction using FABEMD algorithm and Hilbert spiral transform. Opt Express 20(21):23463–23479

54. Riffi J, Mohamed Mahraz A, Tairi H (2013) Medical image registration based on fast and adaptive bidimensional empirical mode decomposition. IET Image Proc 7(6):567–574

55. Bhuiyan SMA, Adhami RR, Khan JF (2008) Fast and adaptive bidimensional empirical mode decomposition using order-statistics filter based envelope estimation. Eurasip J Adv Signal Process

56. Bhuiyan SMA, Adhami RR, Khan JF (2008) A novel approach of fast and adaptive bidimensional empirical mode decomposition. In: 2008 IEEE international conference on acoustics, speech and signal processing, Vols 1–12, pp 1313–1316

57. Du S, Huang D, Wang H (2015) An adaptive support vector machine-based workpiece surface classification system using high-definition metrology. IEEE Trans Instrum Meas 64 (10):2590–2604

58. Du S, Liu C, Xi L (2014) A selective multiclass support vector machine ensemble classifier for engineering surface classification using high definition metrology. J Manuf Sci Eng 137 (1):011003

59. Wang M, Ken T, Du S, Xi L (2015) Tool wear monitoring of wiper inserts in multi-insert face milling using three-dimensional surface form indicators. J Manuf Sci Eng 137(3):031006

60. Du S, Fei L (2015) Co-Kriging method for form error estimation incorporating condition variable measurements. J Manuf Sci Eng

61. Shao Y, Du S, Xi L (2017) 3D machined surface topography forecasting with space-time multioutput support vector regression using high definition metrology. In: Proceedings of the ASME 2017 international design engineering technical conferences & computers and information in engineering conference

62. Huang DL, Du SC, Li GL, Wu ZQ (2017) A systematic approach for online minimizing volume difference of multiple chambers in machining processes based on high-definition metrology. J Manuf Sci Eng Trans Asme 139(8)

63. Nguyen HT, Wang H, Tai BL, Ren J, Hu SJ, Shih A (2016) High-definition metrology enabled surface variation control by cutting load balancing. J Manuf Sci Eng Trans Asme 138(2)

Chapter 5
Surface Classification

5.1 A Brief History of Surface Classification

The surface appearance is sensitive to change in the manufacturing process and is one of the most important product quality characteristics. The classification of workpiece surface patterns is critical for quality control, because it can provide feedback on the manufacturing process. There are many methods to solve the classification problems and some classic classification methods are presented as follows.

(1) Decision Tree. The decision tree method uses the information gain of the information theory to find the attribute field with the largest amount of information in the database, establishes a node of the decision tree and a branch of the tree according to different values of the attribute field and repeatedly establishes the lower nodes and branches of the tree in each subset of branches. The main decision tree methods include ID3 [1], CART [2], PUBLIC [3], SLIQ [4], SPRINT [5], etc.

(2) Rough set. Rough set is a mathematical tool that records incomplete and uncertain data without requirement of prior knowledge. This method can effectively find hidden knowledge from incomplete information, combines with various classification technologies to classify incomplete data. The rough set theory links the classification ability and knowledge and uses the equivalence relationship to formally represent the classification. Some applications of rough set for classification can refer to [6–8].

(3) Fuzzy logic. The classification is inseparable from the calculation of the degree of vector similarity, and the fuzzy classification also requires calculation of the vector fuzzy similarity coefficient. In the fuzzy classification method, a fuzzy similarity matrix should be established to represent the degree of fuzzy similarity of the objects. Fuzzy classification method can well deal with the ambiguity of objective object class attributes, and some applications of fuzzy logic for classification can refer to [9, 10].

© Springer Nature Singapore Pte Ltd. 2019
S. Du and L. Xi, *High Definition Metrology Based Surface Quality Control and Applications*, https://doi.org/10.1007/978-981-15-0279-8_5

(4) Ensemble learning. Ensemble learning is a machine learning method. It attempts to obtain different base learners by successively invoking a single learning method, and then combines the learners according to the rules to solve the same problem, which can significantly improve the generalization ability of the learning system. The voting (weighted) methods are used to combine multiple base learners and common methods include Bagging [11, 12] and Boosting [13]. Since ensemble learning uses a voting average method to combine multiple classifiers, it is possible to reduce the error of a single classifier and obtain a more accurate representation of the spatial model of the problem, thereby improving the classification accuracy of the classifier.

(5) Bayesian. Bayesian classification is a statistical classification method, which uses the knowledge of probability statistics for classification. It predicts the likelihood of class membership when the prior probability and the conditional probability are known. Typical Bayesian classifiers include naive Bayes [14], tree augmented Bayes network [15], Bayesian network classifier [16], etc. Bayesian classifiers can be applied to large databases with simple method, high classification accuracy, and high speed. Since Bayes theorem assumes that the effect of an attribute value on a given class is independent of other attribute values, this hypothesis is often inconsistent in the actual situation, so the classification accuracy may decrease.

(6) Neural networks. Neural network is one of the important methods in classification technology. Artificial neural network (ANN) is a mathematical model for information processing using a structure similar to the synaptic connection of the brain. Neural networks usually require training. The process of training is the process of network learning. Training changes the connection weights of network nodes to make them classified. The trained network can be used for object identification. Common neural network models used for classification include backpropagation (BP) neural network [17], radical basis function (RBF) network [18], Hopfield network [19], etc. Current neural networks still have the disadvantages of slow convergence rate, large amount of calculation, long training time, and inexplicability.

(7) Support vector machine. Statistical learning theory (SLT) is a class of machine learning theory for small samples. The theory establishes a new theoretical framework for statistical problems such as small sample classification. This theory not only takes into account the asymptotic performance but also seeks to achieve optimal results with available information. Vapnik and others have been working on this field of research since the 1960s and 1970s [20]. With the gradual development of the theory and the slow progress in machine learning methods such as neural networks, statistical learning theory has been increasingly developed and more attention is paid.

The theory of statistical learning has a relatively solid theoretical basis, and it gives a unified framework for a limited sample of learning problems. Based on this theory, a new learning method, support vector machine (SVM) [20], was developed. Support vector machines solve a range of machine learning problems such as classification, regression, and prediction. It takes statistical learning Vapnik–Chervonenkis (VC) dimension theory and structural risk minimization principle as the theoretical basis, and seeks the best compromise between the model's learning ability and complexity based on limited sample information, so as to obtain the best generalization ability. The main advantages of SVM are given as follows: (a) SVM is used for a finite sample and it seeks to obtain an optimal solution under existing information rather than an optimal solution when the sample tends to infinity. (b) SVM solves the problem by performing quadratic optimization to obtain the global optimality and overcome the local extremum problem that the neural network can hardly avoid. (c) SVM solves the problem of dimension disaster by using the kernel function, so the method complexity is not affected by the sample dimension.

5.2　A Selective Multiclass Support Vector Machine Ensemble Classifier for Engineering Surface Classification

5.2.1　Introduction

The condition of a workpiece surface has great impact on the functional performance of the product, and the classification of the workpiece surface patterns play an important role in quality control due to its function in providing feedback on the manufacturing process [21]. Since it is not appropriate to directly use the raw data collected from the workpiece surfaces to classify, features need to be extracted first to represent a given workpiece surface before classification. Conventionally, some numerical surface parameters are used for surface texture characterization, such as amplitude parameters like average roughness, maximum profile peak height, spacing parameters like mean peak spacing and high spot count, and shape parameters like slope and kurtosis [22–24]. But with the development of 3D measuring techniques, especially faster optimal methods, a lot of data can be obtained from engineering surfaces in quite a short time [25, 26], and it is inappropriate to simply use one-dimensional parameters to characterize surface textures. Consequently, more surface characterization approaches have been proposed to obtain abundant information about 3D surfaces, such as a 3D parameter set [24, 27–29], gray-level co-occurrence matrix [30], two-dimensional autocorrelation function, spectral analysis [31], and a 3D Monte Carlo model [32]. In recent years, some approaches that first used in signal processing areas have been adopted to extract features of engineering surfaces, such as Gabor filter banks [33], Gaussian filter banks [34], and wavelet packets [35–37], and after filtering, some numerical surface parameters are calculated for each sub-band to represent a given surface.

The goal of feature extraction is to improve the effectiveness and efficiency of classification. Namely, the performance of feature extraction method can affect the accuracy of classification directly. So, it is of great importance to select an appropriate approach to extract features. Wavelet filters can provide multi-scale/ orientation analysis, which makes it a powerful tool in feature extraction and is superior to traditional filters [36, 38]. There are many kinds of wavelets, and discrete wavelet transform (DWT) is the most widely used one in the analysis of workpiece surfaces in previous studies [36, 39, 40]. However, DWT mainly has two disadvantages: lack of shift-invariance and poor directional selectivity for diagonal features, which impair its application in engineering surface analysis. This sub-section adopts dual-tree complex wavelet transform (DT-CWT), which is a type of wavelet filter proposed by Kingsbury [41–44] in recent years, to extract features of engineering surfaces. It turns out that both of the above two problems can be solved effectively by DT-CWT; furthermore, DT-CWT also has some other nice properties as follows: (1) perfect reconstruction, (2) limited redundancy, and (3) efficient order-N computation. Due to the above merits, DT-CWT performs well in feature extraction.

Support vector machine (SVM), proposed by Vapnik [20], is a new machine learning method developed on the basis of statistical learning theory. It can solve a variety of problems such as classification, regression, and prediction [45–48]. Support vector machine produces good generalization abilities by constructing a best hyperplane as the decision surface to maximize the margin between the data of two classes. For nonlinear classification problems, support vector machine uses a kernel function to map the original dataset into a high-dimensional space, in which the data is linear and separable, so it overcomes the "dimension disaster" problem, which is faced by many traditional classification methods.

Classifier combination is now an active area of research in machine learning and pattern recognition. Many studies, both theoretically and empirically, demonstrate that combining classifiers instead of using a single one can improve generalization performance of SVMs and increase the classification accuracy [49–52]. An ensemble of classifiers consists of several basic classifiers created by using several strategies on the training data and aggregates their outputs to classify the testing samples. Classifiers ensemble can deal with some problems that are intractable for a single classifier due to its higher generalization abilities compared with a single one. Though it is well known that SVM ensembles have been applied in many different areas, e.g., face membership authentication [53], text-independent speaker recognition [54], bankruptcy prediction [55], and process mean shifts classification [46], few researches have been done to apply SVMs ensemble technique into classifying the engineering surfaces in the manufacturing areas.

Therefore, this subsection developed a novel method called modified matching pursuit optimization based selective multiclass support vector machine ensemble classifier (MPO-SVME) to solve the classification problem of engineering surfaces. MPO-SVME is a selective ensemble multiclass SVMs, which can improve the classification performance by creating an ensemble consisting of several multiclass SVMs and selecting the optimal ones from the ensemble. The significance of the present method is that it can improve the classification accuracy without bringing in too much computational cost, which means that it can be applied to a practical manufacturing system.

5.2.2   The Proposed Method

This subsection provides an overview of the developed method which consists of DT-CWT and selective ensemble SVMs, and the architecture is shown in Fig. 5.1. To be specific, in Module 1, high-resolution images of engineering surfaces are filtered to extract details in different directions and scales using 2D DT-CWT, and then the means and standard variances of the coefficients of the wavelet sub-bands are calculated as feature vectors to represent a given surface. In Module 2, random subspace algorithm is first used to create several component multiclass SVM classifiers and then MPO algorithm is conducted to select some optimal classifiers from the available ones.

5.2.2.1   Feature Extraction with DT-CWT

(1) The dual-tree complex wavelet transform

The DT-CWT is a multi-scale transform tool that uses a dual tree of wavelets filters to obtain the real and imaginary parts of complex wavelets coefficients. It not only retains the merits of traditional wavelet transform but also provides some nice properties such as approximate shift-invariance and improved directional

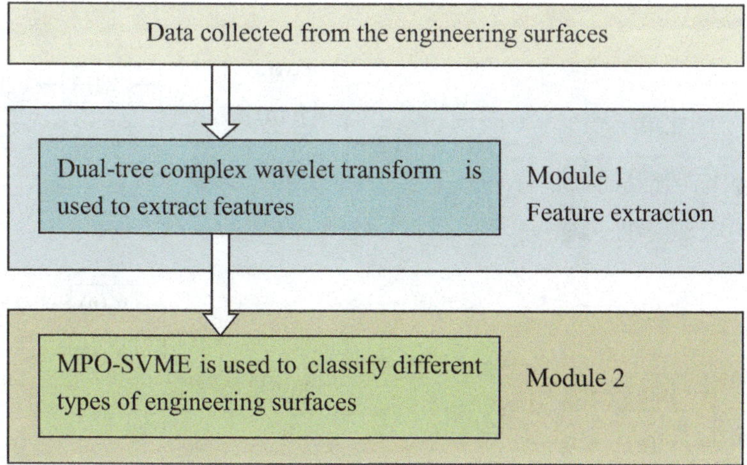

Fig. 5.1 The architecture of the proposed feature extraction and classification method. Reprinted from Ref. [111], copyright 2014, with permission from ASME

selectivity. The one-dimensional DT-CWT is implemented using two filter banks in parallel to operate on the input data $f(x)$, and it is composed of two parallel fully decimated trees of filters, tree a and tree b. Tree a produces the real parts of the complex coefficients while tree b produces the imaginary parts. A four-level decomposition implemented by one-dimensional DT-CWT is illustrated in Fig. 5.2, where $h_0(n)$ and $h_1(n)$ are the low-pass and high-pass filters of tree a, and $g_0(n)$ and $g_1(n)$ are the low-pass and high-pass filters of tree b.

The high-dimensional DT-CWT not only maintains all the attractive properties of the one-dimensional DT-CWT but also provides directional selectivity, which makes it more suitable to extract features from engineering surfaces. Figure 5.3a shows the 2D discrete wavelet filters, note that the first two wavelets are oriented in only three directions (vertical, horizontal, and diagonal). Figure 5.3b shows the 2D dual-tree complex wavelet filters, where the first row illustrates the real part of each complex wavelet, the second row illustrates the imaginary part, and the third row illustrates the magnitude. Note that the 2D dual-tree complex wavelet filters are oriented in six directions ($\pm 15°$, $\pm 45°$, $\pm 75°$), so it does better in reflecting the change of images in different orientations.

(2) Feature extraction

Feature extraction is related to the quantification of surface characteristics, and its quantitative results are known as feature vectors. The selection of these descriptive parameters is important, because it will influence the subsequent classification performance to some degree. A K-level 2D DT-CWT decomposition is implemented to extract features. Near-symmetric 13, 19 tap filters are selected as level 1 filters and Q-shift 18 tap filters for other levels, which has been proved by Kingsbury [42] to be the best combination of filters when compared with other

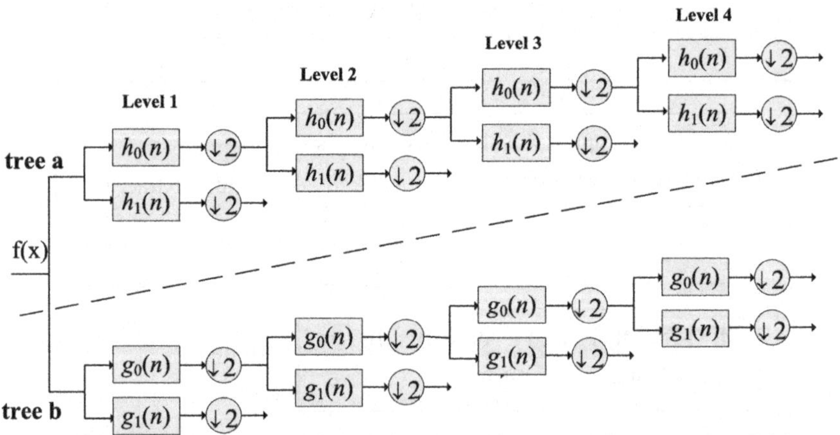

Fig. 5.2 Framework of the dual-tree complex wavelet transform (DT-CWT). Reprinted from Ref. [111], copyright 2014, with permission from ASME

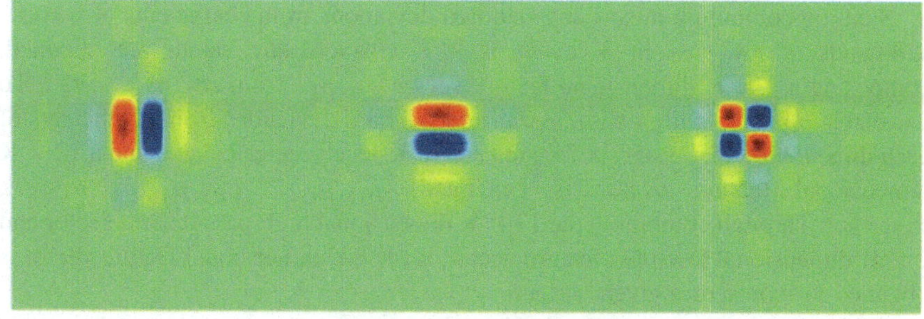

(a) 2D discrete wavelet filters

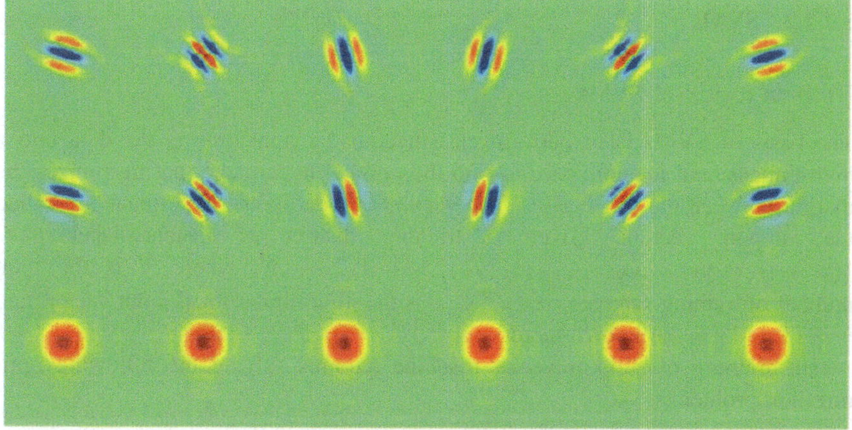

(b) 2D dual-tree complex wavelet filters

Fig. 5.3 The two sets of wavelet filters. Reprinted from Ref. [111], copyright 2014, with permission from ASME

combinations of DT-CWT filters. Feature vectors are generated by computing means and standard deviations from coefficients of wavelet sub-bands [56].

$$\mu = \frac{1}{M \times N} \sum_{i=1}^{M} \sum_{j=1}^{N} |C(i,j)| \tag{5.1}$$

$$\sigma = \sqrt{\frac{1}{M \times N - 1} \sum_{i=1}^{M} \sum_{j=1}^{N} |C(i,j) - \mu|^2} \tag{5.2}$$

As shown in Eqs. (5.1) and (5.2), M denotes the number of points along x-axis, N denotes the number of points along y-axis, $C(i,j)$ denotes the coefficient matrix of wavelet sub-bands, and μ and σ, respectively, denote the means and standard deviations of the sub-bands' coefficients.

Next, by computing means and standard deviations from coefficients of wavelet sub-bands of low-pass in K-levels, $2 \times K$ characteristic values are obtained, composing a $2 \times K$ dimensional feature vector $(\mu_1, \mu_2, \ldots, \mu_K, \sigma_1, \sigma_2, \ldots, \sigma_K)$. To wavelet sub-bands of high-pass, means and standard deviations from coefficients of each direction ($\pm 15°$, $\pm 45°$, $\pm 75°$) in each of K-levels are calculated and a $12 \times K$ dimensional feature vector is generated $(\mu_{11}, \mu_{12}, \ldots, \mu_{K5}, \mu_{K6}, \sigma_{11}, \sigma_{12}, \ldots, \sigma_{K5}, \sigma_{K6})$. Through combining the $12 \times K$ dimensional high-pass feature vector and $2 \times K$ dimensional low-pass feature vector, a $14 \times K$ dimensional feature vector is obtained to represent a given surface.

5.2.2.2 SVMs

(1) Brief review of SVMs

The basis of SVMs is to perform classification by transforming the data into a high-dimensional feature space to find the optimal hyperplane that maximizes the margin between the two classes. The vectors that define the hyperplane are called the support vectors. Given a training dataset of instance-label pairs $\{(x_1, y_1), (x_2, y_2), \ldots (x_i, y_i), \ldots (x_N, y_N)\}$, $i = 1, 2, \cdots, N$, where N is the total number of training samples, $x_i \in R^d \subset R$ is the ith d-dimensional input vector, and $y_i \in \{1, -1\}$ is the known target.

The training of support vector machine requires solving the following optimization problem:

$$\text{Minimize } \frac{1}{2}\mathbf{w}^T\mathbf{w} + C\sum_{i=1}^{N} \xi_i \tag{5.3}$$

$$\text{Subject to } y_i(\mathbf{w}^T \phi(x_i) + b) \geq 1 - \xi_i, \quad \xi_i \geq 0 \tag{5.4}$$

where ξ_i denotes slack variables, measuring the degree of misclassification of the sample x_i, C is a penalty parameter, which is used to penalize training errors, the bias b is a scalar, representing the bias of the hyperplane, $\mathbf{w}$ is the vector of hyperplane coefficients, defining a direction perpendicular to the hyperplane, the index i labels the N training cases, and the map function ϕ is a nonlinear transformation used to map the input vectors into a high-dimensional feature space (as shown in Fig. 5.4). The optimization problem is to determine the trade-off between margin size and training error.

For linearly separable problems, $\xi_i = 0$ and the separating hyperplane that creates the maximum distance between the plane and the nearest data is the optimal one. For the nonlinear classification tasks, a mapping function is often employed to map the training samples from the input space into a higher dimensional feature space, and then the nonlinear classification problem will become a linear one. Any function that satisfies Mercer's theorem [48] can be used as a kernel function.

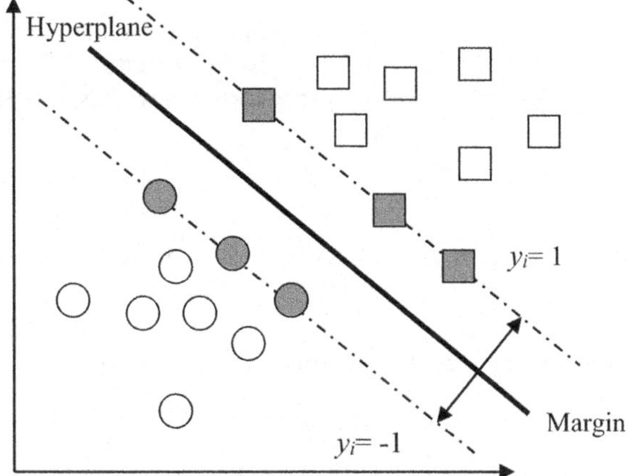

Fig. 5.4 A geometric interpretation of binary classification of SVM. Reprinted from Ref. [111], copyright 2014, with permission from ASME

Though new kernels are being proposed by researchers, the most widely used kernels are as follows:

Linear function:	$K(\mathbf{x}_i, \mathbf{y}_j) = \mathbf{x}_i \times \mathbf{y}_j$
Gaussian radial basis function (RBF:)	$K(\mathbf{x}_i, \mathbf{y}_j) = \exp(-\gamma \|\mathbf{x}_i - \mathbf{y}_j\|^2)\,\gamma > 0$
Polynomial function with degree d:	$K(\mathbf{x}_i, \mathbf{y}_j) = ((\mathbf{x}_i \times \mathbf{y}_j) + \gamma)^d\,\gamma > 0$
Sigmoid:	$K(\mathbf{x}_i, \mathbf{y}_j) = \tanh(\gamma \mathbf{x}_i \times \mathbf{y}_j + r)\,\gamma > 0$

(2) Multiclass SVMs

SVM is in nature a tool for binary classification [20], which is not suitable for classifying engineering surfaces because there are usually several different cases needed to be identified. Thus, it is necessary to extend the binary classifiers in order to solve multiclass problems. The traditional way to implement multiclass classification is to combine several binary SVMs, in light of this, three popular methods, "one-against-one," "one-against-all," and direct acyclic graph (DAG) are proposed to solve multiclass problems. A comprehensive comparison of these three methods conducted by Hsu et al. [57] suggested that the one-against-one method is most suited for practical use than other methods. Therefore, "one-against-one" method is used to classify workpiece surface patterns.

Figure 5.5 shows how to construct a component multiclass SVM classifier using "one-against-one" strategy. For L-class event, the "one-against-one" method constructs $M = C_L^2 = L(L-1)/2$ binary classifiers, each of which is trained by binary-class data. For training data from the ith and the jth class, the following binary classification problem needs to be solved:

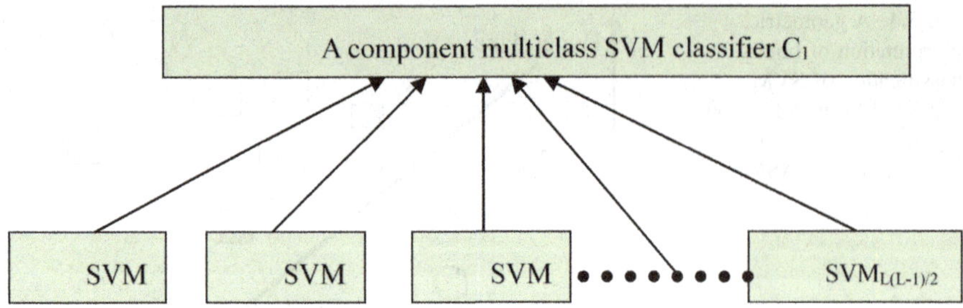

Fig. 5.5 A component multiclass SVM classifier combined by binary SVMs. Reprinted from Ref. [111], copyright 2014, with permission from ASME

$$\min \tfrac{1}{2}\left(\mathbf{w}^{ij}\right)^{T}\mathbf{w}^{ij} + C\sum_{t}\xi_{n}^{ij}\left(\mathbf{w}^{ij}\right)^{T}$$

$$\text{s.t.}\begin{cases} \left(\mathbf{w}^{ij}\right)^{T}K(x_{n}) + b^{ij} \geq 1 - \xi_{n}^{ij} & if \quad y_{n} = i \\ \left(\mathbf{w}^{ij}\right)^{T}K(x_{n}) + b^{ij} \leq \xi_{n}^{ij} - 1 & if \quad y_{n} = j \\ \xi_{n}^{ij} \geq 0 \end{cases} \tag{5.5}$$

where, similarly to binary classification SVMs, $K(x_{n})$ is the kernel function, (x_{n}, y_{n}) is the ith or jth training sample, $\mathbf{w} \in R^{N}$ and $b \in R$ are the weighting factors, ξ_{n}^{ij} is the slack variable, and C is the penalty parameter.

Next, voting strategy is used to output the results after all the $L(L-1)/2$ classifiers are constructed: If $\left(\left(\mathbf{w}^{ij}\right)^{T}K(x_{n}) + b^{ij}\right)$ predicts x is in the ith class, the vote for the ith class increased by one; otherwise, the jth class added by one. Then x is finally predicted in the class with the largest vote. The voting method described above is also called the "Max Wins" strategy.

5.2.2.3   The Proposed MPO-SVME Classifier

(1) Framework of MPO-SVME

The main steps of MPO-SVME are shown in Fig. 5.6.

Step 1: Creation of several training subsets using ensemble strategies. Random subspace algorithm is adopted to create several different training subsets $S_{1}, S_{2},..., S_{T}$.

Step 2: Creation of several component multiclass SVM classifiers using "one-against-one" method. Based on training subsets created in Step 1, several multiclass SVM classifiers are constructed and used as component multiclass SVMs classifiers for optimal selection in Step 3, and they are denoted by $C_{1}, C_{2}, \cdots C_{T}$.

Step 3: Selection of optimal classifiers. Instead of trying to ensemble all the independent component multiclass SVM classifiers created in Step 2, the optimal subset is selected from the component multiclass SVM classifiers through matching pursuit optimization (MPO) algorithm. And the final prediction of the selective

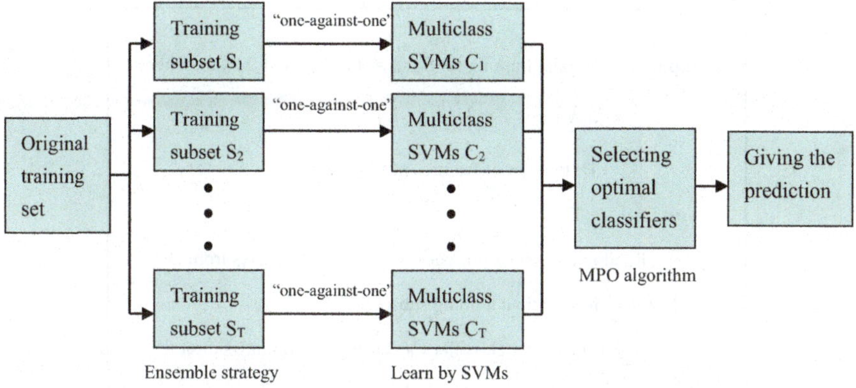

Fig. 5.6 The framework of the proposed selective ensemble classifiers. Reprinted from Ref. [111], copyright 2014, with permission from ASME

ensemble classifiers is given by the aggregation of the predictions of selected component classifiers while taking their weights into consideration.

(2) Ensemble strategy

An ensemble of classifiers is a collection of several classifiers (usually trained on different training subsets) whose individual decisions are combined in some way to classify the test samples. It is known that an ensemble often shows better performance than the individual classifiers that make it up. Therefore, many approaches of constructing SVM ensembles have been proposed, such as bagging [58], boosting [59], and random subspace [60]. In bagging, some new training sets are generated from the original training set via a bootstrap method and then used to train several corresponding classifiers. The main idea behind boosting is to construct a composite classifier by training classifiers sequentially while putting more and more emphasis on misclassified cases. Random subspace is adopted to ensemble learning.

Similar to bagging, the random subspace algorithm also benefits from bootstrapping. However, other than bootstrapping training samples as used in bagging, random subspace performs the bootstrapping in the feature space. It is well known that overfitting is likely to happen when the size of training set is small while the dimensionality of the feature vector is relatively high, and this problem is pretty common in texture or surface classification due to the fact that the size of feature vector is often larger than the number of training samples. In order to avoid overfitting, random subspace method randomly selects a small subset of features to reduce the discrepancy between the training data size and the feature vector length. Using a random sampling method, we construct a set of SVMs. Then these SVMs are combined to construct a more powerful classifier to solve the overfitting problem. The random subspace algorithm is shown in Fig. 5.7.

Input: The original training set $S_{\text{training}}=\{(x_i, y_i)\}$, i=1,2,...,m, where

$$x_i \in X \,,\, y_i \in Y \subset \{l_1,l_2,...,l_k\}$$

D: number of features of training set;

For t=1,..., T

      Randomly select a subspace with d (d<D) features from D

features, and obtain a training subset S_t;

      Construct a classifier $C_t(x)$ using the training subset S_t;

End

Output: The final prediction $\quad P(x_i) = \arg\max_{y \in Y} \sum_{t;C_t(x_i)=y} 1$

Fig. 5.7 The random subspace algorithm. Reprinted from Ref. [111], copyright 2014, with permission from ASME

(3) Selective ensemble strategy

Zhou et al. [61] proposed a concept which has been proved theoretically and empirically in his article that selecting some suitable classifiers from the available classifiers may yield better accuracy than combing all classifiers simultaneously, and this opinion can be deemed as "Many could be better than all." In this subsection, "ensemble all" means ensemble the output results of all the component multiclass SVM classifiers created by random subspace algorithm. The value of selective ensemble lies in reducing the size of ensemble without worsening its performance or even getting a better result. However, it is really a tough task to exclude the underperformed classifiers and retain the optimal subsets since the subspace of possible subsets is very large (2^T-1) for a size of T (T is the total number of component SVM classifiers). Therefore, it is not practical to use the exhaustive method to search for optimal subset. A large number of very diverse selective ensemble methods have been proposed over the past decade, such as search-based methods like Gasen [61] and ensemble pruning via reinforcement learning (EPRL) [62], clustering-based methods like hierarchical agglomerative clustering (HAC) [63], and ranking-based method like "Pruning in Ordered Bagging Ensembles" [64]. So far, there is still no agreement about which is the most appropriate method. In this subsection, we adopt a novel selective ensemble approach named matching pursuit optimization ensemble classifiers (MPOEC), which is proposed by Mao et al. [65] and has been proven to have better performance than some existing methods.

The main idea of MPOEC is to determine the ideal balance between the diversity and the individual accuracy of the ensembles. MPOEC adopts a greedy iterative algorithm of matching pursuit to search for an optimal combination of entire classifiers, and abandons some similar or poor classifiers by giving zero coefficients. In MPOEC, each classifier of the ensemble is deemed as a basis function and the labels of training samples as the target function. The coefficient of each classifier is obtained by minimizing the residual between the target function and the linear combination of individual component classifiers.

The set of all component classifiers $\{C_t(x)\}(t = 1, \ldots, T)$ is regarded as the basis function dictionary, and every component classifier can obtain a coefficient α_t by minimizing the residual R. When $\alpha_t \neq 0$, the classifier corresponding to α_t is selected to ensemble, and when $\alpha_t = 0$, the classifier will be excluded. However, it should be noted that the MPOEC algorithm in [65] can only solve the two-class classification problem, which is not suitable for classification of various workpiece surfaces. So we modify this algorithm to solve the multiclass classification problem, and since SVM is adopted as basic classifiers, the modified algorithm is then called MPO-SVME. The description of MPO-SVME algorithm is shown in Fig. 5.8.

5.2.3   Case Study

Engineering surfaces, which are represented by point cloud data collected by a device called Coherix ShaPix [66], are used to validate the effectiveness of the proposed method on real-world data. ShaPix uses a tunable-wavelength laser to gather up to 1 million data points about the part in its 300 by 300 mm field of view in less than a minute. The basic height (Z) accuracy of it is 1 μm while the lateral (X, Y) resolution is around 0.3 mm.

Six different top surfaces of engine cylinder blocks (as shown in Fig. 5.9) processed by a major domestic car manufacturer are studied here, and it is worth noting that engineering surfaces 1–3 are finished by one machine while surfaces 4–6 by another. Coherix ShaPix is used to scan these six surfaces and gather more than 700,000 data (three-coordinate data) from each surface. Two samples of the color-coded measurement results from each dataset are shown in Fig. 5.10.

It can be seen that parts in set A have a slope problem (the height decreases from bottom left to upper right) while parts in set B do not, and this problem will lead to some adverse effects on the performance of the product. So it is worthwhile to distinguish these problematic parts from the normal ones by classification.

5.2.3.1   Feature Extraction

Twenty square areas are designated from the entire surface with the size of 128 × 128 pixels each. Centroids of those square areas are located in the intersections of the grids, as shown in Fig. 5.11. One of the square areas rather than the entire

Input: The original training set $S_{training}=\{(x_i, y_i)\}$, $i=1,2,\ldots,m$, where $x_i \in X$ $x_i \in X$, testing set $S_{testing}=\{(x_i, y_i)\}$, $i=1,2,\ldots,m^*$

T: number of component SVM classifiers

loop: the iterative number

λ: a threshold error

For t=1 to T

S_t: training subset gained from $S_{training}$ using an ensemble strategy

C_t : the SVM classifier learning for S_t

P_t : the prediction labels for $S_{training}$ by C_t

P_t^* : the prediction labels for $S_{testing}$ by C_t

End for

※ Search for the optimal combination of the T SVM classifiers

The initial residual R_0 is equal to the class labels $\{y_1, y_2,\ldots,y_m\}$;

Obtain the initial α_{ini} with the formula: $\alpha_{ini} = \langle \vec{P}_1, \vec{R}_0 \rangle / \left\| \vec{P}_1 \right\|^2$;

$\alpha_{max} = \alpha_{ini}$;

The number of iteration: $j=1$;

While $(j \le loop \,\&\, error > \lambda)$

  For $t=1,\ldots,$T

  Compute α_t : $\alpha_t = \langle \vec{P}_t, \vec{R}_{j-1} \rangle / \left\| \vec{P}_t \right\|^2$

  If $(\left\| \alpha_t \right\| > \left\| \alpha_{max} \right\|)$

  Note α_t and update α_{max} : $\alpha_{max} = \alpha_t$;

  Else

  Give a zero value to α_t : $\alpha_t = 0$;

  End if

Fig. 5.8 The MPO-SVME algorithm. Reprinted from Ref. [111], copyright 2014, with permission from ASME

Note $P_j(x_i) : P_j(x_i)$ is the prediction label given by the ensemble of classifiers for the ith training sample in jth iteration.

$$P_j(x_i) = \arg\max_{y \in Y} \sum_{t; P_t^*(x_i)=y} \alpha_t, (t = 1, 2, ..., T)$$

Compute the error e_j:

For i=1 to m

If $y_i - P_j(x_i) = 0$ then $e_{ji} = 0$, else $e_{ji} = 1$

End

$$e_j = (\sum_{i=1}^{m} e_{ji}) / m$$

Update the number of iteration: $j=j+1$;

End while

Gain the coefficients of the component classifiers $\{\alpha_1, \alpha_2, ..., \alpha_T\}$ and select the classifiers of the nonzero α_t to ensemble.

Output: The combination of the optimal component classifiers:

$$P_{opt}(x_i) = \arg\max_{y \in Y} \sum_{t; P_t^*(x_i)=y} \alpha_t, (t = 1, 2, ..., T_{opt}, T_{opt} \leq T)$$

Fig. 5.8 (continued)

surface is used to extract features because the amount of data representing the entire surface is too large, and it is too time and memory expensive to handle this data for computers. In addition, although only a small area of the surface is used to represent the entire surface, the classification accuracies are sufficiently high in most of the cases when using these 20 square areas for classification purposes, respectively (discussed in Sect. 5.2.3.2).

To be specific, six small surfaces from the same position (surface area 16 is chosen as example here) of each surface are chosen with the size of 128 × 128 pixels, which are used as the input of 2D DT-CWT. Figure 5.12 shows the six surfaces chosen for classification.

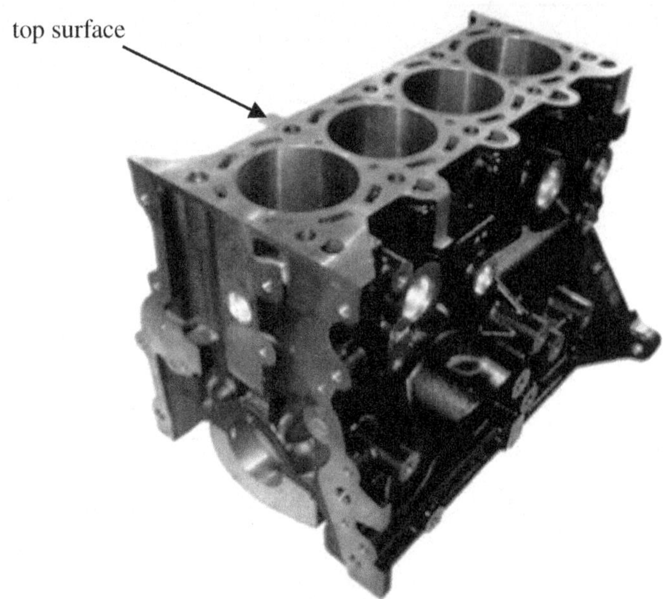

Fig. 5.9 Engine cylinder blocks processed by a major domestic car manufacturer. Reprinted from Ref. [111], copyright 2014, with permission from ASME

It is apparent that the first three surfaces look alike and so do the last three. This is easy to explain, because the first three surfaces are derived from one set while the last three from another set. It is also apparent that surfaces from different sets look different, due to the fact that parts of set A and set B are machined under two different conditions.

Next, each of the surfaces is partitioned into $4 \times 4 = 16$ nonoverlapping small surfaces with size of 32×32 pixels and those stem from the same original surface are considered of the same class (six classes totally). Therefore, 96 samples are obtained for classification and each class has 16 samples. Then, four-level 2D DT-CWT decomposition is implemented on those height data, respectively, feature vectors are generated by computing mean and standard deviation from coefficients of wavelet sub-bands, and then each of the 96 samples is represented by a 56-dimensional feature vector ($D = 56$).

5.2.3.2   Classification Results

Eight surface samples are randomly selected as training samples and the remaining eight samples as testing samples for every class (as seen in Fig. 5.13, Nos. 1–16 represent 16 surface samples, numbers within the circles represent the randomly selected training samples, and the rest represent testing samples), and then 48

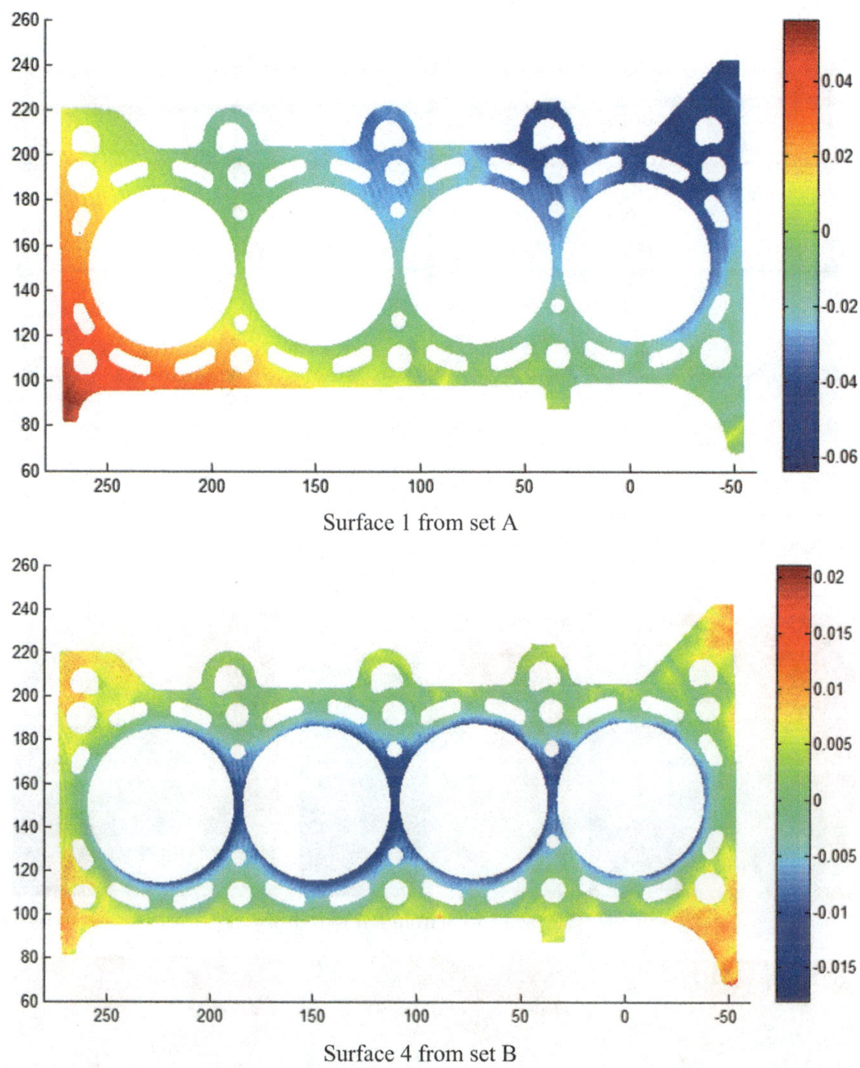

Fig. 5.10 Two samples of the color-coded measurement results from datasets A and B. Reprinted from Ref. [111], copyright 2014, with permission from ASME

training samples and 48 testing samples are obtained finally. Gaussian radial basis function (RBF:$K(\mathbf{x}_i, \mathbf{y}_j) = \exp(-\gamma\|\mathbf{x}_i - \mathbf{y}_j\|^2)$) is adopted. The training of classifiers requires the adequate definition of the kernel parameter γ and the penalty parameter C, for the use of inadequate parameter values is likely to result in a less accurate classification. Usually, the kernel parameters are determined by a grid search using n-fold cross-validation. Potential combinations of C and γ are tested in a user-defined range and the best combinations of C and γ are selected based on the

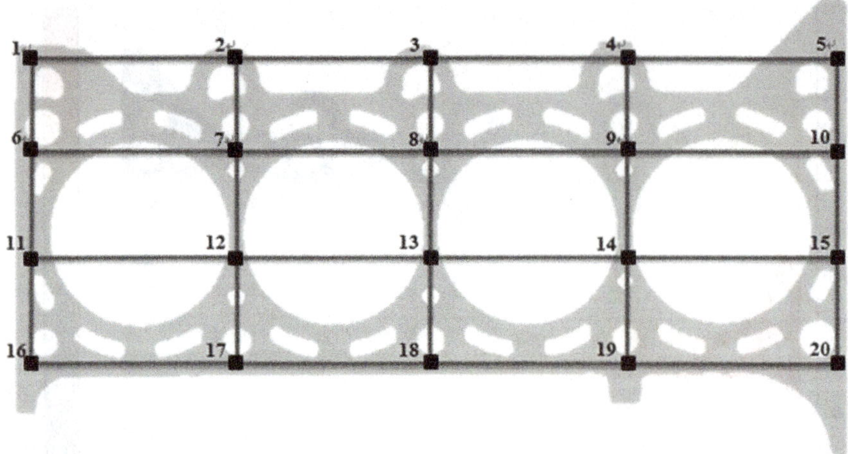

Fig. 5.11 Selection of small surface samples through a grid chart. Reprinted from Ref. [111], copyright 2014, with permission from ASME

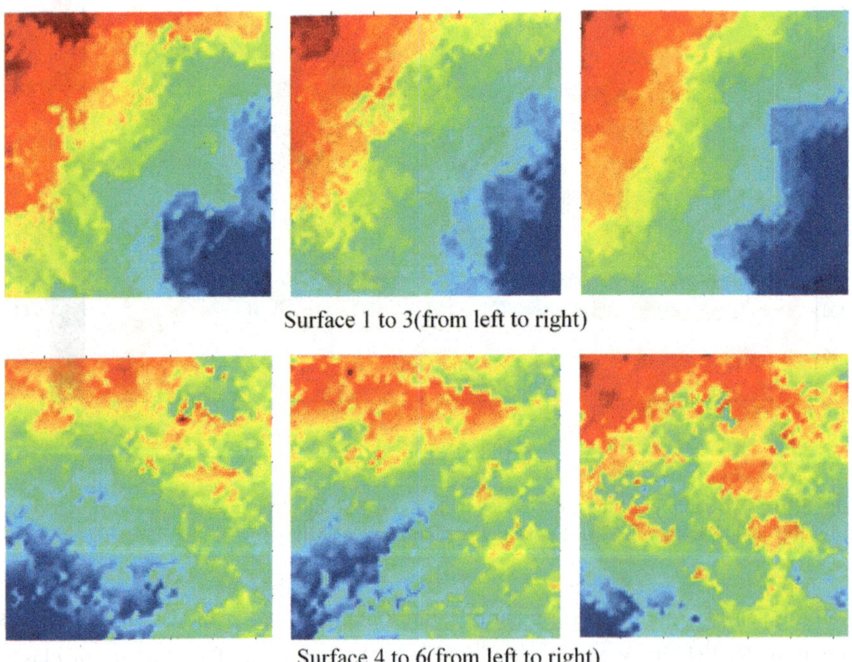

Surface 1 to 3(from left to right)

Surface 4 to 6(from left to right)

Fig. 5.12 Six surfaces selected from set A and set B. Reprinted from Ref. [111], copyright 2014, with permission from ASME

Fig. 5.13 The generation of training samples and testing samples. Reprinted from Ref. [111], copyright 2014, with permission from ASME

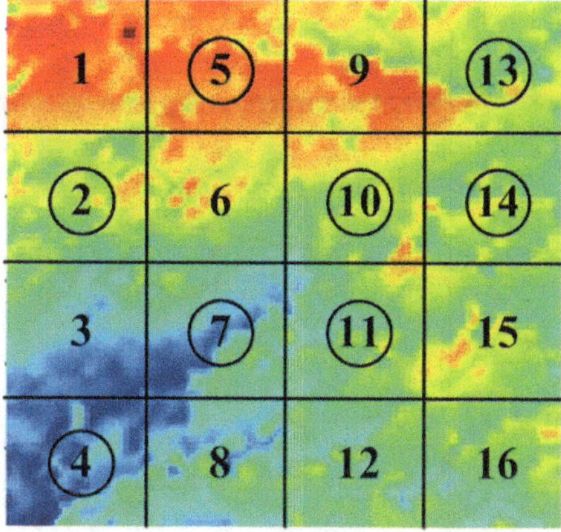

results of the cross-validation. The optimal value of kernel parameter γ and penalty parameter C is searched by 10-fold cross-validation experiment, and the parameters with the highest cross-validation accuracy are selected (here $\gamma = 8$, C = 64).

To study the influence of different decomposition levels on the classification accuracy, levels from 2 to 5 are adopted, respectively. And for comparison purposes, 2D DWT is also implemented to extract features using the same dataset. The classification is implemented using the LIBSVM toolbox [67] in a MATLAB environment and MPO-SVME is adopted as the selective ensemble strategy (two-thirds of the features are randomly selected at each level $(d = \frac{2}{3}D)$ and the number of component multiclass SVM classifiers is 20 (T = 20)), and the final classification results are shown in Fig. 5.14 (CCP(%) means "Correct Classification Percentage," each of the classification results is the average of 30 times).

As Fig. 5.14 shows, the classification result of 2D DT-CWT is much better than that of 2D DWT at each level, which demonstrates that the approximate shift-invariance and good directional selectivity properties are vital to feature extraction. As to the relationship between the value of levels and classification accuracy, it shows that a relatively high CCP is obtained when the number of decomposition levels is 4, no matter using DWT or DT-CWT. So four-level 2D DT-CWT decomposition is adopted.

Table 5.1 presents classification results of single multiclass SVMs, patterns 1–6 represent six different surfaces (1–3 from one set and 4–6 from another). Note that misclassification only occurs within the same set, which means that products machined under the same conditions are quite similar and hard to distinguish.

The small surfaces from other locations (as shown in Fig. 5.11, 20 locations totally) of each of the six surfaces are used to classify, and the results are shown in Fig. 5.15. Here the upper polygonal line denotes the results of two-class classification identifying each surface sample from set A or from set B, while the lower polygonal line denotes the results of six-class classification that solving the problem of determining which surface each small surface sample originates from. It can be

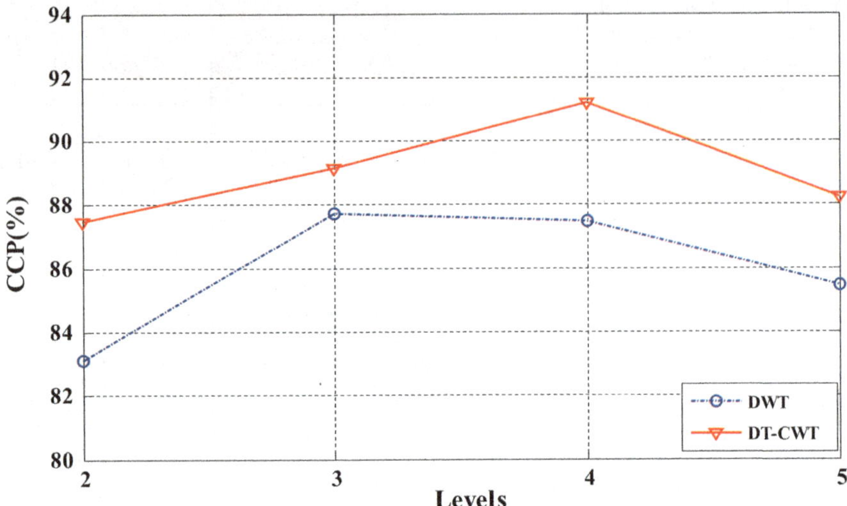

Fig. 5.14 The CCPs using DT-CWT and DWT with different levels. Reprinted from Ref. [111], copyright 2014, with permission from ASME

Table 5.1 CCPs (%) of single multiclass SVMs

| | Classification result | | | | | | CCP |
Pattern	1	2	3	4	5	6	
1	**8**	0	0	0	0	0	100
2	0	**8**	0	0	0	0	100
3	0	**1**	**7**	0	0	0	87.5
4	0	0	0	**5**	**3**	0	62.5
5	0	0	0	0	**7**	**1**	87.5
6	0	0	0	0	0	**8**	100
Average							89.58

seen that the accuracies of two-class classification are higher than 90% in most of the cases and 100% accuracy is achieved in 11 of the 20 cases, which demonstrates that using a small surface to represent a certain kind of surface is feasible. It is a practical conclusion, since, in most of the cases, quality engineers put the greatest emphasis on key positions of the entire workpiece surface, for the quality of these key areas will heavily influence the function of products. For example, the leakage problem is likely to occur in places near the cylinder bore, for the height of these places is relatively low, as the dark blue areas in Surface 4 (see Fig. 5.10). So it is acceptable to identify workpiece surfaces based on the classification results of the interested areas. By contrast, the accuracies of six-class classification problems are relatively low, so the effectiveness of using MPO-SVME to increase the classification accuracy is demonstrated in the following subsections.

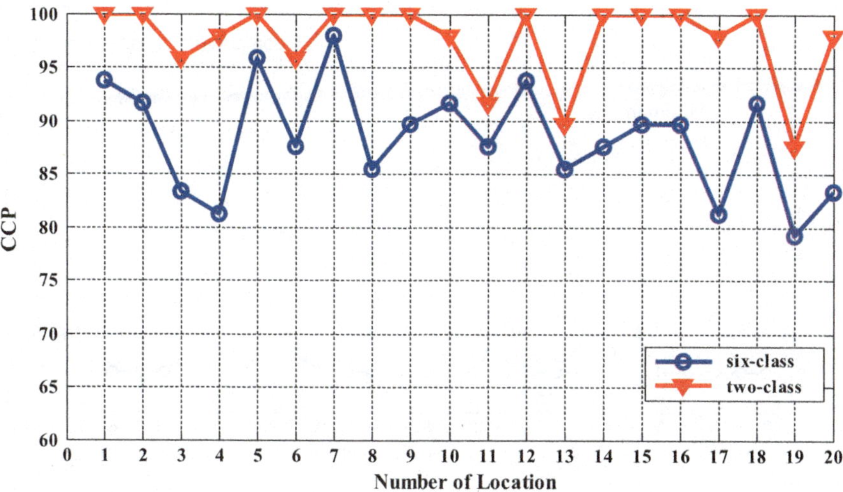

Fig. 5.15 The CCPs of single multiclass SVMs using surface samples at 20 different locations of each surface. Reprinted from Ref. [111], copyright 2014, with permission from ASME

Figure 5.16 presents classification results of MPO-SVME for the six engineering surfaces (likewise, surface area 16 is selected from each surface). In order to get reliable results, the experiment is repeated 30 times ($d = 40$ and $T = 20$). The average correct classification rate is 91.74%, which is 2.41% higher than that of single multiclass SVMs. Figure 5.17 presents the number of selected component

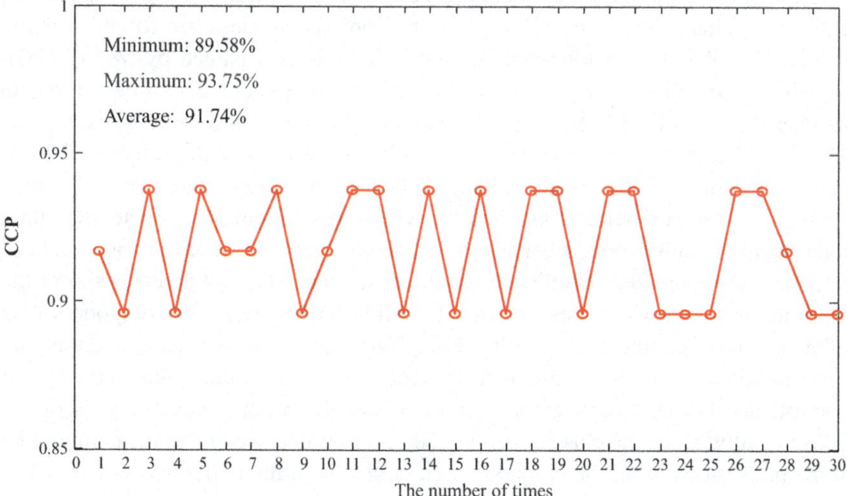

Fig. 5.16 Classification results of MPO-SVME at each time (30 times totally). Reprinted from Ref. [111], copyright 2014, with permission from ASME

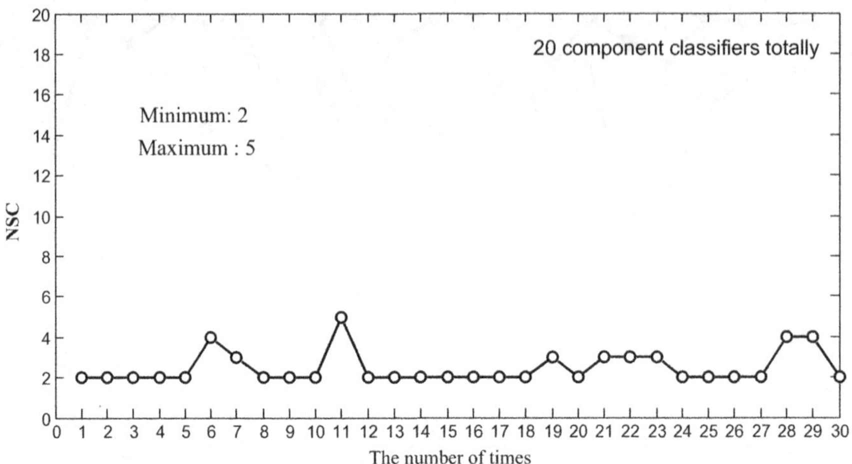

Fig. 5.17 The number of selected component classifiers (NSC) at each time. Reprinted from Ref. [111], copyright 2014, with permission from ASME

classifiers at each time (T = 20), the average is 2.47 each time. The two figures demonstrate that MPO-SVME can gain a better performance than single multiclass SVMs in classification without using too many classifiers. And this property is very useful, because when the quantity of unclassified samples is very large, using too many classifiers will consume a large amount of time, which is not feasible in practical application.

The process of selecting classifiers by MPO-SVME is shown in Fig. 5.18, and 30 component multiclass classifiers are generated by random subspace algorithm. The upper chart shows the CCP of every component classifier for the original training samples, the middle chart shows the coefficients gained by MPO-SVME, and the lower chart shows the CCP of every component classifier for testing samples. In this selective ensemble, the correct classification percentage is 93.75%, which is higher than that of the single multiclass SVM. In the upper and lower charts, the points within black rectangles denote that these component classifiers obtain positive coefficients; the points within red rectangles denote that these component classifiers obtain negative coefficients, and the rest of the points indicate classifiers that gain zero coefficients in the ensemble. The middle chart shows that six component classifiers are selected by MPO-SVME from 30 component classifiers to ensemble (the 2nd, 7th, 9th, 11th, 14th, and 22nd classifiers), and the 14th, 22nd component classifiers obtain greater coefficients than others, due to their better performance in the classification of training samples when compared with others. Consequently, since the classification result of ensemble is equal to the combination of the predictions of all the component classifiers with their coefficients α_t(t = 1, 2, …,T), the selected component classifiers are helpful for the better performance of the ensemble due to their higher coefficients. From the lower chart, it can be seen that there are nine classifiers whose CCPs are lower than 89.58%. If all the

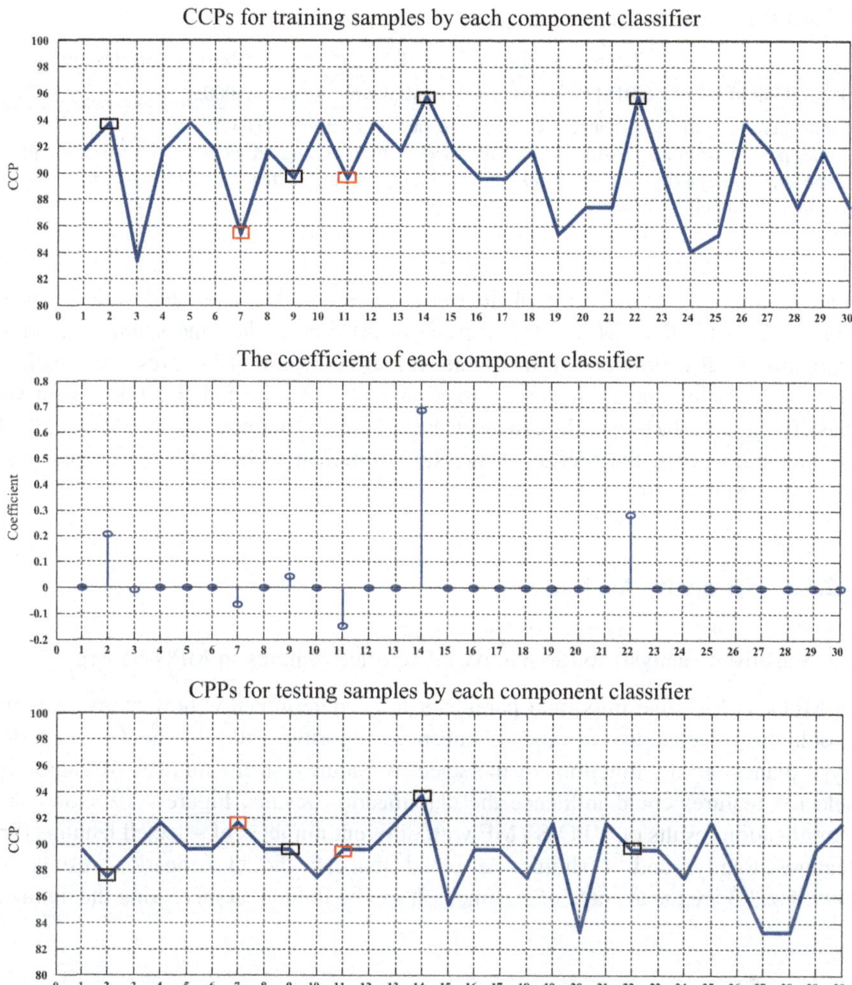

Fig. 5.18 The process of selective classifiers by MPO-SVME (T = 30, d = 30). Reprinted from Ref. [111], copyright 2014, with permission from ASME

component classifiers, including these poor classifiers, are treated equally in the classification of testing samples, the performance of the ensemble is likely to be encumbered by these poor classifiers. Hence, MPO-SVME improves the classification performance compared with general ensemble strategies.

Since the proposed surface classification method includes several different algorithms in its three steps (feature extraction, training of classifiers, and classification of new surfaces), it is necessary to analyze its time complexity. So the presented method was implemented in a MATLAB R2012a programming environment on a PC with Intel Core2 Duo E6550 CPU running Windows 7. The

Table 5.2 Computational time of the proposed methodology

	Feature extraction[a]	Training[b]	Classification (testing)[c]
Time (Unit s)	0.0111	0.0046	0.0003

[a]Time spent on extracting features of a surface sample using 2D DT-CWT
[b]Time spent on training classifiers with MPO-SVME (the training set includes 48 surface samples)
[c]Time spent on classifying an unlabeled surface sample (input is feature vector) with the selected classifiers obtained by training

training time and testing time of the presented method in case study are given in Table 5.2, both of which are the average of 30 times. The time spent on feature extraction is also provided in this table. It can be seen that the presented method costs very little time in feature extraction and training, and the time spent on classifying a surface sample can even be ignored. Therefore, from viewpoint of computational cost, the presented method is efficient to be applied to a practical manufacturing system.

5.2.3.3   Sensitivity Analysis

(1) Sensitivity analysis to the number of selected features in MPO-SVME

In MPO-SVME, one important parameter to be determined is how many features should be selected in the random subspace algorithm, due to the fact that the hyperplanes are the functions of the selected features, so the number of randomly selected features could influence the classification results. Figure 5.19 shows the classification results of MPO-SVME with different numbers of selected features (56 features totally, and it should be noted that all the classification results in Sect. 5.2.3.3 are the average of 30 times). It can be found that increasing the number

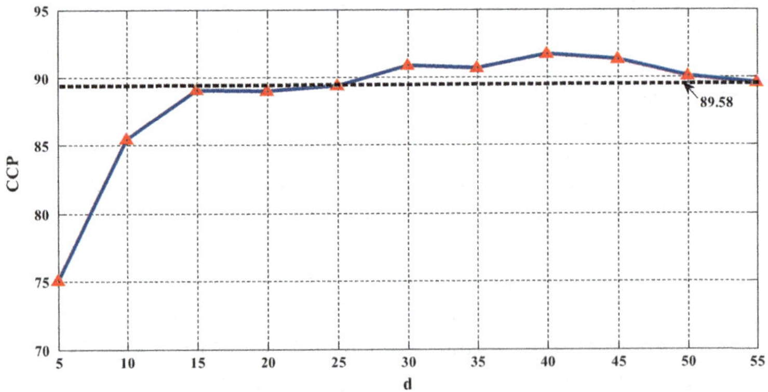

Fig. 5.19 Comparison of CCPs with different numbers of random features (T = 20). Reprinted from Ref. [111], copyright 2014, with permission from ASME

of features can substantially improve the performance of MPO-SVME when d is relatively small (less than 15). As d continues to increase, the correct classification percentage will grow slowly and reach the maximum when d is 40. Hence, selecting about 40 features from the total 56 features is appropriate to get the optimal result.

(2) Sensitivity analysis to the number of component classifiers T

The number of component multiclass SVM classifier T influences the performance of SVM ensemble. Figure 5.20 presents CCPs for different numbers of component multiclass SVM classifier T. It shows that increasing the number of T can improve the performance of MPO-SVME up to a certain level of accuracy. This may be explained by the fact that it is more likely to generate some preferable component classifiers that are available for selection when T is larger. But this is not to say bigger is better, since any further increase of the number of T after reaching such a limit will not obviously improve the performance of MPO-SVME, and a larger T will cost more time in training SVMs and decreases the effectiveness of classifying engineering surfaces as quickly as possible. Therefore, 25 component multiclass SVM classifiers are enough to get the optimal results.

(3) Sensitivity analysis to the size of training set

As 96 surface samples are generated in case study and 16 samples for each class, 30 surface samples are randomly selected as testing samples (5 training samples from each class), then different numbers of training samples are randomly selected from the remaining ones and the number of the selected training samples is the same among each class. The experiment is repeated 30 times (d = 40 and T = 20). From Fig. 5.21, it can be found that increasing the training samples can improve the performance of MPO-SVME up to a certain level of accuracy. This could be explained by the fact that there is a better chance of true representation of a problem space with enough training sets. However, any further increase of the training size

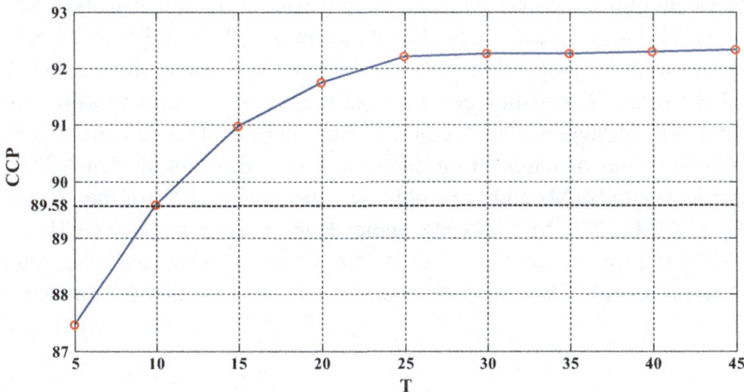

Fig. 5.20 Comparison of CCPs for different multiclass SVM classifiers T (n = 40). Reprinted from Ref. [111], copyright 2014, with permission from ASME

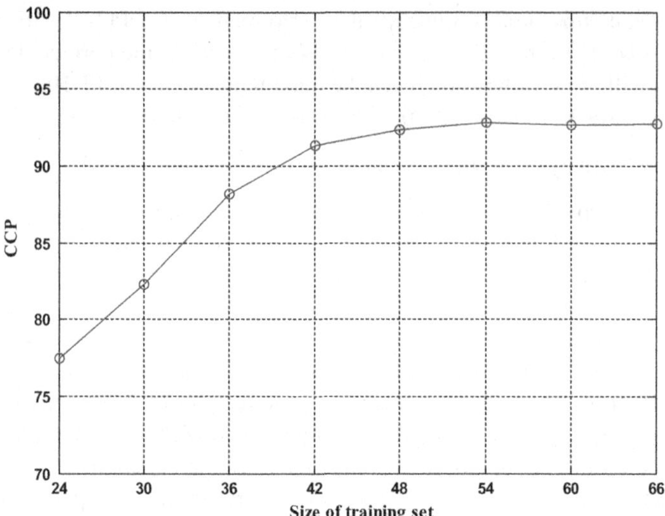

Fig. 5.21 CCPs for different sizes of training set. Reprinted from Ref. [111], copyright 2014, with permission from ASME

after reaching such limits will not significantly improve the performance of MPO-SVME. Moreover, the larger training set results in higher time on training classifiers.

(4) Comparison analysis for MPO-SVME, single SVMs, and Ensemble All

In order to further evaluate the performance of MPO-SVME, it is compared with other SVMs strategies including single SVMs and Ensemble All (i.e., ensemble all the component multiclass classifiers $C_1, C_2, C_3, \ldots, C_T$). And the classification results of each strategy are shown in Fig. 5.22 and Table 5.3 (in Ensemble All strategy (including bagging [58], random subspace [60], and Adaboost.M2 [68]), $T = 25$; in MPO-SVME, $T = 25$, $d = 40$). The boxplots in Fig. 5.22 reflect the distributions of the classification results of each method, for all the methods that are repeated 30 times. It can be seen that random subspace outperforms other two "Ensemble All" strategies, which denotes that random subspace is more applicable to the classification of engineering surfaces. It is worth noting that MPO-SVME outperforms single SVMs and Ensemble All strategies in most of the 30 cases, and according to Table 5.3, MPO-SVME achieves the highest average CCPs in all the methods. This proves that MPO-SVME has better performance when compared with those commonly used methods such as single SVMs and Ensemble All.

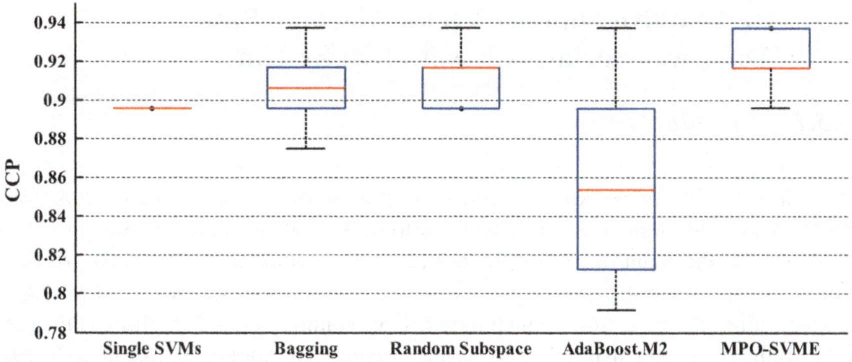

Fig. 5.22 Boxplots of the classification results using different strategies. Reprinted from Ref. [111], copyright 2014, with permission from ASME

Table 5.3 CCPs of different strategies (average of 30 times)

Strategy	Single SVMs	Ensemble all			Selective ensemble
		Bagging	Random subspace	AdaBoost.M2	MPO-SVME
CCP(%)	89.58	90.69	90.97	85.56	92.22

5.2.4   Conclusions

A method consists of feature extraction and classification is proposed to classify engineering surfaces using high-definition metrology. In this two-step method, DT-CWT is used to extract features and MPO-SVME is used for the purpose of classification. A case study is conducted, and it first demonstrates the effectiveness of using DT-CWT for feature extraction. The results indicate that DT-CWT outperforms DWT in feature extraction, and this can be explained by the fact that DT-CWT has approximate shift-invariance and good directional selectivity properties whereas DWT does not. Then the performance of MPO-SVME in classification is validated using real-world data. The results demonstrate that MPO-SVME shows improved generalization performance by assembling some accurate and diverse component multiclass SVMs. Furthermore, the influences of some key parameters of MPO-SVME upon its performance are analyzed, which aims to find the suitable parameters for constructing MPO-SVME. In general, the proposed method based on DT-CWT and MPO-SVME is effective and efficient in classifying engineering surfaces, especially surfaces with clear surface patterns.

5.3   An Adaptive Support Vector Machine Based Workpiece Surface Classification System

5.3.1   Introduction

The classification of workpiece surface patterns is considered as an essential element in understanding the functional performance of the product and providing feedback on the manufacturing process [21]. As traditional measurement devices (such as coordinate measuring machines) measure only a few scattered points or profiles due to economic constraints, they cannot sample high-density data describing three-dimensional (3D) surface spatial variation patterns in industrial applications. Recently, high-definition metrology (HDM) [69, 70] provides opportunities for product quality control since high-density data collected by an HDM device can precisely characterize the surface and reflect the impact of manufacturing processes on surface quality of a machined workpiece. HDM is superior to traditional measurement since it can generate a surface height map of millions of data points within seconds [25] and help to understand and improve surface quality control strategies in high-precision manufacturing [71], process control [72], and process improvement [26, 73]. Though lots of researches have been done for image classification, the research work about workpiece surface classification using high-density data points collected by HDM is sparse.

The classification tasks using HDM can be divided into two main steps: feature extraction and classification. Since it is not appropriate to directly use the raw data collected from the workpiece surfaces to classify, features should be extracted to represent a given workpiece surface before classification. Many surface characterization methods have been proposed to obtain abundant information about 3D surfaces, such as a 3D parameter set [25, 27], gray-level co-occurrence matrix [30], and two-dimensional autocorrelation function and spectral analysis [31]. In recent years, some methods that are first used in signal processing areas have been adopted to extract features of workpiece surfaces, such as Gabor filter banks [33], Gaussian filter banks [34], and wavelet packets [35–37], and after filtering, some numerical surface parameters are calculated for each sub-band to represent a given surface. Feature extraction is implemented by non-subsampled contourlet transform (NSCT) developed by [74], which possesses the properties of full shift-invariant, multi-scale, and multi-direction.

Automatic and accurate classifications are important tasks in industry practice and lots of methods are developed for different classification purposes [75–79]. Among many statistical learning-based classification methods, support vector machine (SVM) proposed by Vapnikin [20] is gaining popularity due to its many attractive features and promising generalization performance. Unlike other classification methods (such as artificial neural networks), SVM has a high capacity for generalization [47] and is used as a relatively novel statistical learning tool [80, 81]. SVM can effectively solve such practical classification problems of small samples, nonlinear, and high dimensions [45, 46, 48, 82]. The performance of SVM has a close relationship with the parameters of kernel function and penalty coefficient. Therefore, it is imperative to study the optimization method of SVM parameters.

Optimization of parameters (such as kernel function and penalty coefficient) in an SVM classifier faces challenges. Previous parameter optimization method is adopted by experiences or experiments [83], not only leading to problems such as intensive computation and low efficiency but also limiting the application of SVM since the selected parameters are usually not optimal. In order to overcome the drawback, some researchers propose many methods based on different search algorithms such as grid search [84], genetic algorithm (GA) [85], and particle swarm optimization (PSO) algorithm [86, 87]. These methods have been proven effective by experiments in the corresponding articles. However, grid search has heavy burden of computation with low learning accuracy [88], and implementation of GA is complex and needs to design different ways in creating crossover or mutation [89].

Though PSO algorithm is a viable optimization with good robustness [86, 87, 90], it has the problem of premature convergence into local extreme points and poor local search capability [91, 92]. In a standard PSO algorithm, the inertia weight cannot well balance the ability of particles between global search and local search as it is a fixed value [93]. Inertia weight is used to characterize the extent of the impact on particle's previous velocity relative to its current speed. Therefore, inertia weight affects the balance between particles' ability of global search and local search. The adjustment methods to inertia weight currently include mainly linear change, fuzzy adaptive, and random changes. The most widely used method among them is linearly decreasing strategy proposed by Shi [94, 95], but this method is likely to trap the solutions in a local optimum when solving multimodal function problems. To overcome these drawbacks, this subsection proposes a novel adaptive SVM-based classification method. An adaptive PSO (APSO) algorithm is proposed to control the update of inertia weight in every iteration for each particle, which is easy to implement and helps in preventing the solution stuck at a local minimum. Then a varied step-length pattern search (VSPS) algorithm is proposed to make use of its good performance in local search, leading the entire particle swarm to search potential optimal solution more efficiently.

5.3.2   The Framework of the Proposed Classification System

This subsection presents an overview of the proposed classification system with components of NSCT and an SVM classifier based on APSO-VSPS algorithm. To be specific, in feature extraction, high-resolution images are reflected by 3D coordinates of workpiece surfaces, which are filtered to extract details in different directions and scales by NSCT. The means and standard deviations of the

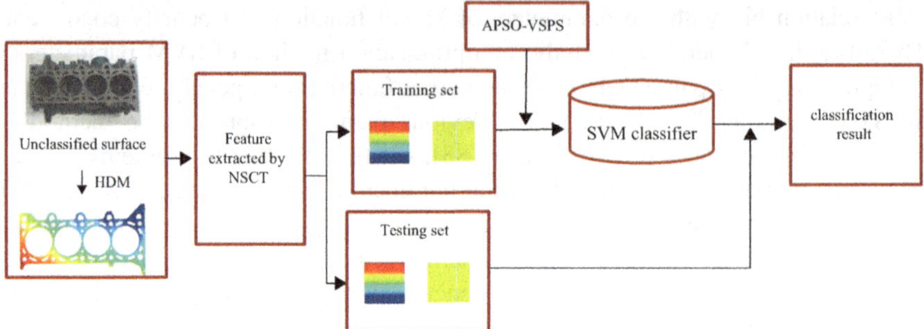

Fig. 5.23 The framework of the workpiece surface classification system. Reprinted from Ref. [112], copyright 2015, with permission from IEEE

coefficients of the wavelet sub-bands are calculated as feature vectors to represent a given surface. In the proposed classification method, the APSO algorithm is the main algorithm to select the optimal combination of parameters and the VSPS algorithm is nested in the main algorithm to improve the efficiency and accuracy of the search by its good performance in local search. The framework of the proposed system is shown in Fig. 5.23.

The procedure involves the following steps:

Step 1: HDM is employed to measure and collect three-dimensional coordinates from workpiece surfaces.

Step 2: NSCT is used to extract features from height maps of the three-dimensional coordinates. The extracted feature dataset is divided into two subsets: (1) training dataset and (2) testing dataset.

Step 3: The training dataset is input for the proposed APSO-VSPS algorithm to build an SVM classifier.

Step 4: The testing dataset is classified by the SVM classifier (built in Step 3) and the classification result is the final output.

5.3.3 Feature Extraction Using NSCT

The original data gained by HDM should not be directly used for classification since these data are just 3D coordinates that cannot be used as features. Therefore, a feature exaction procedure is necessary. An NSCT algorithm is used to exact features because it contains many merits of few constraints, good frequency selectivity, and good decomposition of image sub-band.

5.3.3.1 The Non-subsampled Contourlet Transform

NSCT combines non-subsampled pyramids (NSP) and non-subsampled directional filter bank (NSDFB) and it has many properties such as multi-scale, multi-direction, and shift-invariance, which can effectively eliminate pseudo-Gibbs phenomenon [96] in signal processing, and is sensitive to texture feature extraction [97].

The non-subsampled pyramid filter bank (NSPFB or NSP) is a two-channel non-subsampled filter bank [98, 99]. The process of NSP is similar to non-subsampled wavelet transform (NSWT) [74] based on à trous algorithm [100] in achieving the multi-scale decomposition of an image. The building block of a non-subsampled directional filter bank (NSDFB) is also a two-channel non-subsampled filter bank. The NSDFBs are iterated to obtain finer directional decomposition. The NSPFB provides multi-scale decomposition and NSDFB provides directional decomposition.

Figure 5.24 shows the block diagram of NSCT that contains two parts of structure figure and frequency division result. It can be seen that the transformation can be divided into two shift-invariant parts including a non-subsampled pyramid structure to ensure NSCT in multi-scale characteristics and a non-subsampled directional filter under a given direction. The input is decomposed into high-pass parts and low-pass parts by the NSPFB, and then the high-pass sub-band is decomposed into several directional bands by NSDFB. The scheme is iterated repeatedly on the low-pass sub-band.

5.3.3.2 Feature Extraction

Feature extraction is related to quantification of surface characteristics and its quantitative results are known as feature vectors. Optimization of these descriptive parameters is important because these parameters will influence the subsequent classification performance to some degree. A K-level NSCT decomposition is implemented to extract features. In order to reflect the degree of image texture, the mean and standard deviation of the coefficient matrix extracted by NSCT for decomposition transform are used to constitute an eigenvector $f = [\mu, \delta]$. Mean μ and standard deviation δ are calculated as

$$\mu = \frac{1}{A \times B} \sum_{x=1}^{A} \sum_{y=1}^{B} |P(x,y)| \tag{5.6}$$

$$\delta = \sqrt{\frac{1}{A \times B - 1} \sum_{x=1}^{A} \sum_{y=1}^{B} [P(x,y) - \mu]^2} \tag{5.7}$$

In Eqs. (5.6) and (5.7), A denotes the number of points along x-axis, B denotes the number of points along y-axis, $P(x,y)$ denotes the coefficient matrix of wavelet

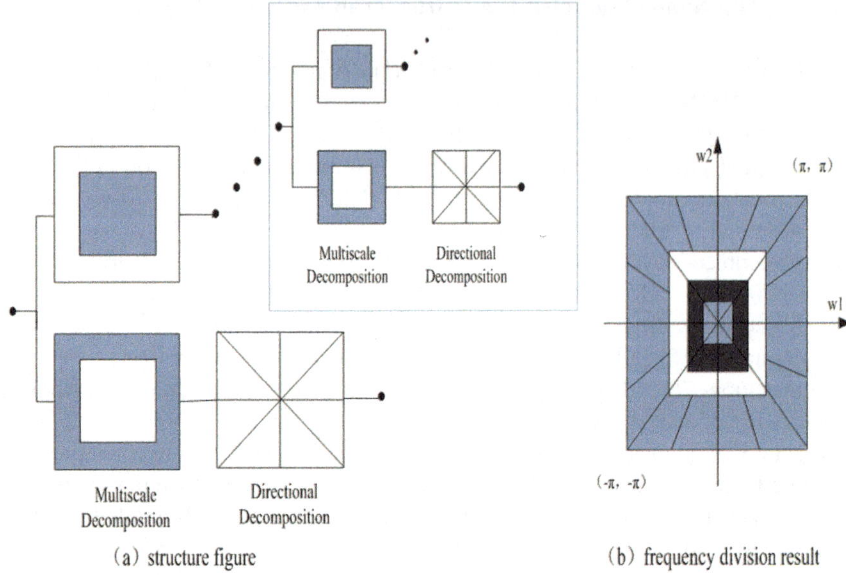

Fig. 5.24 Non-subsampled contourlet transform. Reprinted from Ref. [112], copyright 2015, with permission from IEEE

sub-bands, and μ and δ denote the mean and standard deviation of the sub-bands' coefficients, respectively.

Next, by computing means and standard deviations from coefficients of wavelet sub-bands of low-pass in K-levels, $2 \times K$ characteristic values are obtained to compose a $2 \times K$ dimensional feature vector $(\mu_1, \mu_2, \ldots, \mu_K, \delta_1, \delta_2, \ldots, \delta_K)$. The decomposition by NSCT just outputs the last level's coefficients of wavelet sub-bands of low-pass. Therefore, it is necessary to find another K-1 levels of coefficients of wavelet low-pass' sub-bands since these coefficients affect the classification results. To wavelet sub-bands of high-pass, means and standard deviations are calculated from coefficients matrix in each direction $([a_1, a_2, \ldots, a_K])$ of K-levels and a $2 \times \sum_{i=1}^{K} 2^{a_i}$ dimensional feature vector is generated $(\mu_1, \mu_2, \ldots, \mu_{\sum_{i=1}^{K} 2^{a_i}}, \delta_1, \delta_2, \ldots, \delta_{\sum_{i=1}^{K} 2^{a_i}})$. By combining the $2 \times \sum_{i=1}^{K} 2^{a_i}$ dimensional high-pass feature vector and $2 \times K$ dimensional low-pass feature vectors, a $2 \times \left(\sum_{i=1}^{K} 2^{a_i} + K\right)$ dimensional feature vector is obtained to represent a given surface.

5.3.4 The Proposed Adaptive SVM Classifier

Since SVM is currently one of the most popular classification tools, there is no need to discuss it in much detail. For more details about SVM, please refer to the work of

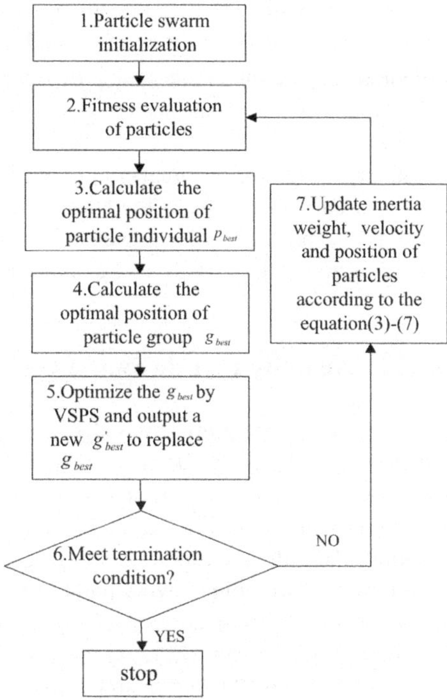

Fig. 5.25 Flow diagram for adaptive parameter optimization algorithm. Reprinted from Ref. [112], copyright 2015, with permission from IEEE

[47]. Two parameters of C (controls the trade-off between function complexity and training error of SVM classifier) and σ (denotes the width parameter of the Gaussian radial basis function (RBF) and controls the radial scope of function) have a great impact on the classification performance of SVM classifier. An APSO algorithm and a VSPS algorithm are proposed to jointly optimize the two parameters. The flow diagram (shown in Fig. 5.25) is the process of the proposed APSO-VSPS algorithm, which is APSO algorithm nested by the VSPS algorithm.

The APSO algorithm is an adjustment strategy adopting different inertia weights to update the particle swarm in the same generation, which accelerates the convergence speed and can jump out of local optimal solution. In this way, it provides a better initial value for VSPS, which is conducive to obtain the final optimal solution.

The proposed algorithm can be described as follows:

Step 1: Initialize all particles. Set initial position and velocity of particles within the allowable range randomly. The position of a particle is used to indicate a potential solution of the optimization problem in the search space. Set initial position of each particle as its local optimum p_{best} and set the best value in the set of p_{best} as g_{best}.

Step 2: Evaluate the fitness value of each particle and calculate objective function value of each particle according to the fitness function.

Step 3: Calculate the best position of each particle p_{best}.

Step 4: Calculate the best position of all particles g_{best}.

Step 5: The g_{best} is set as the initial search point in VSPS algorithm and the output is g'_{best} which replaces the previous g_{best}. In the vicinity of the better point for local search by VSPS, the search should continue until the result meets the accuracy requirements.

Step 6: Check the terminal condition (the optimal solution stagnates and changes within a range of $\pm 2\%$). If the termination is met, then stop. If not, go to Step 7.

Step 7: Update the inertia weight, velocity, and position of the current particle according to the Eqs. (5.8)–(5.12) and back to Step 2.

5.3.4.1 Adaptive Particle Swarm Optimization (APSO) Algorithm

PSO is an iterative optimization algorithm which is initialized to a group of random particles. The flying direction and distance of each particle may be controlled by its own velocity, while the pros and cons of each particle can be evaluated by a fitness value function. In each iteration, the algorithm is mainly to update each particle by tracking individual extreme point p_{best} and global extreme point g_{best}. When searching the two points, every particle updates its speed and position according to the following equations:

$$v_t = \omega v_{t-1} + c_1 r_1 \left(p_{best} - x_{t-1}\right) + c_2 r_2 \left(g_{best} - x_{t-1}\right) \tag{5.8}$$

$$v_t = \begin{cases} v_{max}, & v_t > v_{max} \\ -v_{max}, & v_t \leq v_{max} \end{cases} \tag{5.9}$$

$$x_t = x_{t-1} + v_t \tag{5.10}$$

where ω is inertia weight, c_1 and c_2 are acceleration constraints, $r_1, r_2 \in [0, 1]$ are random values, x_t is current position of the particle that represents the current parameter value of C and σ in SVM, g_{best} is the best position of group in history records, and p_{best} is the best position of current particle in history records.

To accelerate the convergence speed while not being trapped in the local optima, an adaptive ω-strategy and the concrete equations are adopted as follows:

$$\omega(i, x) = u(d_i, x) * \left[\omega_1 + (\omega_2 - \omega_1) * \frac{CurIter}{MaxIter}\right] \tag{5.11}$$

where MaxIter is the maximum number of iteration, CurIter is the current number of iteration, and ω_1, ω_2 are initial value and final value of ω, respectively.

$$u(d_i, x) = \begin{cases} 1 + \theta, & d_i \leq S_1 M \\ 1, & S_1 M < d_i < S_2 M \\ 1 - \varsigma, & d_i \geq S_2 M \end{cases} \tag{5.12}$$

where M is population size, S_1, S_2 are control parameters, and θ, ς are adjustment parameters, and satisfy the following conditions: $S_1 < S_2 < 1, \theta > 0, \varsigma > 0$.

5.3.4.2   Varied Step-Length Pattern Search (VSPS) Algorithm

In this subsection, VSPS algorithm is embedded within APSO algorithm due to its good performance in local search. Hooke and Jeeves [101] first proposed PS in 1961, and then Torczon [102] completed the convergence proof of the PS algorithm for unconstrained problems and gave a framework for a class of pattern search method which contains coordinate search method, Hooke–Jeeves (H–J) method, and Powell conjugate direction method [103]. The H–J PS algorithm [101] alternates implementation of two searches from the initial point: axial search and PS. The first type of search is an exploratory move (also called axial search explained in the subsequent Step 2) designed to determine the new base point and the direction conducive to decrease the values of fitness function, and move in the direction of m-axis. The second type of search is a pattern move designed to utilize the information acquired in the exploratory moves, and accomplishes the actual minimization of the function by moving in the conjunction direction of two adjacent base points.

The H-J PS algorithm is a constant nonlinear function optimization method that is widely used in many fields. It is a simple iteration as it does not require the objective function to be derivable. Besides, the H-J PS algorithm has good performance in local convergence and is sensitive to initial value in consideration of local convergence. However, in the process of the H-J PS algorithm, the same step length for search in all directions restricts the convergence speed and accuracy and if the search point is a local maximum point of the function, such an exploratory move may lead to direction of selecting an inferior point rather than the better one. Therefore, a VSPS algorithm is proposed to avoid these drawbacks, which consists of a variable step change strategy, accelerating factors, and a new way for exploratory moving. The flow diagram of the proposed VSPS algorithm is shown in Fig. 5.26.

To describe the two modes of axial search and pattern search, the procedure is presented as follows. With respect to $\min f(x)$ (where min is to minimize and $f(x)$ is objective function), let $e_i = (0, \ldots, 0, 1, \ldots, 0)^T, i = 1, 2, \ldots, m$ and e_i denote unit vector and m denote m-direction of axis. $y^i (i = 1, 2, \ldots, m)$ is used to indicate a starting point searching in the direction of the ith axis e_i.

Step 1: Select initial point x^1, initial step length $\chi > 0$ (here χ is an m-dimensional vector rather than a numerical value), acceleration factor $\eta \geq 1$, increase rate of step length $\lambda > 1$, decrease rate of step length $\beta \in (0, 1)$. ε is the accuracy requirement. Let C_{pso}, σ_{pso}, the search loop step number $k = 1. x^k$ is used to denote the kth base point.

Step 2: Start axial search mode. In every axial search, there are two search directions: one is along the positive direction $(+e_i)$ of the axis and the other is the

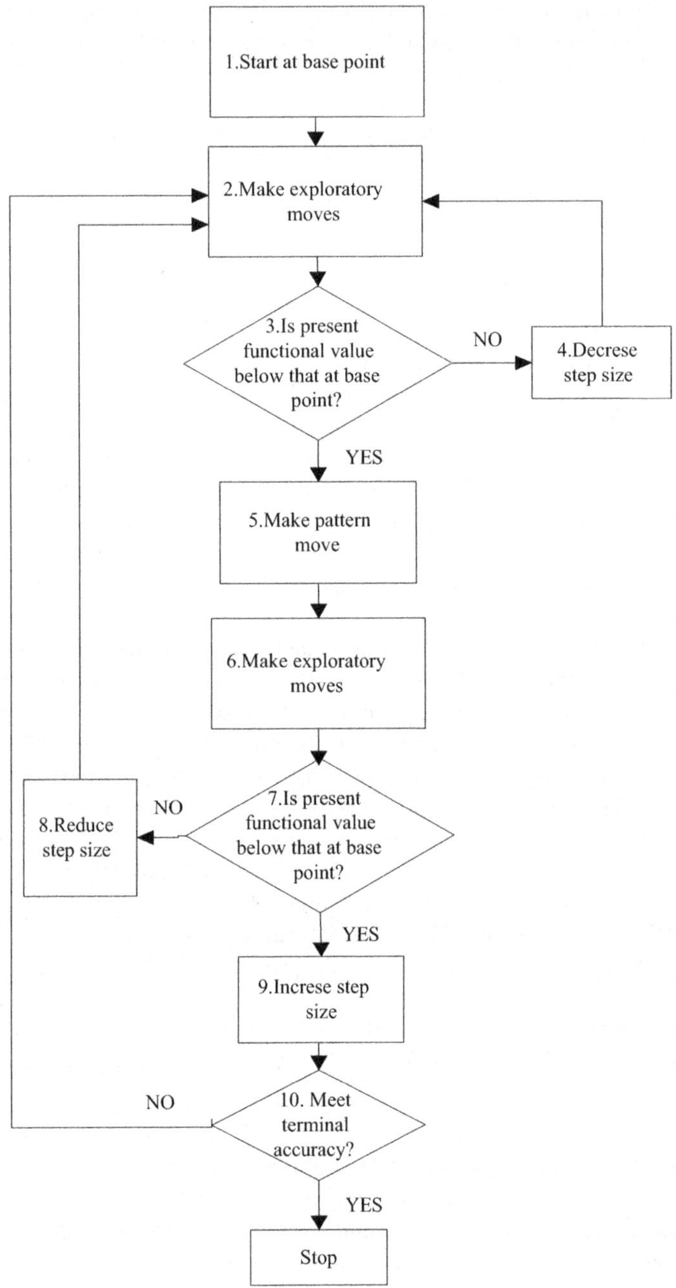

Fig. 5.26 Flow diagram of the proposed VSPS algorithm. Reprinted from Ref. [112], copyright 2015, with permission from IEEE

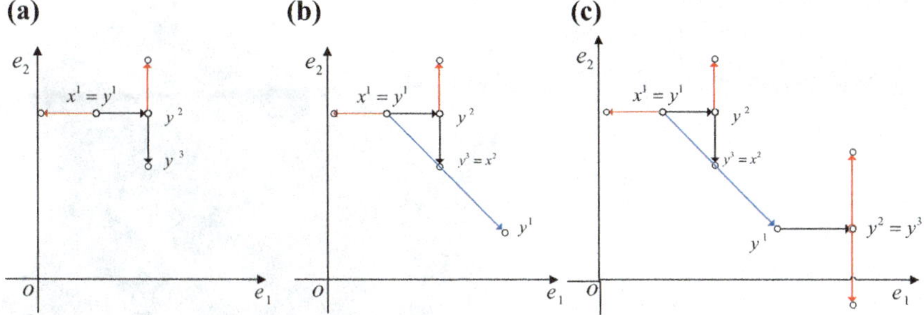

Fig. 5.27 Axial search and pattern search in two-dimensional coordinate system. Reprinted from Ref. [112], copyright 2015, with permission from IEEE

opposite $(-e_i)$. $i = 1, 2, \ldots, m$, compare the value of $f(y^i + \chi(i)e_i) = f_{i1}$ (positive direction), $f(y^i - \chi(i)e_i) = f_{i2}$ (negative direction), and $f(y^i) = f_{i3}$. If $f_{i1} < f_{i3} \leq f_{i2}$ or $f_{i1} < f_{i2} < f_{i3}$, let $y^{i+1} = y^i + \chi(i)e_i$ and $\chi = \chi \times \lambda$ (the search along the positive direction of the axis is effective); if $f_{i2} < f_{i3} \leq f_{i1}$ or $f_{i2} < f_{i1} < f_{i3}$, let $y^{i+1} = y^i - \chi(i)e_i$ and $\chi = \chi \times \lambda$(the search along the negative direction of the axis is effective); else let $y^{i+1} = y^i$ and search along another axis. The search should be along m-axes and a value of $f(y^{m+1})$ is obtained. An example is shown in Fig. 5.27a in two-dimensional coordinate system. In Fig. 5.27, x^1, x^2 are base points, y^1, y^2, y^3 are points found by axial search and pattern search, e_1, e_2 are unit vectors, and O is the origin of the coordinate.

Step 3: If $f(y^{m+1}) \geq f(x^k)$ (it means axial search mode fails and that process should be repeated until it succeeds), go to Step 4. If $f(y^{m+1}) < f(x^k)$, go to Step 5.

Step 4: Decrease step size $\chi = \chi \times \beta$ and back to Step 2.

Step 5: Start pattern search mode. Let $x^{k+1} = y^{m+1}$ and the direction of $(x^{k+1} - x^k)$ is probable to decrease the function value. Therefore, pattern search will be along that direction, which is $y^1 = x^{k+1} + \eta \times (x^{k+1} - x^k)$ (shown in Fig. 5.27b in two-dimensional coordinate system).

Step 6: Do the same as Step 2 (shown in Fig. 5.27c in two-dimensional coordinate system).

Step 7: Let $k = k + 1$. If $f(y^{m+1}) < f(x^k)$ (it means pattern search mode is successful and another pattern search mode can begin with acceleration at the point of y^{m+1} obtained in Step 6), go to Step 9. If $f(y^{m+1}) \geq f(x^k)$ (it means pattern search mode is failed and another iteration of VSPS can begin at the point of y^{m+1} obtained in Step 2), go to Step 8.

Step 8: Reduce step size $\chi(\chi = \chi \times \beta, \eta = \eta \times \beta)$, and back to Step 2.

Step 9: Increase step size $\chi = \chi \times \lambda$, and go to Step 10.

Step 10: If $\chi < \varepsilon, i \in [1, m]$, then stop. If not, back to Step 2.

The VSPS algorithm improves the optimization process of detecting move by enabling detecting step length in each direction to change differently according to the variations of objective function value in all directions, which makes the search

Fig. 5.28 Engine cylinder block. Reprinted from Ref. [112], copyright 2015, with permission from IEEE

direction closer to the optimal pattern of descent direction. Acceleration factor is shrinking with the search conducted, which is helpful to search more in detail and is not easy to skip the optimal point.

5.3.5 Case Study

5.3.5.1 Case 1: Engine Block Top Surface

This subsection validates the proposed classification system based on measurement data from engine block top surfaces. The measurement is conducted by a laser holographic interferometer (ShaPix Detective [66]) which is capable of gathering up to 1 million data points over a surface area of 300×300 mm^2 within 1 min. The basic height (Z) accuracy of it is 1 μm while the lateral (X, Y) resolution is around 0.3 mm.

Six surface samples are selected from different engine blocks (shown in Fig. 5.28) manufactured in different conditions. It is apparent that surface samples 1–3 (called set A) are machined in one condition while surface samples 4–6 (called set B) are machined in another condition. ShaPix is used to scan these six surfaces and collect 3D coordinate data from each surface. Two samples of the color-coded measurement results from each dataset are shown in Fig. 5.29.

It is obvious that the height of both sets decreases from left to right while the former set has a larger fluctuation range in surface than the latter. As the car engine block is a critical functional part, low precision will lead to poor engine performance such as leakage or bore distortion. Therefore, it is imperative to identify the defective blocks from the normal ones through classification.

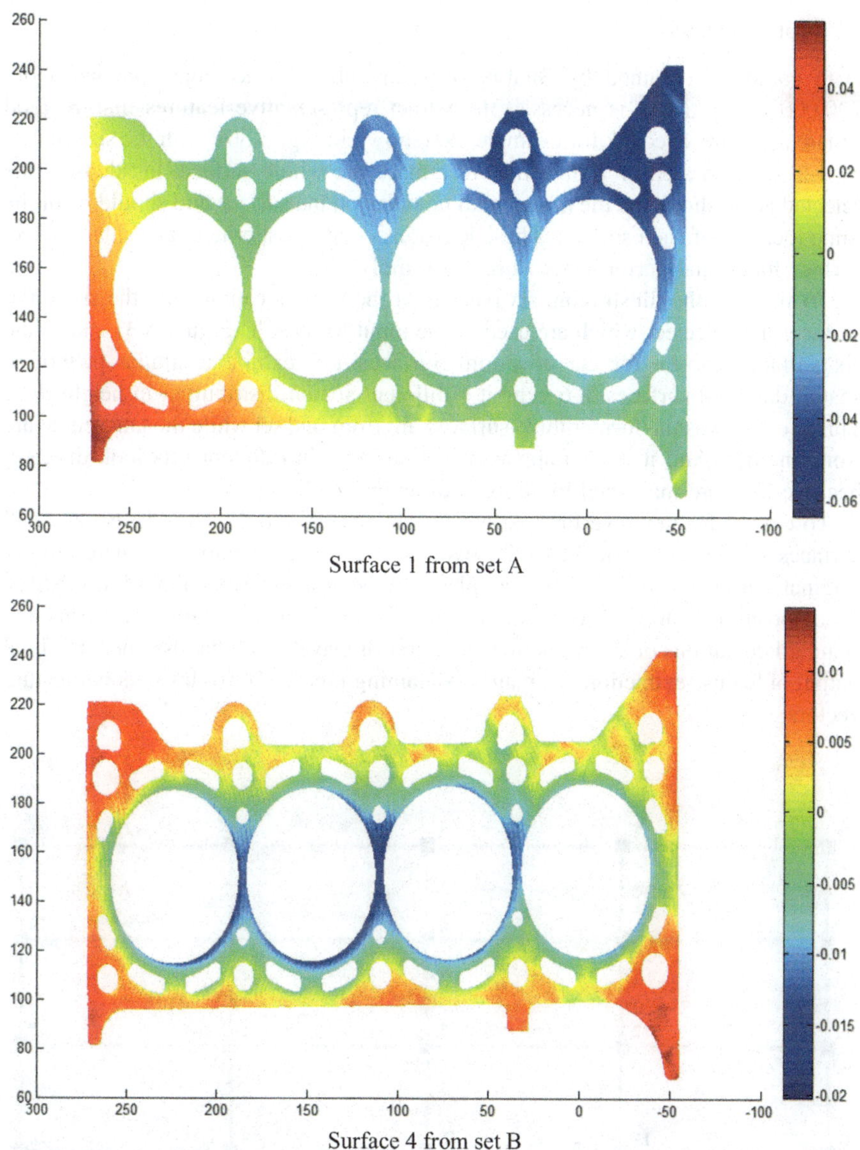

Fig. 5.29 Two samples of the color-coded measurement results from datasets A and B. Reprinted from Ref. [112], copyright 2015, with permission from IEEE

(1) Feature extraction

Data could be obtained by ShaPix over an entire surface that contains about 790,000 points and it is necessary to extract representative features first. Several square areas are selected, for example, 20 (shown in Fig. 5.30), with the size of 128 by 128 over an area of 16384 mm^2 to replace the whole surface. In addition, the selected areas should be the places with no holes on the surface and should be of the same location of each surface. These locations cover most critical areas that impact surface functional performance during assembly.

To simplify the illustration, six patches of the same location from the six initial surfaces are selected, which are used as the input for NSCT. Figure 5.31 shows the six surfaces chosen for classification. Surfaces 1–3 exhibit a similar pattern in spatial data and surfaces 4–6 exhibit a different spatial distribution in height data. This is evident as the former three surfaces are from one set while the latter three are from another. And it is also apparent that surfaces in different sets look different because they are machined by tools in different conditions.

To extract features, we first partition each surface into $4 \times 4 = 16$ equal small surfaces with the size of 32 by 32 over an area of 1024 mm^2. As there are six original surfaces, $16 \times 6 = 96$ samples can be gotten. Then three-level NSCT decomposition is applied to extract feature vectors, which contains the means and standard deviations of every coefficient matrix of wavelet sub-bands. Then, the final output of feature extraction is a matrix containing ninety-six 56-dimensional feature vectors.

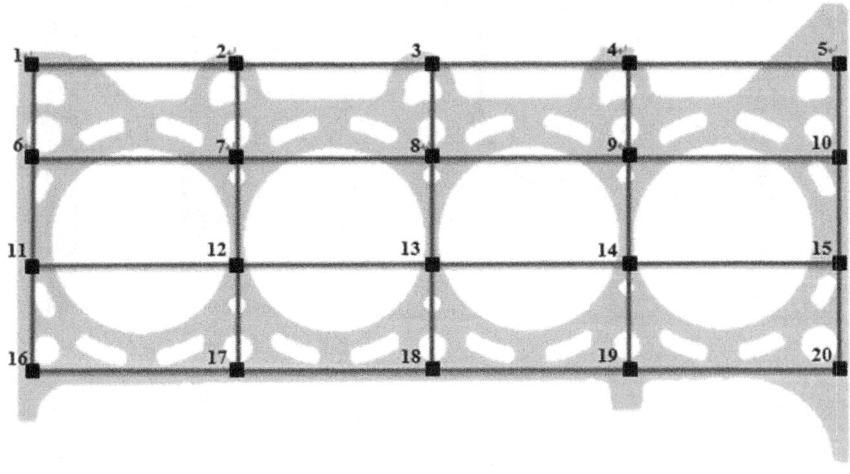

Fig. 5.30 Optimization of small surface samples through a grid chart. Reprinted from Ref. [112], copyright 2015, with permission from IEEE

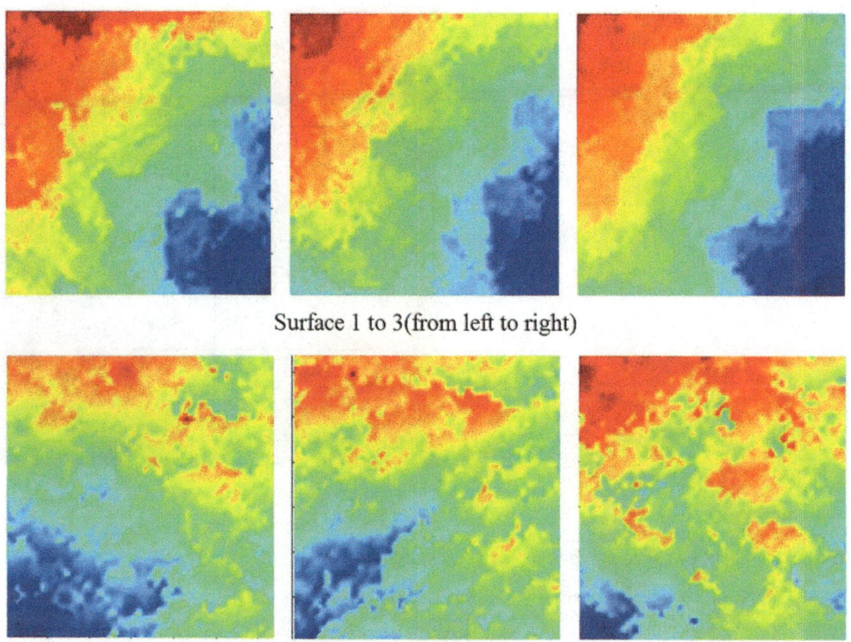

Surface 1 to 3(from left to right)

Surface 4 to 6(from left to right)

Fig. 5.31 Six surfaces selected from set A and set B. Reprinted from Ref. [112], copyright 2015, with permission from IEEE

(2) Classification results

In Fig. 5.32, Nos. 1–16 represent 16 surface samples. In terms of each original surface sample, eight circled surface samples are randomly selected as training samples and the other eight ones are testing samples.

Since there are six surfaces in this case, the total number of training samples and testing samples is $6 \times 8 = 48$. As to the selection of the kernel function, there are four common kernel functions: linear kernel function, polynomial kernel function, RBF, and sigmoid kernel function. In fact, RBF is the most widely used kernel function and the reasons are given as follows: (1) RBF can realize the nonlinear mapping. (2) The number of parameters that affects the model complexity in RBF is few. (3) RBF kernel has less numerical difficulties.

Table 5.4 shows the correct classification percentage (CCP) for the four types of kernel functions. The results represent that the CCP of RBF is the highest and it is reasonable to be chosen as the kernel function of SVM. It is however worth noting that although the kernels presented in Table 5.4 are the most popular, multiple other functions (such as Laplace, Fourier, etc.) can also be used for the classification task as well. This subsection focuses on the optimization of parameters of SVM rather than the selection of kernel functions, and therefore other types of functions are not compared here.

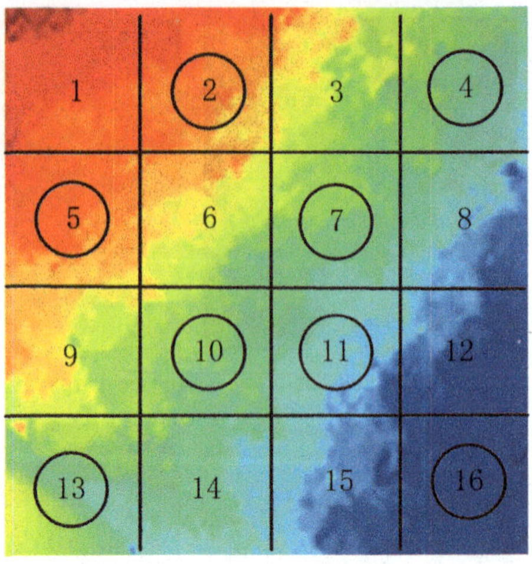

Fig. 5.32 The generation of training samples and testing samples. Reprinted from Ref. [112], copyright 2015, with permission from IEEE

Table 5.4 The CCP of four types of kernel functions

Kernel function	Linear kernel	Polynomial kernel	RBF	Sigmoid kernel
CCP	87.5	91.67	93.75	87.5

The level of NSCT has an effect on the classification accuracy. To study the influence of different decomposition levels on the classification accuracy, levels from 2 to 5 are adopted, respectively. For the purpose of comparison, each level of NSCT implemented to extract features should use the same dataset. The final classification results of each level are shown in Fig. 5.33.

Figure 5.34 shows the classification results for the six engine cylinder block surfaces (surface areas are of the same location from each surface).

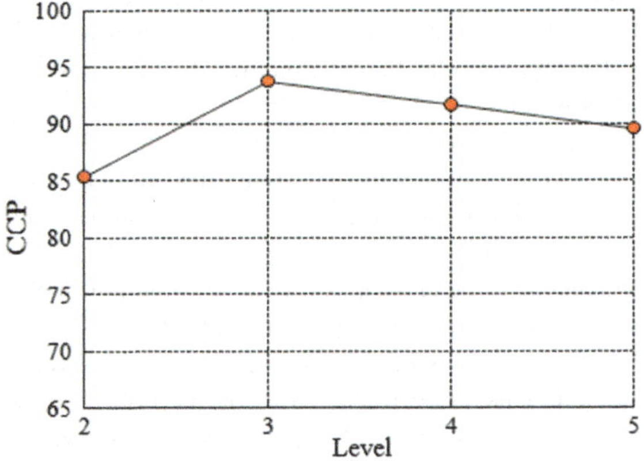

Fig. 5.33 The CCPs using NSCT with different levels. Reprinted from Ref. [112], copyright 2015, with permission from IEEE

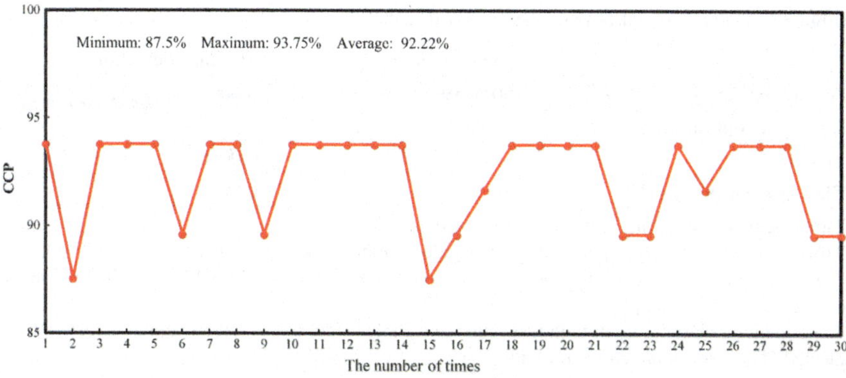

Fig. 5.34 Classification results of the proposed adaptive SVM-based system. Reprinted from Ref. [112], copyright 2015, with permission from IEEE

In order to obtain a more accurate result, the experiment is repeated for 30 times. Correct classification percentage (CCP) is used as an evaluation index. The proposed system finds the CCP of 93.75% repeatedly (19/30 in the experiments), which shows good performance of stability. The average CCP is 92.22%.

Table 5.5 presents the final results of 30 experiments by the proposed adaptive SVM-based classification system, which contains the classification accuracy and corresponding parameter combination. It can be seen that although the accuracies are same, the parameter combinations are quite different. The detailed analysis will be discussed in next subsection.

To analyze the time duration of the proposed algorithm, it was implemented in the MATLAB R2012b programming environment on a PC with Intel Core2 Duo T6600 CPU running Windows 7. The typical time complexity of SVM is between $O(Nsv^3 + LNsv^2 + dLNsv)$ and $O(dL^2)$, in which Nsv is the number of support

Table 5.5 Results of parameters and CCP by the proposed adaptive SVM-based system

Index	1	2	3	4	5	6	7	8	9	10
C	104.03	244.06	216.68	292.57	153.53	105.09	153.53	93.23	108.42	45.64
σ	4.57	9.49	1.15	1.19	1.94	8.96	1.94	4.4	7.87	5.52
CCP	93.75	87.5	93.75	93.75	93.75	89.58	93.75	93.75	89.58	93.75
Index	11	12	13	14	15	16	17	18	19	20
C	75.07	242.3	174.78	257.03	277.36	225.89	120.45	163.66	137.57	28.02
σ	4.66	1.53	1.81	1.85	8.91	6.28	2.01	2.25	2.06	9.71
CCP	93.75	93.75	93.75	93.75	87.5	89.58	91.67	93.75	93.75	93.75
Index	21	22	23	24	25	26	27	28	29	30
C	57.45	150.42	233.79	46.28	40.15	249.95	50.64	263.05	269.68	204.88
σ	5.43	9.78	6.05	6.14	6.1	1.03	5.91	1.41	3.06	3.76
CCP	93.75	89.58	89.58	93.75	91.67	93.75	93.75	93.75	89.58	89.58

Table 5.6 Computational time of the proposed method

	Time (Unit s)	Standard deviation
Feature extraction[a]	0.0459	0.0009
Parameters optimization[b]	7.0856	0.7663
Training[c]	0.0349	0.0060
Classification (testing)[d]	0.0081	0.0012

[a]Time spent on extracting features of a surface sample
[b]Time spent on optimizing parameters of penalty coefficient and kernel function
[c]Time spent on training classifiers with the parameters optimized by APSO-VSPS (the training set includes 48 surface samples)
[d]Time spent on classifying an unlabeled surface sample (input is extracted feature vector) with the selected classifiers obtained by training

vectors, L is the number of training samples, and d is the dimensional number (original dimensional number that has not been mapped to high-dimensional space) of each sample. The time complexity of APSO-VSPS is $\max\{O_{APSO}, O_{VSPS}\} = O_{APSO} = N*m*T_T$, in which N is the number of particles, m is the maximum number of iterations, and T_T is the computational time of each particle in every iteration.

To evaluate the actual duration of the implemented algorithm, statistical methods including mean and standard deviation values are used. The former determines the most probable execution time, while the latter gives the information about the size of outlayers caused by the operating system's work regime [104]. The training time, testing time, and parameters optimization time of the presented method in this case study are given as follows, both of which are the average of 30 times. The time spent on feature extraction is also provided in Table 5.6. The total implement time cost is about 7.1745 s.

(3) Sensitivity analysis

(a) Sensitivity analysis to the range of parameters in APSO-VSPS

Table 5.7 presents that the optimization of parameters has a great impact on the classification performance.

Parameter of penalty coefficient (C) controls the extent of the punishment for classifying wrong samples and the value of C denotes the degree of punishment for misclassified samples. If value of C tends to infinity, all constraints must be satisfied, which means that all the training samples should be correctly classified and would result in the complexity of classification hyperplane and intensive computation. From Table 5.7, it can be seen that the classification accuracy shows a trend that it first increases and then decreases on the whole with the value of penalty coefficient (C) increasing. The selection of C is a compromise since the optimal number of support vectors and CCP can be obtained when C reaches a certain value, if value of C continues to increase, it will only add the training time and have no improvement for the performance of SVM. The experiments with the σ

Table 5.7 CCP by different parameter combinations of C and σ

σ	C					
	10	20	50	100	200	300
1	87.5	87.5	87.5	87.5	89.58	93.75
2	87.5	87.5	89.58	89.58	93.75	89.58
5	87.5	87.5	93.75	91.67	89.58	89.58
10	87.5	89.58	91.67	89.58	87.5	87.5
50	93.75	89.58	87.5	87.5	85.42	85.42
100	89.58	87.5	87.5	85.42	91.67	89.58
200	87.5	87.5	85.42	91.67	89.8	87.5
300	87.5	87.5	85.42	89.58	87.5	87.5

parameter show that as long as it tends to be zero and the vector of Lagrange multiplier is nonnegative, all training samples are support vectors and SVM classifier classifies the whole training set correctly. If σ tends to be infinite, which indicates that more training samples will be in the interval (between two different classes of data) or be misclassified, the classification accuracy of training samples by SVM is zero and SVM classification performance will drop substantially since all the samples are classified as the same class. If value of σ is appropriate, the number of support vectors can decrease and the ability to classify test samples will be greatly enhanced. Therefore, choosing the appropriate parameter combination together can construct a classifier with high performance. Besides, for datasets obtained from different kinds of workpiece surfaces, the parameter combination should be reoptimized because the ranges are not the same.

(b) Sensitivity analysis to the step length of pattern search in APSO-VSPS

If the length of the step size during the normal pattern search is the same in all directions, the deviation between the direction of pattern movement and optimal descent direction is large, limiting the convergence speed and accuracy. Thus, a non-monotonic and varied step-length pattern search method is adopted to change the step length and make the direction of pattern move closer to the optimal descent direction. However, the initial step length should be selected appropriately as large step length may lead to a skip out of the optimal point and small step length may lead to a stuck in local convergence that is not globally optimal. Table 5.8 represents frequency of CCP by APSO-VSPS with different lengths of step size (operation for 30 times totally).

(c) Comparison analysis for APSO, APSO-PS, and APSO-VSPS

Table 5.9 presents the comparison of APSO, APSO-PS, and APSO-VSPS in the aspects of maximum, minimum, average, and standard deviation of CCP.

From Table 5.9, it can be seen that CCP of APSO-VSPS is much higher than that of APSO since PS (pattern search) performs well in local search, which does good to the performance of global search by APSO. Besides, it also can be seen that although the standard deviation, maximum and minimum CCP of APSO-PS are

Table 5.8 Frequency of CCP by APSO-VSPS algorithm

Times of value number		Combination of initial step length in VSPS					
		(0.1, 0.1)	(1, 1)	(1, 9)	(10, 2)	(10, 10)	(20, 3)
CCP	87.5	14	8	5	6	4	2
	89.58	4	10	3	8	8	7
	91.67	4	4	3	5	1	2
	93.75	8	8	19	11	17	19

Table 5.9 Comparison of APSO, APSO-PS, and APSO-VSPS

CCP	Maximum (%)	Minimum (%)	Average (%)	Standard deviation
APSO	91.67	85.42	87.43	0.027
APSO-PS	93.75	87.5	89.65	0.021
APSO-VSPS	93.75	87.5	92.22	0.021

equal to that of APSO-VSPS, the average CCP of APSO-VSPS is higher than that of APSO-PS because APSO-VSPS adopts varied step-length pattern search, which is uneasy to step out of the primal point. Therefore, APSO-VSPS can perform better. According to the comparisons, APSO-VSPS is superior to APSO and APSO-PS.

(4) Comparisons for the proposed adaptive SVM method and other methods

(a) Comparison for standard SVM and the proposed adaptive SVM method

Weka [105] with LIBSVM toolbox [67] is chosen as software tool for computation. The standard SVM is used to classify the six patterns which represent six different surfaces (1–3 belong to Set A while 4–6 belong to Set B) in detail. Here the default parameters of $C = 1, \sigma = 0$ are selected for standard SVM. Tables 5.10 and 5.11 show the results of CCP of standard SVM and the proposed adaptive SVM, respectively.

According to Table 5.10, classifications of several patterns are entirely correct and the final classification result is the average CCP of six patterns. There also

Table 5.10 CCP of standard SVM

Pattern	Classification result						CCP
	1	2	3	4	5	6	
1	**8**	0	0	0	0	0	100
2	0	**8**	0	0	0	0	100
3	0	2	**6**	0	0	0	75.0
4	0	0	0	**5**	3	0	62.5
5	0	0	0	0	**7**	1	87.5
6	0	0	0	0	0	**8**	100
Average							87.5

Table 5.11 CCP of the proposed adaptive SVM method

Pattern	Classification result						CCP
	1	2	3	4	5	6	
1	**8**	0	0	0	0	0	100
2	0	**8**	0	0	0	0	100
3	0	**1**	**7**	0	0	0	87.5
4	0	0	0	**7**	**1**	0	87.5
5	0	0	0	0	**7**	**1**	87.5
6	0	0	0	0	0	**8**	100
Average							93.75

exists a phenomenon that misclassification is more likely to occur in the same set as they are too similar to differentiate.

Comparing Tables 5.10 and 5.11, it can be seen that the classification result of the proposed adaptive SVM method is 6.25% higher than that of standard SVM and the proposed adaptive SVM method can increase the classification accuracy of misclassified patterns in Table 5.10 to some extent. It is also apparent that the proposed adaptive SVM method can improve the CCP of patterns with much misclassification (pattern 3 with two misclassifications and pattern 4 with four misclassifications), but it cannot improve the CCP of misclassified patterns to 100 percentages.

(b) Comparison for the proposed adaptive SVM method and other classification methods

There are several other classic classification methods such as bagging-SVM [58], random subspace-SVM [60], naive Bayesian [106], Adaboost.M2-SVM [68], logistic regression [107], neural network [108], fuzzy logic [109], and rough sets [110]. The performances of some methods are greatly affected by their parameters and architectures. For instance, rough sets require setting discretization and reduction methods, and neural network has several specific architectures (multi-layered perceptron, RBF network, etc.) to choose. Due to limited space, the comparisons with the rough sets and neural network methods are not provided in this subsection. The classification results of the other strategies are shown in Table 5.12.

According to Table 5.12, it can be seen that the proposed adaptive SVM method outperforms other methods in most of the 30 cases and the proposed adaptive SVM method achieves the highest average CCPs in all the methods. This proves that the proposed adaptive SVM method has a better performance when compared to those commonly used methods. Due to a large production volume in engine production, an improvement of 1% in CCP could lead to significant savings of cost incurred by potential scraps.

Table 5.12 CCP of different strategies (average of 30 times)

Strategy		CCP
Other methods	Bagging-SVM	90.69
	Random Subspace-SVM	90.97
	AdaBoost.M2-SVM	85.56
	Naive Bayesian	87.50
	Logistic regression	87.50
	Fuzzy logic	89.50
Proposed method	The proposed adaptive SVM method	93.75

5.3.5.2 Case 2: Cylinder Head Cover Surface

The second workpiece surface is the surface of a cylinder head that is made of aluminum alloy. The height map of the two different (new and used) surfaces is shown in Fig. 5.35 and the units of X-axis, Y-axis, and height are mm. Eight locations are selected (locations 1–8, shown in Fig. 5.36) from the surface with the same size and then eight small surfaces are obtained.

To classify the cylinder head cover surfaces, the proposed adaptive SVM-based classification method is adopted and the same process should be done as shown in Case 1. The classification accuracy is 95.83%, which is higher than that of Case 1 since the samples in this case are more apparent in difference. An optimal set of parameters is $C = 214.28$, $\sigma = 6.91$. Table 5.13 presents classification results of the proposed adaptive SVM-based classification system, where patterns 1–6 represent six different surfaces (1–3 from one set and 4–6 from another).

5.3.5.3 Case 3: Pump Valve Plate Top Surface

The third workpiece surface is the surface of a cam carrier. The height map of the two different (new and used) surfaces is shown in Fig. 5.37 and the units of X-axis, Y-axis, and height are mm. Eight locations are selected (locations 1–8, shown in Fig. 5.38) from the surface with the same size and then eight small surfaces are obtained.

To classify the cam carrier cover surfaces, the proposed adaptive SVM-based classification system is adopted and the same process should be done as shown in Case 1. The classification accuracy is 93.75% and an optimal set of parameters is $C = 91.54$, $\sigma = 4.05$. Table 5.14 presents classification result of the proposed adaptive SVM-based classification system, where patterns 1–6 represent six different surfaces (1–3 from one set and 4–6 from another).

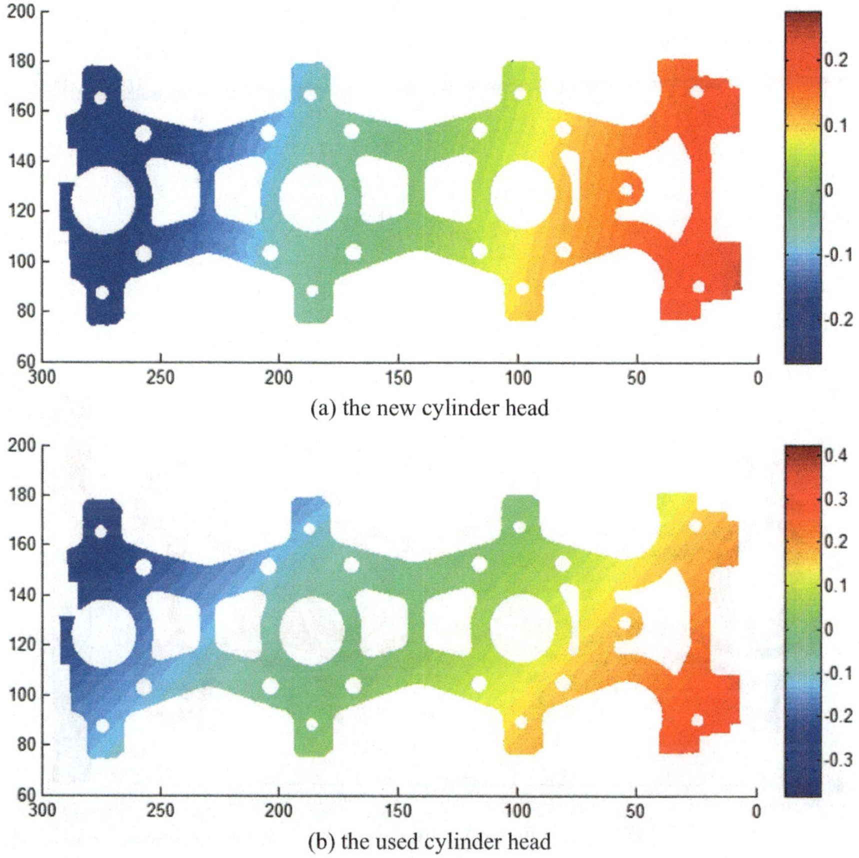

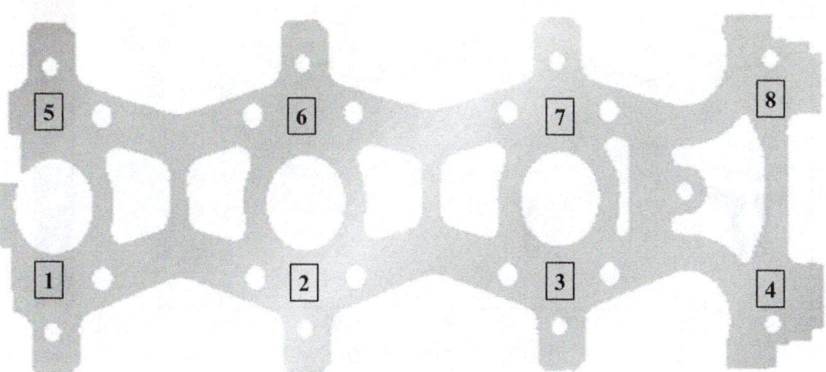

Fig. 5.35 The height map of two different cylinder heads. Reprinted from Ref. [112], copyright 2015, with permission from IEEE

Fig. 5.36 Eight locations selected from the surface. Reprinted from Ref. [112], copyright 2015, with permission from IEEE

Table 5.13 CCP of the proposed adaptive SVM method

Pattern	Classification result						CCP
	1	2	3	4	5	6	
1	**8**	0	0	0	0	0	100
2	0	**8**	0	0	0	0	100
3	0	**1**	7	0	0	0	87.5
4	0	0	0	7	**1**	0	87.5
5	0	0	0	0	**8**	0	100
6	0	0	0	0	0	**8**	100
Average							95.83

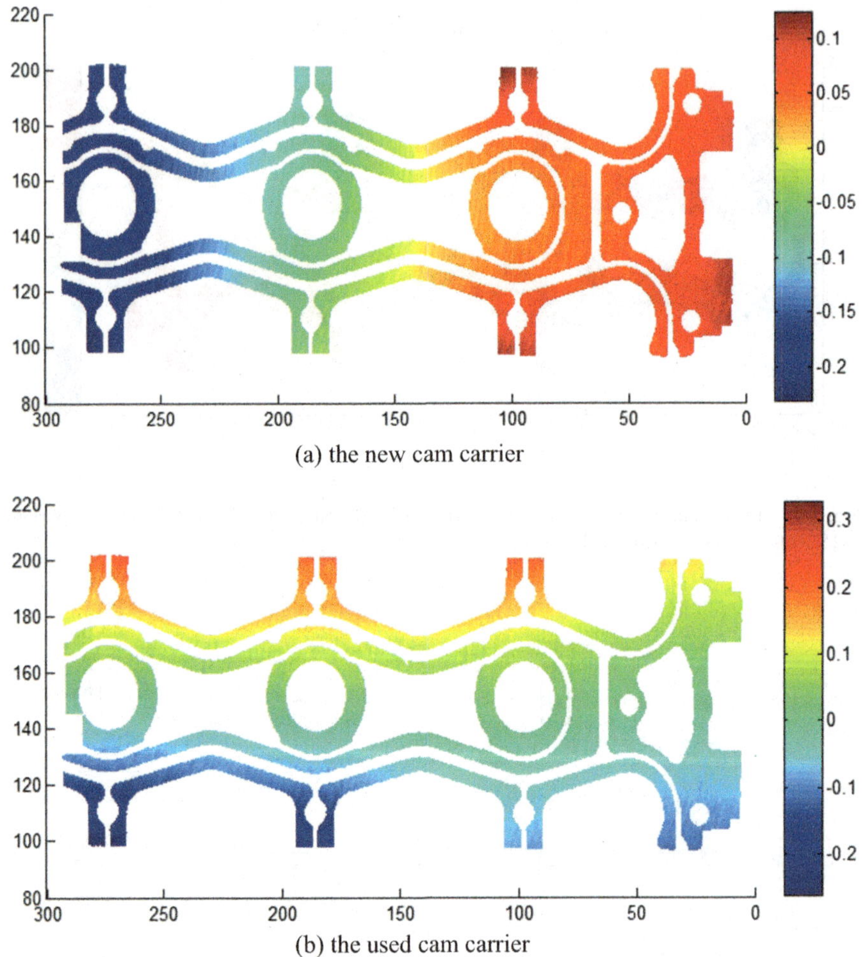

(a) the new cam carrier

(b) the used cam carrier

Fig. 5.37 The height map of two different cam carriers. Reprinted from Ref. [112], copyright 2015, with permission from IEEE

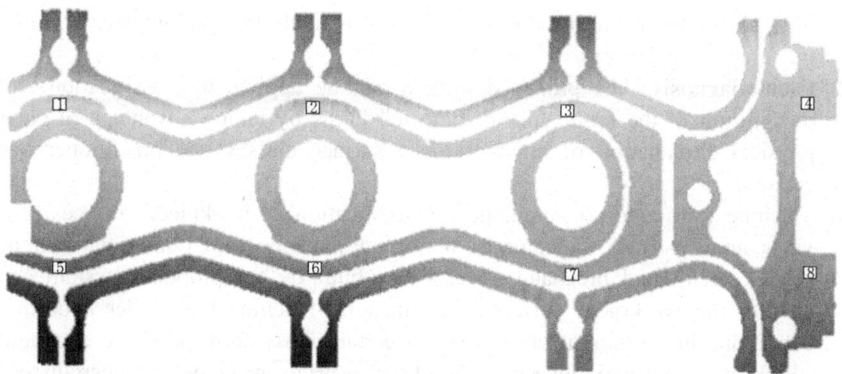

Fig. 5.38 Eight locations selected from this surface. Reprinted from Ref. [112], copyright 2015, with permission from IEEE

Table 5.14 CCP of the proposed adaptive SVM method

Pattern	Classification result						CCP
	1	2	3	4	5	6	
1	**8**	0	0	0	0	0	100
2	0	**8**	0	0	0	0	100
3	0	2	**6**	0	0	0	75
4	0	0	0	**7**	1	0	87.5
5	0	0	0	0	**8**	0	100
6	0	0	0	0	0	**8**	100
Average							93.75

5.3.6 Conclusions

A novel adaptive system based on SVM is proposed to classify workpiece surfaces using HDM. In this classification system, NSCT is first used to extract features and an adaptive SVM-based classification method is proposed which consists of selection of appropriate parameters C and σ to build an SVM classifier for the purpose of classification. Three case studies are conducted to evaluate the classification performance. The results demonstrate that the proposed adaptive SVM-based classification system can not only enhance the capability of APSO algorithm in global search but also reduce the likelihood that the identified solution being trapped in local optimum. The proposed adaptive SVM-based classification system is effective and efficient in classification of workpiece surfaces.

The proposed adaptive SVM-based classification system can be applied in the following three aspects:

(1) Process monitoring. The proposed system shows a good classification performance in workpiece surface. If the categorized defective workpieces account

for a high proportion of the total amount, it is likely that problems exist in the manufacturing process.

(2) Fault diagnosis. The proposed system can be applied to classification and recognition of the fault types of detected faults by taking advantage of clear physical explanation of the workpiece surface classes one practitioner has created.

(3) Machine tool condition prediction. Classification of workpiece surfaces provides an important indicator of abnormal machine tool condition such as chatter, wear, and breakage. A classifier built by the proposed system can classify the workpiece surfaces machined by machine tool under abnormal conditions into different classes and a certain class corresponds to a certain condition of machine tool. Once a workpiece surface machined by a certain tool is classified as one class with known conditions, the machining tool condition can be predicted for diagnosis.

References

1. Quinlan JR (1983) Learning efficient classification procedures and their application to chess end games. Mach Learn 463–482
2. Breiman L, Friedman JH, Olshen R, Stone CJ (1984) Classification and regression trees. Encycl Ecol 40(3):582–588
3. Rastogi R, Shim K (2000) PUBLIC: a decision tree classifier that integrates building and pruning. Data Min Knowl Disc 4(4):404–415
4. Mehta M, Agrawal R, Rissanen J (1996) SLIQ: a fast scalable classifier for data mining. In: International conference on extending database technology, pp 18–32
5. Shafer JC, Agrawal R, Mehta M (1996) SPRINT: a scalable parallel classifier for data mining. In: International conference on very large data bases, pp 544–555
6. Khoo LP, Tor SB, Zhai LY (2003) A rough-set-based approach for classification and rule induction. Int J Adv Manuf Technol 15(6):438–444
7. Greco S, Matarazzo B, Slowinski R (1998) A new rough set approach to multicriteria and multiattribute classification. In: International conference on rough sets and current trends in computing, pp 60–67
8. Bazan JG, Nguyen HS, Nguyen SH, Synak P (2000) Rough set algorithms in classification problem. In: Rough set methods and applications, pp 49–88
9. Suganthi L, Iniyan S, Samuel AA (2015) Applications of fuzzy logic in renewable energy systems–a review. Renew Sustain Energy Rev 48:585–607
10. Melin P, Castillo O (2014) A review on type-2 fuzzy logic applications in clustering, classification and pattern recognition. Appl Soft Comput J 21(5):568–577
11. Zhu X, Yang Y (2008) A lazy bagging approach to classification. Pattern Recogn 41 (10):2980–2992
12. Abellán J, Masegosa AR (2012) Bagging schemes on the presence of class noise in classification. Expert Syst Appl 39(8):6827–6837
13. Kai MT, Zheng Z (1998) Boosting trees for cost-sensitive classifications. In: European conference on machine learning, pp 190–195
14. Mccallum A (1998) A comparison of event models for naive bayes text classification. In: AAAI-98 workshop on learning for text categorization, vol 752, no 1, pp 41–48

15. Zheng F, Webb GI (2011) Tree augmented naive Bayes. In: Encyclopedia of machine learning. Springer US, pp 990–991.

16. Pernkopf F (2005) Bayesian network classifiers versus selective-NN classifier. Pattern Recogn 38(1):1–10

17. Liu Y (2010) The BP neural network classification method under Linex loss function and the application to face recognition. J Northeast Norm Univ 42(1):592–595

18. Chen S, Cowan CFN, Grant PM (1991) Orthogonal least squares learning algorithm for radial basis function networks. IEEE Trans Neural Netw 2(2):302–309

19. Jacyna GM, Malaret ER (2002) Classification performance of a Hopfield neural network based on a Hebbian-like learning rule. IEEE Trans Inf Theory 35(2):263–280

20. Vapnik V (1999) The nature of statistical learning theory. Springer

21. Whitehouse DJ (1999) Surface metrology. Comput Stand Interfaces 21(9):955–972

22. Whitehouse DJ (1982) The parameter rash—is there a cure? Wear 83(1):75–78

23. ISO 4287 (1997) Geometrical product specifications (GPS)-surface texture: profile method-terms, definitions and surface texture parameters. International Organization for Standardization, Geneva

24. ASME B46.1, Surface Texture (Surface Roughness, Waviness, and Lay)

25. Leopold J, Günther H, Leopold R (2003) New developments in fast 3D-surface quality control. Measurement 33(2):179–187

26. Hai TN, Wang H, Hu SJ (2013) Characterization of cutting force induced surface shape variation in face milling using high-definition metrology. J Manuf Sci Eng 135(4):041014-1-12

27. Blunt L, Jiang X, Scott PJ (2003) Future developments in surface metrology. In: Advanced techniques for assessment surface topography: development of a basis for 3D surface texture standards "Surfstand", pp 339–347

28. Graziano AA, Ganguly V, Schmitz T, Yamaguchi H (2014) Control of lay on cobalt chromium alloy finished surfaces using magnetic abrasive finishing and its effect on wettability. J Manuf Sci Eng 136(3):031016-1-8

29. Grzesik W, Zak K (2014) Characterization of surface integrity produced by sequential dry hard turning and ball burnishing operations. J Manuf Sci Eng 136(3):031017-1-9

30. Ramana KV, Ramamoorthy B (1996) Statistical methods to compare the texture features of machined surfaces. Pattern Recogn 29(9):1447–1459

31. Dong WP, Sullivan PJ, Stout KJ (1993) Comprehensive study of parameters for characterizing three-dimensional surface topography II: Statistical properties of parameter variation. Wear 167(1):9–21

32. Satoh G, Yao YL, Huang X, Ramirez AG (2012) Characterization and prediction of texture in laser annealed NiTi shape memory thin films. J Manuf Sci Eng 134(5):051006-1-11

33. Tsa DM, Wu SK (2000) Automated surface inspection using gabor filters. Int J Adv Manuf Technol 16(7):474–482

34. Zhang M, Levina E, Djurdjanovic D, Ni J (2008) Estimating distributions of surface parameters for classification purposes. J Manuf Sci Eng 130(4):031010-1-9

35. Fu S, Liu X, Muralikrishnan B, Raja J (2003) Engineering surface analysis with different wavelet bases. J Manuf Sci Eng 125(4):844–852

36. Liao Y, Stephenson DA, Ni J (2012) Multiple-scale wavelet decomposition, 3D surface feature exaction and applications. J Manuf Sci Eng 134(1):011005-1-13

37. Li Y, Ni J (2011) B-spline wavelet-based multiresolution analysis of surface texture in end-milling of aluminum. J Manuf Sci Eng 133(1):011014-1-11

38. Yu J (2012) Machine tool condition monitoring based on an adaptive gaussian mixture model. J Manuf Sci Eng 134(3):031004-1-13

39. Raja J, Muralikrishnan B, Fu S (2002) Recent advances in separation of roughness, waviness and form. Precis Eng 26(2):222–235

40. Badashah SJ, Subbaiah P (2011) Image enhancement and surface roughness with feature extraction using DWT. In: International conference on sustainable energy and intelligent systems, pp 754–759

41. Selesnick IW, Baraniuk RG, Kingsbury NC (2005) The dual-tree complex wavelet transform. IEEE Signal Process Mag 22(6):123–151
42. Kingsbury N (2001) Complex wavelets for shift invariant analysis and filtering of signals. Appl Comput Harmon Anal 10(3):234–253
43. Kingsbury N (1999) Image processing with complex wavelets. Philos Trans Math Phys Eng Sci 357(1760):2543–2560
44. Kingsbury N (1998) The dual-tree complex wavelet transform: a new technique for shift invariance and directional filters. In: 8th IEEE digital signal process workshop
45. Malak RJ, Paredis CJJ (2010) Using support vector machines to formalize the valid input domain of predictive kodels in systems design problems. J Mech Des 132(10):101001-1-14
46. Du S, Lv J, Xi L (2012) On-line classifying process mean shifts in multivariate control charts based on multiclass support vector machines. Int J Prod Res 50(22):6288–6310
47. Brereton RG, Lloyd GR (2010) Support vector machines for classification and regression. ISIS Tech Rep 14(1):5–16
48. Andrew AM (2002) An introduction to support vector machines and other kernel-based learning methods. Kybernetes 32(1):1–28
49. Wang SJ, Mathew A, Chen Y, Xi LF, Ma L, Lee J (2009) Empirical analysis of support vector machine ensemble classifiers. Expert Syst Appl 36(3):6466–6476
50. Pal SK, Pal A (2001) Pattern recognition: from classical to modern approaches. World Scientific Publishing Company, pp 427–451
51. Kim HC, Pang S, Je HM, Kim D, Bang SY (2003) Constructing support vector machine ensemble. Pattern Recogn 36(12):2757–2767
52. Ho TK (2000) Complexity of classification problems and comparative advantages of combined classifiers. In: International workshop on multiple classifier systems, pp 97–106
53. Pang S, Kim D, Bang SY (2005) Face membership authentication using SVM classification tree generated by membership-based LLE data partition. IEEE Trans Neural Netw 16(2):436
54. Lei Z, Yang Y, Wu Z (2006) Ensemble of support vector machine for text-independent speaker recognition. Int J Comput Sci Netw Secur 5:163–167
55. Hung C, Chen JH (2009) A selective ensemble based on expected probabilities for bankruptcy prediction. Expert Syst Appl 36(3):5297–5303
56. Priya KJ, Rajesh RS (2010) Local fusion of complex dual-tree wavelet coefficients based face recognition for single sample problem. Procedia Comput Sci 2(6):94–100
57. Hsu CW, Lin CJ (2002) A comparison of methods for multiclass support vector machines. IEEE Trans Neural Netw 13(2):415–425
58. Breiman L (1996) Bagging predictors. Mach Learn 24(2):123–140
59. Schapire RE (1989) The strength of weak learnability. In: Proceedings of the second annual workshop on computational learning theory, vol 5, no 2, pp. 197–227.
60. Ho TK (1998) The random subspace method for constructing decision forests. IEEE Trans Pattern Anal Mach Intell 20(8):832–844
61. Zhou ZH, Wu J, Tang W (2002) Ensembling neural networks: many could be better than all. Artif Intel 137(1):239–263
62. Partalas I, Tsoumakas G, Katakis I, Vlahavas I (2006) Ensemble pruning using reinforcement learning. In: Advances in artificial intelligence, pp 301–310
63. Giacinto G, Roli F, Fumera G (2000) Design of effective multiple classifier systems by clustering of classifiers. In: International conference on pattern recognition, vol 2, pp 160–163.
64. Martınez-Munoz G, Suárez A (2006) Pruning in ordered bagging ensembles. In: Proceedings of the 23rd international conference on machine learning, pp 609–616
65. Mao S, Jiao LC, Xiong L, Gou S (2011) Greedy optimization classifiers ensemble based on diversity. Pattern Recogn 44(6):1245–1261
66. http://www.coherix.com
67. Chang CC, Lin CJ (2011) LIBSVM: a library for support vector machines 2(3):1–27

68. Freund Y, Schapire RE (1995) A decision-theoretic generalization of on-line learning and an application to boosting. In: European conference on computational learning theory, vol 904, pp 23–37.

69. Nguyen HT, Wang H, Tai BL, Ren J, Hu SJ, Shih A (2016) High-definition metrology enabled surface variation control by cutting load balancing. J Manuf Sci Eng 138 (2):021010-1-11

70. Wang M, Xi L, Du S (2014) 3D surface form error evaluation using high definition metrology. Precis Eng 38(1):230–236

71. Liu HZ, Shi YS, Yin L, Jiang WT, Lu BH (2013) Roll-to-roll imprint for high precision grating manufacturing. Eng Sci 11(1):39–43

72. Zhou L, Wang H, Berry C, Weng X, Hu SJ (2012) Functional morphing in multistage manufacturing and its applications in high-definition metrology-based process control. IEEE Trans Autom Sci Eng 9(1):124–136

73. Hai TN, Wang H, Hu SJ (2014) Modeling cutter tilt and cutter-spindle stiffness for machine condition monitoring in face milling using high-definition surface metrology. Int J Adv Manuf Technol 70(5–8):1323–1335

74. Da CA, Zhou J, Do MN (2006) The nonsubsampled contourlet transform: theory, design, and applications. IEEE Trans Image Process 15(10):3089–3101

75. Panahandeh G, Mohammadiha N, Leijon A, Händel P (2013) Continuous hidden markov model for pedestrian activity classification and gait analysis. IEEE Trans Instrum Meas 62 (5):1073–1083

76. Li W, Zhang S, He G (2013) Semisupervised distance-preserving self-organizing map for machine-defect detection and classification. IEEE Trans Instrum Meas 62(5):869–879

77. He Z, Wang Q, Shen Y, Jin J, Wang Y (2013) Multivariate gray model-based BEMD for hyperspectral image classification. IEEE Trans Instrum Meas 62(5):889–904

78. He Z, Shen Y, Wang Q, Wang Y (2014) Optimized ensemble EMD-based spectral features for hyperspectral image classification. IEEE Trans Instrum Meas 63(5):1041–1056

79. He S, Li K, Zhang M (2013) A real-time power quality disturbances classification using hybrid method based on s-transform and dynamics. IEEE Trans Instrum Meas 62(9):2465–2475

80. Salgado DR, Alonso FJ (2006) Tool wear detection in turning operations using singular spectrum analysis. J Mater Process Technol 171(3):451–458

81. Du S, Huang D, Lv J (2013) Recognition of concurrent control chart patterns using wavelet transform decomposition and multiclass support vector machines. Comput Ind Eng 66 (4):683–695

82. Du S, Lv J (2013) Minimal Euclidean distance chart based on support vector regression for monitoring mean shifts of auto-correlated processes. Int J Prod Econ 141(1):377–387

83. Cherkassky V, Ma Y (2004) Practical selection of SVM parameters and noise estimation for SVM regression. Neural Netw 17(1):113–126

84. Lin J, Zhang J (2013) A fast parameters selection method of support vector machine based on coarse grid search and pattern search. In: 2013 fourth global congress on intelligent systems (GCIS), vol 3, no 4, pp 77–81

85. Chen PW, Wang JY, Lee HM (2004) Model selection of SVMs using GA approach. In: IEEE international joint conference on neural networks, vol 3, pp 2035–2040

86. Lin SW, Ying KC, Chen SC, Lee ZJ (2008) Particle swarm optimization for parameter determination and feature selection of support vector machines. Expert Syst Appl 35 (4):1817–1824

87. Du S, Xi L (2011) Fault diagnosis in assembly processes based on engineering-driven rules and PSOSAEN algorithm. Comput Ind Eng 60(1):77–88

88. Huang Q, Mao J, Liu Y (2012) An improved grid search algorithm of SVR parameters optimization. In: IEEE international conference on communication technology, pp 1022–1026

89. Huang CL, Wang CJ (2006) A GA-based feature selection and parameters optimization for support vector machines. Expert Syst Appl 31(2):231–240

90. Ho SL, Yang SY, Ni GZ, Wong KF (2007) An improved PSO method with application to multimodal functions of inverse problems. IEEE Trans Magn 43(4):1597–1600
91. Li Y, Peng Y, Zhou S (2013) Improved PSO algorithm for shape and sizing optimization of truss structure. J Civil Eng Manag 19(4):542–549
92. Xiao L, Zhang W, Zhang W (2007) Particle Swarm optimization with adaptive local search. Comput Sci 34(8):199–201
93. Guo WZ (2006) Fuzzy self-adapted particle swarm optimization algorithm for traveling salesman problems. Comput Sci 33(6):161–163
94. Shi Y, Eberhart R (1998) A modified particle swarm optimizer. In: IEEE international conference on evolutionary computation proceedings, pp 69–73
95. Shi Y, Eberhart RC (1998) Parameter selection in particle swarm optimization. In: International Conference on Evolutionary Programming, pp 591–600
96. Coifman RR, Donoho DL (1995) Translation-invariant de-noising. Wavelets Stat 103 (2):125–150
97. Li S, Fu X, Yang B (2008) Nonsubsampled contourlet transform for texture classifications using support vector machines. In: IEEE international conference on networking, sensing and control, pp 1654–1657
98. Yang B, Li S, Sun F (2007) Image fusion using nonsubsampled contourlet transform. In: International conference on image and graphics, pp 719–724
99. Jianping Z, Cunha AL, Do M. N (2005) Nonsubsampled contourlet transform: construction and application in enhancement. In: IEEE international conference on image processing, vol 1, pp 469–472
100. Shensa MJ (1992) The discrete wavelet transform: wedding the a trous and Mallat algorithms. IEEE Trans Signal Process 40:2464–2482
101. Jeeves TA (1961) Direct search solution of numerical and statistical problems. J Assoc Comput Mach 8(2):212–229
102. Torczon V (1993) On the convergence of pattern search algorithms. SIAM J Optim 7(1):1–25
103. Audet C, Dennis JE (2001) Pattern search algorithms for mixed variable programming. SIAM J Optim 11(3):573–594
104. Bilski P, Winiecki W (2012) Methods of assessing the time efficiency in the virtual measurement systems. Comput Stand Interfaces 34(6):485–492
105. http://weka.wikispaces.com
106. Domingos P, Pazzani M (1997) On the optimality of the simple bayesian classifier under zero-one loss. Mach Learn 29(2–3):103–130
107. Dreiseitl S, Ohno-Machado L (2002) Logistic regression and artificial neural network classification models: a methodology review. J Biomed Inform 35(5):352–359
108. Ciresan D, Meier U, Schmidhuber J, and IEEE (2012) Multi-column deep neural networks for image classification. In: 2012 IEEE conference on computer vision and pattern recognition, pp 3642–3649
109. Thakkar M, Bhatt M, Bhensdadia CK (2011) Fuzzy logic based computer vision system for classification of whole cashew kernel. Comput Netw Inf Technol 142:415–420
110. Hu YC (2013) Rough sets for pattern classification using pairwise-comparison-based tables. Appl Math Model 37(12–13):7330–7337
111. Du S, Liu C, Xi L (2014) A selective multiclass support vector machine ensemble classifier for engineering surface classification using high definition metrology. J Manuf Sci Eng 137 (1):011003
112. Du S, Huang D, Wang H (2015) An adaptive support vector machine-based workpiece surface classification system using high-definition metrology. IEEE Trans Instrum Meas 64 (10):2590–2604

Chapter 6
Surface Monitoring

6.1 A Brief History of Surface Monitoring

As an effective tool for quality control, statistical process control (SPC) has been widely used in various industries for special cause identification, removal and variation reduction [1]. Different control charts, including univariate charts and multivariate charts, have been extensively studied in many researches. Recently, control charts for profiles are also drawing more and more attention [2]. With the development of sensing technology, in an advanced process, process or product quality is characterized by two-dimensional (2D) surfaces or three-dimensional (3D) flat surfaces.

There are many researches about SPC methods for 2D surface monitoring. For example, profile monitoring is applied for surface monitoring. Williams et al. [3] extended the use of the T^2 control chart to monitor the coefficients resulting from a parametric nonlinear regression model fit to profile data. Colosimo et al. [4] proposed a profile monitoring method based on combining a spatial autoregressive regression model to multivariate and univariate control charting, which was aimed at extending the proposed method to surface monitoring. Chen et al. [5] proposed a high-dimensional control chart approach for profile monitoring that is based on the adaptive Neyman test statistic for the coefficients of discrete Fourier transform of profiles. Colosimo et al. [6] presented a novel method consisting of modeling the manufactured surface via Gaussian processes models and monitoring the deviations of the actual surface from the target pattern estimated in Phase I. Wang et al. [7] proposed a new chart based on the Gaussian–Kriging model, in which the spatial correlations within the 2D surface profile were represented by a parametric function. More profile monitoring methods can refer to [8]. However, profile monitoring

© Springer Nature Singapore Pte Ltd. 2019
S. Du and L. Xi, *High Definition Metrology Based Surface Quality Control and Applications*, https://doi.org/10.1007/978-981-15-0279-8_6

is based on the assumption that the can be represented by linear, nonlinear, or nonparametric model.

With the development of high-definition metrology (HDM), researches about monitoring 3D flat surface variation based on HDM have also been conducted. Etesami [9] and Xia et al. [10] proposed parameterized methods to monitor high-dimensional data by fitting a parametric surface model to the data and monitoring the parameters. Wang et al. [11] developed a framework for efficient monitoring of spatial variation in HDM data using principal curves and quality control charts. Suriano et al. [12] improved a sequential method for global and localized monitoring of shape variations in HDM data by refining the localized monitoring scheme and applied the method to HDM data collected from an automotive engine head machining process. Suriano et al. [13] also proposed a new methodology for efficiently measuring and monitoring flat surface variations by fusing in-plant multi-resolution measurements and process information.

However, only a few researches focus on monitoring of 3D curved surface based on high-density point cloud data. Quantile–quantile (Q–Q) plot is the mainly used method to monitor 3D point cloud data. Wang et al. [14] used Q–Q plot to characterize a huge sample and transform it into a linear profile and proposed profile monitoring techniques to improve the performance of a conventional control chart. Wells et al. [15] also used Q–Q plots to transform high-density point cloud data into linear profiles that can be monitored by well-established profile monitoring techniques. But the methods using Q–Q plot cannot identify the locations of defective regions from the curved surface in out-of-control (OOC) condition. He et al. [16] proposed a method to transform point clouds into 2D images to benefit from the image analysis and monitoring techniques that were currently being implemented on the shop floor. This method is limited to the application of the cylindrical surface, but spatial correlation will be lost if this method is used for freeform curved surface.

6.2 Tool Wear Monitoring of Wiper Inserts in Multi-insert Face Milling Using 3D Surface Form Indicators

6.2.1 Introduction

Cutting tool wear has a direct impact on product quality, production efficiency, and tool change cost. Therefore, monitoring tool wear is essential to ensure cutting tools in satisfactory condition and reduce downtime and tool costs [17].

Methods for tool wear monitoring have been extensively studied. These methods consist of two categories: direct methods and indirect methods. Direct methods measure actual geometric changes of cutting edge using machine vision [18–22] or scanning electron microscope [23] directly. However, direct methods are influenced by illumination condition, reflectance of tool edge, coolant mist, and efficiency. Indirect methods take advantage of related signals as tool wear indicators, such as cutting force [24, 25], torque [26], vibration [27, 28], drive parameters [29, 30], acoustic emission [31, 32], etc. Among indirect methods, using surface texture as tool wear indicators has long been a matter of interest. Because tool wear affects the surface texture of the final parts [33]. In other words, the texture of the machined surface is a replica of the changing of the cutting tool shape [34]. Furthermore, the primary factor that results in degradation of surface finish and form accuracy is the change of cutting tool shape under stable machining condition [35]. Therefore, surface texture can provide reliable and detectable information for tool wear monitoring.

The surface texture used for tool wear monitoring mainly comes from three kinds of measuring techniques: (a) contacting stylus techniques, (b) machine vision techniques, and (c) noncontacting optical techniques. Contacting stylus techniques can directly measure surface texture, such as roughness and waviness, for tool condition monitoring [36]. However, despite the high measuring accuracy, contacting stylus techniques are highly localized and are difficult for online application. Machine vision techniques extract surface texture parameters from the obtained images. With rapid image capturing speed, machine vision systems are qualified for online tool wear monitoring. A thorough review of image texture analysis methods for tool condition monitoring can be found in [37]. However, extracting surface texture using machine version is not straightforward as surface texture is predicted by captured surface images indirectly. The image quality is influenced by nonuniform illumination, depth of focus, and noises (dirt, cut chips, etc.) in the industrial environment. Noncontacting optical techniques, such as phase-shifting interferometry, coherence scanning interferometry, and atomic force microscopy, can achieve accurate 3D surface measurement directly [38]. Some of these techniques have huge potentials for online tool condition monitoring due to fast measurement speed. However, these techniques usually have a small field of view so that they cannot reveal the subtle textural changes of the entire surface.

Due to a lack of metrology that is able to measure large surfaces with high resolution, tool wear monitoring using 3D surface information of the entire surface has seldom been studied. A recent advancement of noncontact optical technique, high-definition metrology (HDM), gains the ability to measure 3D surface height map of the entire surface at sub-µm level within 40 s [39]. For example, as shown in Fig. 6.1, the engine block surface measured by HDM including about 0.8 million data points covers an area of 320 mm $\times$ 160 mm with 0.15 mm resolution in x–y-direction and 1 µm accuracy in depth direction. Therefore, the 3D surface topography of the entire surface examined by HDM presents a novel platform for online tool condition monitoring. Liao, Stephenson, and Ni [40] used 2D Gaussian filter to extract 3D surface waviness of machined to access flank wear of a single cutting insert. In order to avoid boundary distortion, a minimum evaluation area is required when using 2D fillers. However, discontinuous surfaces like engine block surfaces with cylinder holes, bolt holes, and cooling holes do not always have the continuous large enough area for 2D filtering operations. 3D surface form features introduced in [41] can describe the 3D height and spatial information of the entire discontinuous surface measured by HDM, which serve as a promising tool to monitor the wear of wiper inserts. Nevertheless, for multi-insert face milling with cutting inserts and wiper inserts in the same cutter, surface finish quality depends more on wiper inserts than cutting inserts. Therefore, this research aims to monitor the wear of wiper inserts in multi-insert face milling using 3D surface form indicators.

6.2.2 Measurement of Wiper-Insert Wear

An experiment on engine blocks is performed to study the wear of wiper inserts in a multi-insert face milling. Engine block surfaces are machined by a KNOLL KF400 machining center. The engine block material is FC250 cast iron with the following chemical composition: 3.0–3.4% C, 1.9–2.4% Si, 0.66–0.9% Mn, 0.2–0.5% Cr, 0.3–0.8% Cu, 0.15%S, and 0.10%P. Quaker 370 KLG cutting fluid is used. Cutting parameters are set the same as those parameters for the production line: the depth of cut is 0.5 mm, the cutting speed is 1300 rpm, and feed rate is 3360 mm/min.

The face milling cutter has a diameter of 200 mm with 15 cutting inserts intercalated by three wiper inserts. As shown in Fig. 6.2, inserts 1–5, 7–11, and 13–17 are the cutting inserts and inserts 6, 12, and 18 are the wiper inserts. The cutting tool material for the wiper insert is PCBN. The cutting inserts and wiper inserts are mounted based on the following criterion. The maximum admissible axial deviation between the three wiper inserts is 0.003 mm. The cutting inserts are installed between 0.02 and 0.06 mm below the lowest wiper insert. Table 6.1 shows the cutting angles of the cutting inserts and wiper inserts.

Worn wiper inserts are measured using a Hitachi Scanning Electron Microscope (SEM) S–3400N to identify the predominant wear modes. Figure 6.3 shows the flank face and the rake face of the worn wiper insert. The dominant wear patterns

Fig. 6.1 3D surface height map measured by HDM. Reprinted from Ref. [57], copyright 2015, with permission from ASME

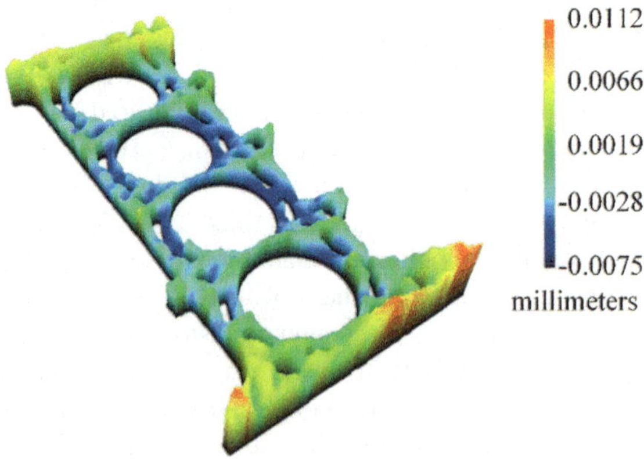

Fig. 6.2 The face milling cutter with 15 cutting inserts and 3 wiper inserts. Reprinted from Ref. [57], copyright 2015, with permission from ASME

Table 6.1 Geometry of cutting inserts and wiper inserts

Tool geometry	Cutting inserts	Wiper inserts
Radial rake angle	0°	−16°
Axial rake angle	9°	0.2°
Cutting edge angle	90°	90°
Minor cutting edge angle	0°	4°
Inclination angle	9°	0.2°
Tip radius	0.8 mm	1 mm

within the normal life period are observed to be micro-cracks and axial lowering of the wiper edge. The micro-cracks, which combine the wear on the flank face and rake face, are adjacent to the wiper edge. This is because the cutting force is concentrated adjacent to the wiper edge as the insert–chip contact length for wiper inserts is short. Adhesion and abrasion are also considered to be the wear mechanism for PCBN wiper inserts [42]. The lowering of the wiper edge in the direction parallel to the axis of the cutter could be caused by plastic deformation or plastic lowering due to high-temperature creep and could be the predominant cause of premature tool failure [43].

However, the inserts had to be dismounted from the cutter to measure the subtle wear of wiper inserts using SEM or tool maker's microscope. The disengagement and engagement of the inserts and the cuter will result in mounting error, which will alter the normal wear process of the wiper inserts in subsequent operations. To tackle this problem, ZOLLER Venturion 600, a tool presetter and measuring machine, is used to measure the axial positions of all inserts without dismounting the inserts from the cutter. Then the lowering of the wiper edge is calculated as the changes of axial positions of wiper edges.

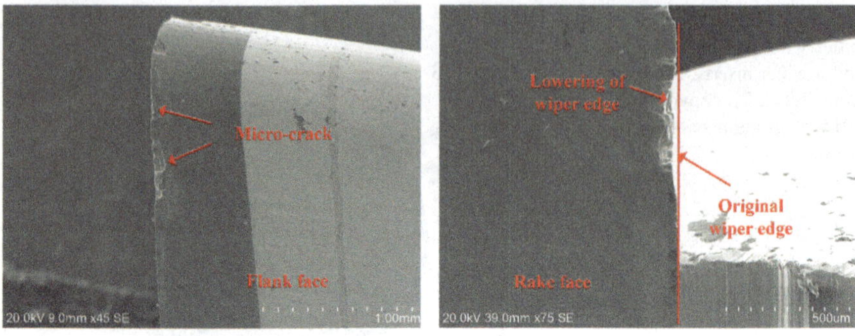

Fig. 6.3 Wear of wiper inserts. Reprinted from Ref. [57], copyright 2015, with permission from ASME

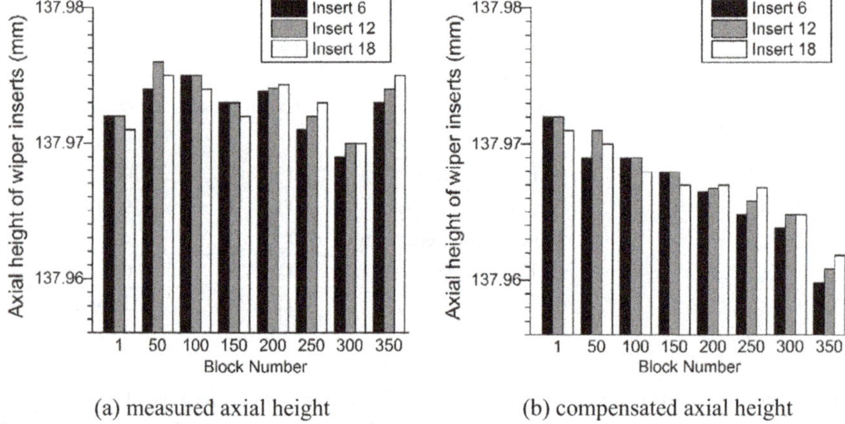

(a) measured axial height   (b) compensated axial height

Fig. 6.4 Axial height of wiper edges versus number of blocks machined. Reprinted from Ref. [57], copyright 2015, with permission from ASME

As we mainly focus on monitoring the tool wear in normal production, the number of machined engine blocks is set at 350, which is the preset tool life span. Axial positions of all inserts were measured every 50 engine blocks. Figure 6.4a shows the measured axial positions of the three wiper inserts 6, 12, and 18. However, we found that the present axial positions could be higher than previously measured axial positions. For example, the axial positions of the wiper inserts after machining 50 engine blocks are higher than the initial axial positions. This phenomenon is not consistent with common scene that the inserts would be shorter due to the wear of cutting edges. This strange phenomenon happens because ZOLLER Venturion 600 introduced reference error for each measurement. This reference error has no effects on tool presetting because it does not affect the relative axial positions of the three wiper inserts and the 15 cutting inserts. However, absolute axial positions of wipers edges cannot be obtained due to the reference error. Therefore, a compensation process is needed to acquire the actual axial positions from the measured results.

Assume that the axial positions of all inserts are lower or equal to the previous positions in the machining process. Therefore, there is a least reference error to make sure that the axial position of each insert is lower or equal to the axial position of the same insert in the previous measurement. The least reference error can be used to compensate for the absolute axial positions of the wiper inserts. The least reference error is calculated as follows:

$$E_i = \begin{cases} 0, i = 1 \\ \max(z_i^j - z_{i-1}^j), i = 2, \ldots, 8, j = 1, \ldots, 18 \end{cases} \tag{6.1}$$

where z_i^j is the ith measured axial position of the jth insert. E_i is the least reference error between the ith measurement and $(i-1)$th measurement. $E_1 = 0$ because there is no reference error for the first measurement. The range of subscript i is from 1 to 8 because the experiments machined a total of 350 workpieces and measured every 50 workpieces. In addition, the range of superscript j is from 1 to 18 representing the 15 cutting inserts and 3 wiper inserts. The compensated axial positions of the inserts are calculated as follows:

$$Z_i^j = z_i^j - E_i, i = 1, \ldots, 8, j = 1, \ldots, 18 \tag{6.2}$$

where Z_i^j is the compensated axial position of the jth insert for the ith measurement. The compensated axial positions for wiper inserts 6, 12, and 18 are shown in Fig. 6.4b.

The axial lowering of each wiper edge is obtained by calculating the difference between the initial axial positions and the compensated axial positions.

$$d_i^j = Z_{i+1}^j - Z_1^j, \; j = 6, \; 12, \; 18 \tag{6.3}$$

The lowering of the wiper edge in this experiment is calculated as the average lowering of the three wiper inserts, and the results are shown in Fig. 6.5. The lowering of the wiper edge increased with the number of machined engine blocks. The magnitude of the lowering of the wiper edge is at the scale of micrometer. The lowering value reaches 10.85 μm after machining 350 blocks.

6.2.3   Extraction of Wear Indicators

This subsection describes how to extract wiper-wear-related indicators from HDM data. The wiper inserts with a wiping flat (also called parallel land) are used to produce featureless surface finish. However, in practical machining situations, wear

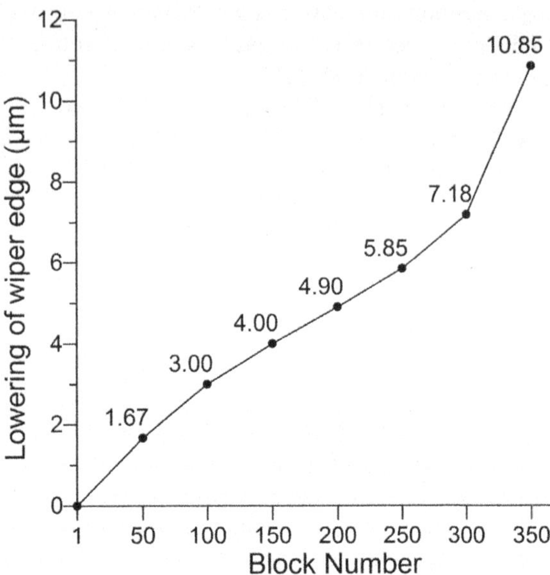

Fig. 6.5 Average axial lowering of the three wiper inserts. Reprinted from Ref. [57], copyright 2015, with permission from ASME

and axial offsets of wiper inserts make rather complicated surface texture. Wear indicators of wiper inserts can be extracted from 3D surface height map measured by HDM data. The extraction process contains three steps. First, HDM data is converted into a height-encoded gray image that contains all the height and spatial information of the entire surface. Second, the curvilinear toolmarks of the converted image are straightened to ensure equivalent distance between two adjacent toolmarks along the same direction. Finally, a modified gray level co-occurrence matrix (GLCM) method is adopted to describe the second-order statistical information of the machined surface, which avoids boundary distortion. 3D surface form features, including entropy and contrast, are calculated from the modified GLCM along the direction perpendicular to the straightened toolmarks as wear indicators. The correlation of these indicators with wear of wiper inserts is studied in experiments conducted in real industrial environment.

6.2.3.1 Gray Image Converting

Raw HMD data contains a large number of data points with noises in both horizontal direction and vertical direction. The HDM data should be preprocessed to a suitable data type for further analysis. Therefore, the HDM data preprocessing method proposed in [41] is introduced to convert HDM coordinates $[X_i\ Y_i\ Z_i]$ into a height-encoded gray pixel $I(m, n)$ with all the height and spatial information of the raw HDM data. The pixel index (m, n) and pixel gray intensity I are calculated as follows:

$$m = \frac{X_i - X_{\min}}{l}, n = \frac{Y_i - Y_{\min}}{l} \tag{6.4}$$

$$I(m,n) = \left[\frac{Z_i - S_L}{S_U - S_L} \times 255 \right] \tag{6.5}$$

where l is the resampling interval, S_L and S_U are the lower and upper limits of Z-coordinates.

Considering lateral resolution of HDM equal to 0.15 mm and flatness specification of engine block surface equal to 0.05 mm, l is set as 0.2 mm and S_L and S_U are set as—0.03 and 0.03 mm. Under the same image converting criterion, i.e., the same l, S_L, and S_U, pixel index (m,n) and pixel gray intensity I have the same quantitative relation with the coordinates $[X_i\ Y_i]$ and $[Z_i]$. In other words, gray images measured from different workpiece can be compared and evaluated at the same level. Figure 6.7a shows an example of the converted gray image of an engine block surface and its partial enlarged detail.

This quantitative relation between measured points and image pixels is an advantage of HDM-converted images compared with images obtained by machine version in [34, 35, 44, 45]. For HDM-converted image, surface texture can be calculated directly using pixel intensities, whereas for machine-version-acquired image, surface texture is predicted indirectly by correlating pixel intensities with point heights.

6.2.3.2   Toolmark Straightening

Surface features related to the wear of wiper inserts are most significant perpendicular to the direction of toolmarks. For face milled surfaces, the directions of toolmarks are changing in one revolution of one insert because the cutting process consists of a translation and a rotation at the same time. Therefore, the distance between two adjacent toolmarks along the same direction is varying at different positions. For example, as shown in Fig. 6.7a, toolmark distance at the edge and center of the surface is different along the same direction. As a result, surface information of profiles along the same direction at two different positions cannot guarantee the consistent results. To tackle this problem, the curvilinear toolmarks are straightened to straight lines. Thus, wear indicators can be extracted along the perpendicular direction of the straightened toolmarks.

The toolmark straightening procedure described in [46] is adopted for face milled surfaces with no back cutting. Figure 6.6 shows one curvilinear toolmark produced by the cutting path of one wiper insert on one revolution. The milling cutter is moving along the cutter path (X-axis) at the feed rate f (mm/s) and the rotational speed ω (rad/s). Assume the cutter center is at point O and the wiper insert starts cutting at point A ($t = 0$). After a quarter revolution, the cutting insert

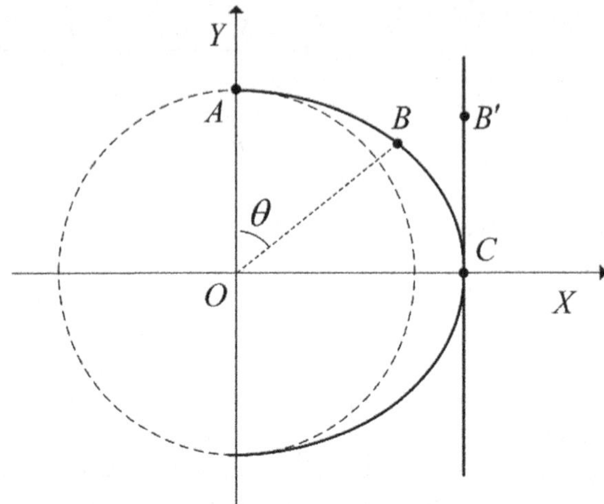

Fig. 6.6 Face milling toolmark straightening. Reprinted from Ref. [57], copyright 2015, with permission from ASME

intersects with the X-axis at point C $(t = t_C)$. Then the curvilinear arc AC is the toolmark of one wiper insert.

The curvilinear toolmark can be projected onto a straight line passing through point C and perpendicular to the X-axis. For an arbitrary point B on arc AC, the corresponding point on the projected line is written as B'. The coordinates of point B are given as follows:

$$x_B = x_O + R\sin\ \omega t_B + ft_B$$
$$y_B = R\cos\ \omega t_B$$

$$(6.6)$$

Therefore, $t_B = \arccos(y_B/R)/\omega$, $0 < t_B < \pi/2\omega$ and $x_O = x_B - ft_B - R\sin\omega t_B$, where R is the cutter radius. The length of $B'C$ should be the same as that of BC. Therefore, the coordinates of point B' are given as follows:

$$x_{B'} = x_C = x_O + R + f\frac{\pi}{2\omega}$$
$$y_{B'} = \text{length}(BC) = \int_{t_B}^{\pi/2\omega} \sqrt{\left(\frac{dx_B}{dt}\right)^2 + \left(\frac{dy_B}{dt}\right)^2}\, dt$$

$$(6.7)$$

The toolmark straightening process is repeated for all the points on the surface until all the toolmarks are straightened to straight lines. Figure 6.7b shows the toolmark-straightened engine block surface. Toolmark distance at the edge and center of the surface is equal along the same direction. Therefore, analyzing surface features from profiles along the same direction at different positions will give consistent results.

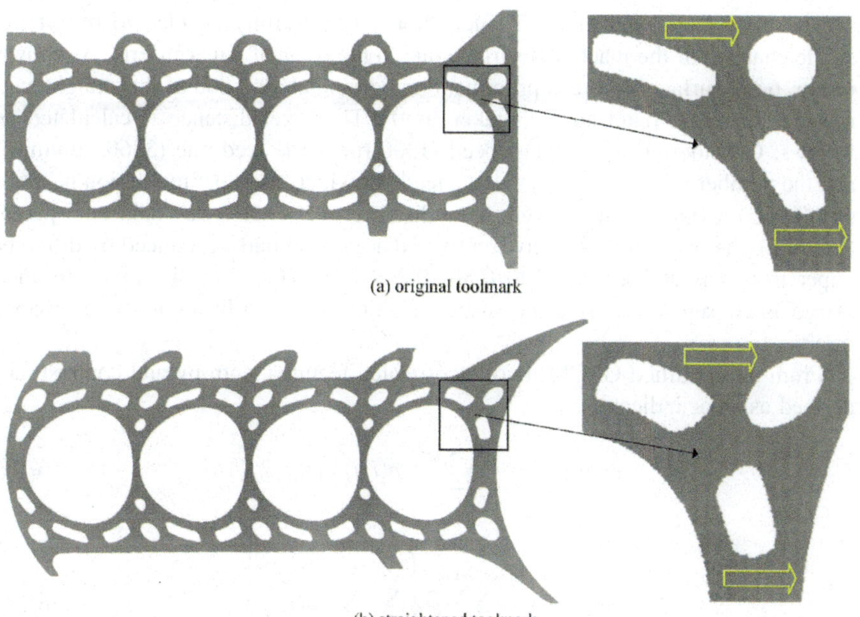

(a) original toolmark

(b) straightened toolmark

Fig. 6.7 Height-encoded gray image converted from HDM data. Reprinted from Ref. [57], copyright 2015, with permission from ASME

6.2.3.3   Surface Feature Extracting

A modified GLCM technique is used to analyze the 3D surface form error using the toolmark-straightened image. GLCM describes the co-occurrence of two pixels for a specified distance and direction [47]. Given the converted gray image $I(x, y)$, the normalized co-occurrence frequencies $p(i, j)$ between gray level i and j are defined as [41]:

$$p(i, j) = P(i, j) \bigg/ \sum_{i=1}^{N_g} \sum_{j=1}^{N_g} P(i, j) \tag{6.8}$$

$$P(i, j) = \#\{I(x_1, y_1) = i, I(x_2, y_2) = j, x_2 = x_1 + D\cos\theta, y_2 = y_1 + D\sin\theta\}, \\ i \neq 0, j \neq 0 \tag{6.9}$$

where N_g is the number of gray levels, i and j are the corresponding gray levels of pixels $I(x_1, y_1)$ and $I(x_2, y_2)$ with the pixel distance D and the pixel angle θ, $\#$ denotes the number of pixel pairs satisfying the conditions, and $i \neq 0, j \neq 0$ eliminate the pixel pairs of the non-measured area.

The pixel angle and pixel distance should be carefully decided to reflect the subtle changes of the machined surface due to the wear of wiper inserts. As shown in Fig. 6.6b, surface feature is prominent on horizontal direction of the straightened image. Therefore, pixel angle is taken as 0°. The pixel distance is calculated as follows: Considering the cutting speed (1300 rpm), the feed rate (3360 mm/min), and the number of wiper inserts (3), the feed per wiper insert per revolution is 3360/(1300 × 3) = 0.862 mm. Given the distance between two adjacent pixels (0.2 mm), the pixel distance between two adjacent toolmarks produced by different wiper inserts is at least equal to 0.862/0.2 = 4.31. Therefore, the pixel distance $D = 5$ is chosen to describe the surface feature produced by two adjacent wiper inserts.

From the modified GLCM, two uncorrelated features, entropy and contrast, are derived as wear indicators.

$$\text{entropy} = -\sum_i \sum_j p(i,j) \log(p(i,j)) \tag{6.10}$$

$$\text{contrast} = \sum_{n=0}^{N_g-1} n^2 \left\{ \sum_{\substack{i=1 \\ }}^{N_g} \sum_{\substack{j=1 \\ |i-j|=n}}^{N_g} p(i,j) \right\} \tag{6.11}$$

Entropy evaluates the randomness of the surface height distribution. Entropy is small for regularly distributed surface height, and it is large for randomly distributed surface height. Contrast measures the degree of surface local variation, such as grooves and ridges, which is the amount of local surface height difference between a pixel and its neighbor over the image. The correlation of these indicators with the wear of wiper inserts is discussed in the next subsection.

6.2.4   Results and Discussion

Along with the measurement of the wear of wiper inserts, three successive engine blocks were measured every 50 workpieces by HDM. The modified GLCM is calculated with pixel angle equal to 0° and pixel distance equal to 5. Entropy and contrast are the averages of the three engine blocks. The averaged entropy and contrast versus number of blocks machined are shown in Fig. 6.8a and Fig. 6.8b, respectively.

One of the most consistent observations is that entropy increased with the number of blocks machined and the lowering of the wiper edge. The correlation coefficient between entropy and the lowering of the wiper edge is 0.947, which illustrates that entropy is a good indicator of the lowering of the wiper edge. The high correlation coefficient between entropy and the lowering of the wiper edge is reasonable. On a geometric view, wiper edges become duller along with the

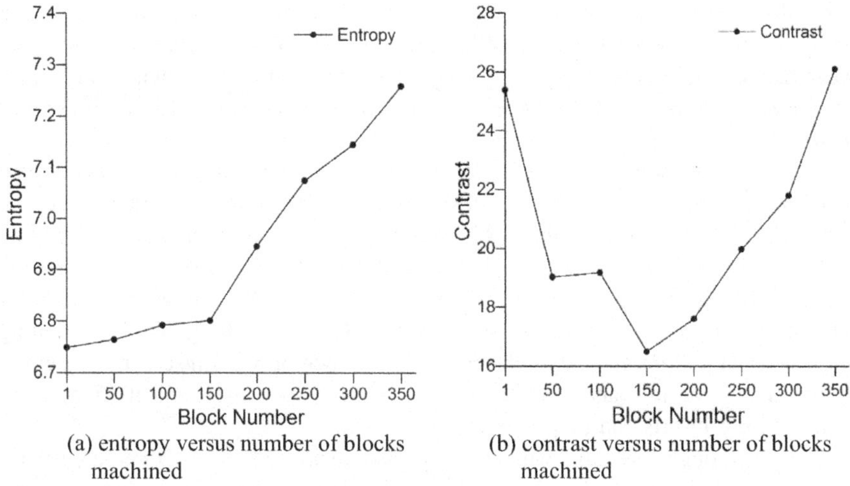

(a) entropy versus number of blocks machined

(b) contrast versus number of blocks machined

Fig. 6.8 The averaged entropy and contrast versus number of blocks machined. Reprinted from Ref. [57], copyright 2015, with permission from ASME

lowering of the wiper edge. Thus, the surface randomness and nonuniformity increase, and entropy goes in the same direction. The strong correlation between entropy and lowering of the wiper edge can be used for the prediction of tool life and control of machining accuracy, allowing the optimum tool change before tool failure.

Knowledge about the lowering rate of wiper inserts verses the machined workpieces can be also used for online tool position compensation. The milling process uses axial position of the wiper edge as tool position reference. Along with the machining progress, the wiper edge becomes lower. The lowing of wiper edge shortens the milling tool and gradually shortens the depth of cut in the axial direction. The shortening of the depth of cut could lead to the surface dimension out of tolerance. If the tool position can be compensated based on the knowledge of lowering rate of the wiper edge, surface dimension would be improved and cutting tools can reach its maximum capability.

An unexpected finding is that contrast first decreases with machined parts before machining 150 engine blocks and then increases after machining 150 engine blocks. This phenomenon can be explained as follows. As described in Sect. 6.2.3, contrast describes the degree of surface local variation. Because the surface local variation is caused by alteration of the depth of cut of the three wiper inserts due to various axial offsets, contrast is likely to be associated with the evolution of axial offset of the three wiper inserts in the milling process. The axial offset of wiper inserts is determined by the initial axial position and individual wear rate for each wiper insert. As can be seen in Fig. 6.4b, at the beginning of machining, wiper inserts 6 and 12 are higher than insert 18. As wiper inserts 6 and 12 wear more quickly than wiper insert 18, the three wiper inserts tend to have equal axial positions, i.e., small axial offset. Therefore, the local variation caused by axial offset of the three wiper inserts become smaller and contrast become lower. After machining 150 parts, the wear rate of wiper insert 6 and 12 is still higher than that of wiper insert 18. Therefore, axial offset of the three wiper inserts become larger again, so contrast increases.

In addition to monitoring the cutting tool wear, the experimental results can also help increase the cutting tool life. As shown in Fig. 6.4b, the wear rates of all the three wiper inserts are different, which could cause the early changing of cutting tool due to the failure of one particular wiper insert. Therefore, tool life can be increased by balancing the different wear rates of wiper inserts. In general, the wear rate of wiper inserts would result from the different cutting loads or cutting allowance. Therefore, balancing the cutting allowance of wiper inserts by presetting the relative axial and radial positions between the wiper insert and its five prece- dence cutting inserts can help increase the tool life.

Flatness, roughness, and waviness are also measured to test their ability for monitoring the wear of wiper inserts in multi-insert face milling. Flatness is aver- aged from the HDM data of the three sampled engine blocks for each measurement. As can be seen in Fig. 6.9a, flatness shows a tendency to increase with the number of blocks machined except at the initial cutting period. Therefore, flatness does not always provide consistent values that can be related to the wear of wiper inserts. Roughness Rz and waviness Wt are measured at eight different positions for the three sampled engine blocks using a Hommel T8000 instrument. The results shown in Fig. 6.9b are an average of the measured values. As can be seen, roughness Rz and waviness Wt are not always reliable tool wear indicators. This is because roughness and waviness are affected by multiple factors including radial and axial

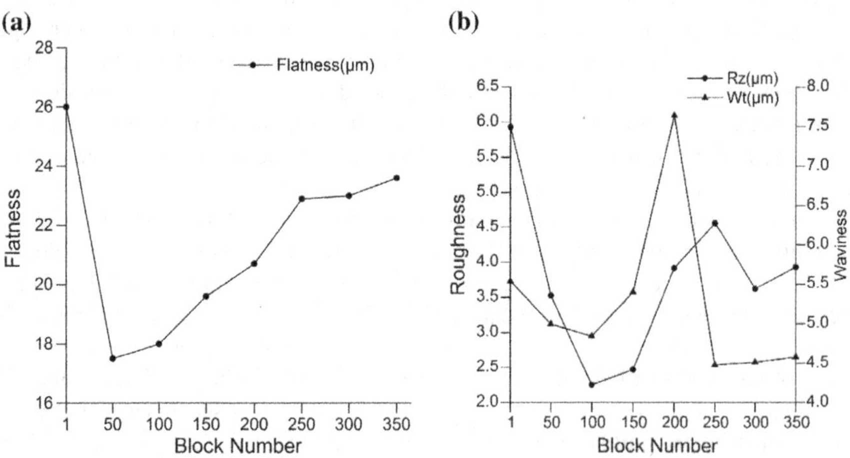

Fig. 6.9 **a** Flatness versus number of blocks machined, and **b** Roughness and waviness versus number of blocks machined. Reprinted from Ref. [57], copyright 2015, with permission from ASME

offsets between inserts, wear, runout, mechanical impact, transient thermal stresses, and back cutting. Even for the same workpiece, the measurement of roughness and waviness can vary substantially for different measuring positions and directions.

Besides face milling process, other surface machining processes such as end milling and grinding can be monitored using the proposed method. For certain machined surfaces with no clear toolmarks, surface features could be extracted from the HDM-converted image using the modified GLCM technique without toolmark straightening. At the same time, it would be possible to define a direction locally for extracting GLCM features, given that the tangent to the surface height gradient is readily found. Extensive experiments are required to establish the correlation between the extracted surface features and the tool wear condition.

6.2.5   Conclusions

This work presents a wear monitoring method for wiper inserts in multi-insert face milling using 3D surface form as tool wear indicators. The 3D surface form indicators including entropy and contrast are extracted from height-encoded and toolmark-straightened gray images that contain the entire 3D surface topography information. The wear of wiper inserts is represented using the axial lowering of the wiper edge, which is measured using a tool presetter and measuring machine without dismounting wiper inserts from the cutter. Experimental results indicate that entropy showed a strong correlation with the axial lowering of the wiper edge and contrast is related to the evolution of axial offset between multiple wiper inserts. These findings can be used to control multi-insert milling accuracy by compensating tool positions and increase tool life through proper inserts setting. In future work, it can be focused on the repeatability study of various feed rate, cutting speed, and insert mounting patterns and on monitoring marked wear of wiper inserts.

6.3   Detection and Monitoring of Defects on Three-Dimensional Curved Surfaces Based on High-Density Point Cloud Data

6.3.1   Introduction

The surface quality of three-dimensional (3D) curved surfaces is very important for product performance. For instance, the inner surface of an engine cylinder head combustion chamber is a 3D curved surface. If the inner surface is defective, the volume of the combustion chamber will be affected and in turn one of the significant performances of an engine, namely, compression ratio, will be affected

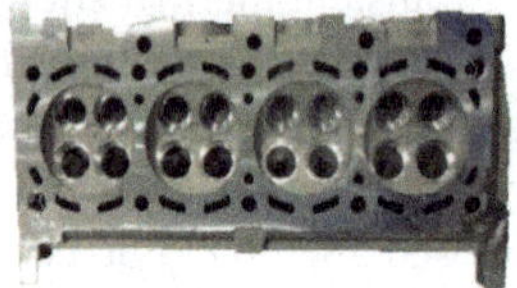
(a) An engine cylinder head

(b) A combustion chamber

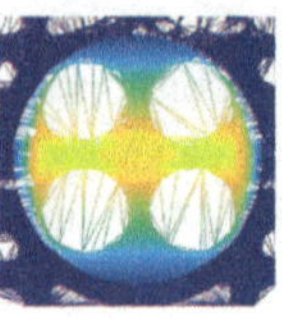
(c) High-density point cloud data

Fig. 6.10 The high-density point cloud data of a curved surface of an engine cylinder head. Reprinted from Ref. [101], copyright 2018, with permission from Elsevier

greatly. The traditional quality control method of the inner surface is applied based on coordinate measurement machine (CMM). Although CMM is high in accuracy, the number of measured points is small, and the entire surface contour cannot be fully reflected. With the development of high-definition metrology (HDM) technologies, high-density point cloud data is measured for reflection of the entire curved surface contour (see Fig. 6.10), which provides great opportunities for quality control of 3D curved surfaces. About one million measurement points are collected from a cylinder head by HDM system [48].

Many researches have been conducted on detection and monitoring of defects on flat surfaces. Wang et al. [49] employed the fused least absolute shrinkage and selection operator (LASSO) algorithm to identify potentially shifted sites on wafer surfaces and proposed a variable selection-based statistical process control (SPC) method for monitoring two-dimensional (2D) data maps. He et al. [50] proposed a multivariate generalized likelihood ratio (MGLR) control chart for monitoring and detecting numerous number of fault clusters per image in industrial applications. Sullivan [51] proposed a method based on profile monitoring which is effective in detecting single or multiple shifts and/or outliers and described the algorithm and an effective stopping rule that controlled the false detection rate. Woodall et al. [52] viewed the monitoring methods of process and product profiles (representation of quality characteristics) in statistical process control and exhibited comparisons of the monitoring methods for linear calibration relationships, change-point methods for simple linear regression profile data, and the use of generalized linear models to represent profiles. However, profile monitoring is based on the assumption that the can be represented by linear, nonlinear or non-parametric model.

With the development of measurement technologies, several researches about controlling flat surface variation based on HDM have also been conducted. Du et al. [53–55] proposed a shearlet-based method and support vector machine-based methods to separate and extract different surface components using HDM. Du et al. [56] also presented a co-Kriging method based on multivariate spatial statistics to estimate surface form error. Wang et al. [41, 57, 58] developed a modified gray level co-occurrence matrix to extract features from the images converted from face milled surface measured by HDM. Suriano et al. [13] proposed a new methodology for efficiently measuring and monitoring flat surface variations by fusing in-plant multi-resolution measurements and process information. Nguyen et al. [59] presented a method to reduce flat surface variation in face milling processes based on HDM. Wells et al. [60] proposed an adaptive generalized likelihood ratio (AGLR) technique to monitor the planar surface defects by transforming high-density data into a grayscale image.

However, the above researches based on HDM only focus on the flat surface. It is desired to develop a systematic method for detection and monitoring of defects on 3D curved surface based on high-density point cloud data. Since the number of the curved surface point cloud data is large, it is unrealistic to directly monitor the entire surface quality by monitoring the condition of each point. Wang et al. [14] used quantile–quantile (Q–Q) plot to characterize a huge sample and transform it to a linear profile, and proposed profile monitoring techniques to improve the performance of a conventional control chart. Wells et al. [15] also used Q–Q plots to transform high-density point cloud data into linear profiles that can be monitored by well-established profile monitoring techniques. But the methods using Q–Q plot cannot identify the locations of defective regions from the curved surface in out-of-control (OOC) condition. Colosimo et al. [6] developed a method consisting of modeling the manufactured surface via Gaussian processes models and monitoring the deviations of the actual surface from the target pattern estimated in Phase I. The proposed method is limited to monitoring surfaces (for example, a cylindrical surface) characterized by parametric models.

In this subsection, the curved surface is divided into hundreds of small sub-regions, and each small sub-region contains dozens of 3D points. The defects of the small sub-region are characterized by two explored evaluation indexes: the nonrandom distribution of abnormal points (NDAP) of a sub-region and the plane direction deviation (PDD) of a sub-region. The NDAP is calculated by the wavelet packet entropy and the PDD is calculated by the normal vector. The wavelet packet coefficients are used to extract the NDAP feature of a sub-region, which are quantified by the information entropy. That is, the wavelet packet entropy is used as the NDAP feature of each sub-region. In order to accurately extract the PDD features of the divided sub-regions, the normal vector of each sub-region is also calculated. Wavelet packet entropies and normal vectors of the modulus curved surface are calculated as the criteria to evaluate whether the sub-regions of the measured curved surface are out of limit (OOL). When the OOL sub-regions are identified, three quality parameters that represent quality characteristics of the measured curved surface are calculated based on the clusters of OOL sub-regions. Three individual control charts are made to monitor the three quality parameters.

6.3.2 The Proposed Method

6.3.2.1 Framework

This subsection presents an overview of the proposed method for monitoring 3D curved surface quality of a workpiece based on high-density point cloud data. The framework of the proposed method is shown in Fig. 6.11. The procedure involves the following steps.

Step 1: Region division of curved surfaces. HDM is employed to collect 3D coordinates from workpiece surfaces. The curved surface (represented by the measured point cloud data) is divided into multiple sub-regions through remove of outlier, boundary recognition, and sub-region division. A sub-region is a small point cloud (usually consisting of dozens or hundreds of 3D points) that represents a part of the curved surface. In order to evaluate the quality of each divided sub-region, the modulus point cloud data (i.e., product design specification) of the curved surface is processed by boundary recognition and sub-region division.

Step 2: Feature evaluation of each sub-region. Two new evaluation indexes (NDAP and PDD) based on wavelet packet entropies and normal vectors are explored to represent the features of sub-regions of the measured curved surface. Wavelet packet entropies and normal vectors of sub-regions of the modulus curved surface also need to be calculated as the criteria.

Step 3: Quality parameters calculation of the curved surface. When the OOL sub-regions of the measured curved surface are recognized, three quality parameters that represent quality characteristics of the measured curved surface are calculated by clusters of the OOL sub-regions.

Step 4: Monitoring the three quality parameters. Each quality parameter is monitored by an individual control chart. If any quality parameter of a curved surface is out of the control range, the manufacturing process is OOC and the correction should be conducted.

6.3.2.2 Region Division of Curved Surfaces

(1) Algorithm to remove outliers

Due to the interference of the measurement environment and the surface reflection of the measured workpiece, etc., there are mainly outliers in the measured point cloud data (the measured curved surface). There are three types of outliers: the outliers generated by ambient light interference, the outliers caused by the reflective properties of the curved surface of a workpiece, and the outliers that are not obvious. An algorithm including three steps is proposed to remove the corresponding type of outliers.

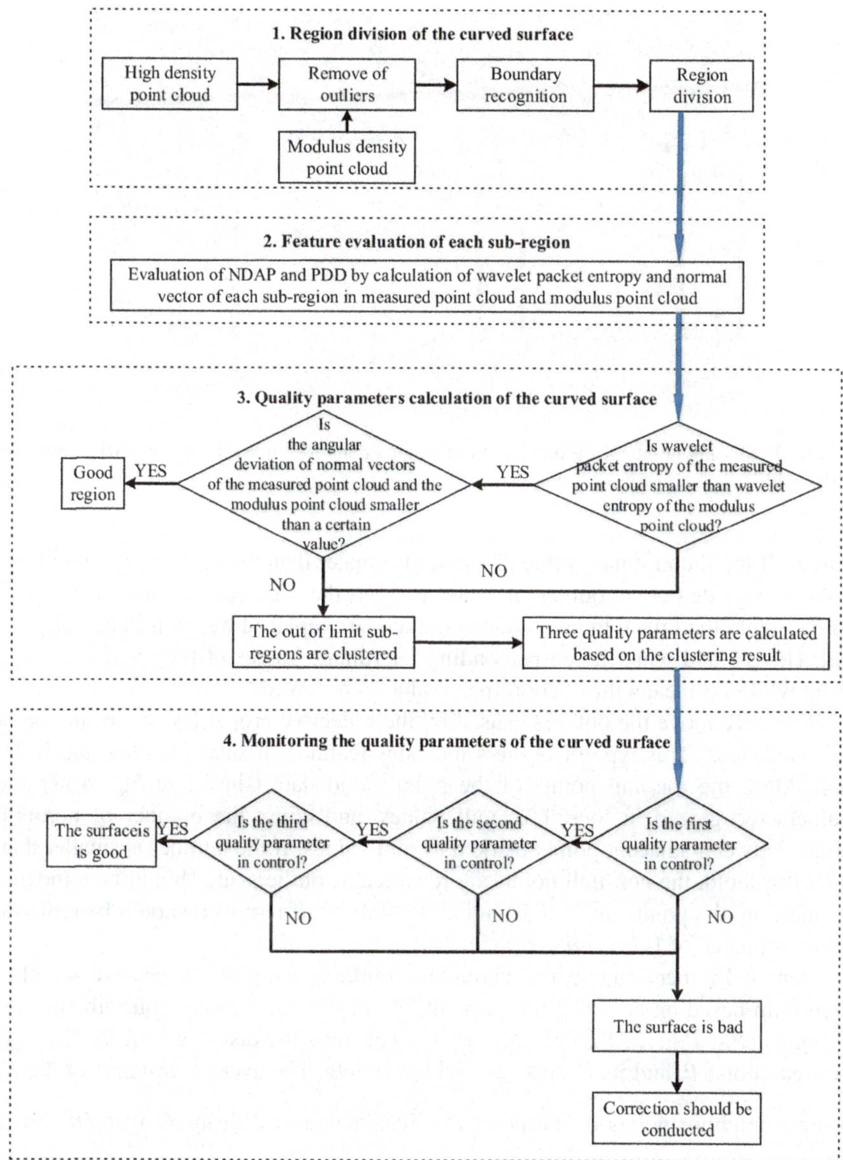

Fig. 6.11 Framework of the proposed method. Reprinted from Ref. [101], copyright 2018, with permission from Elsevier

Step 1: Remove the outliers generated by ambient light interference. Since the maximum distance from the surface to the bottom of a workpiece is known, the maximum distance is set as the distance threshold to determine whether a point is an

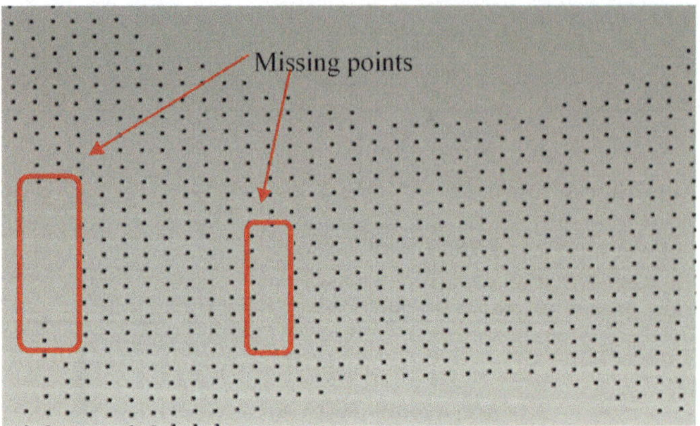

Fig. 6.12 Missing points in the measured point cloud data. Reprinted from Ref. [101], copyright 2018, with permission from Elsevier

outlier. If the Z-coordinate value of a point is greater than the distance threshold, the point is regarded as an outlier. In order to keep the ordered structure of the measured point cloud data, the recognized outliers are replaced by coordinates of [x; y; na]. Here, x and y are the corresponding coordinate values of the recognized outliers, while na means the Z-coordinate value is removed.

Step 2: Remove the outliers caused by the reflective properties of the surface of the workpiece. This type of outliers indicates features of small number and isolation. Mark the missing points of the point cloud data (shown in Fig. 6.12) and outliers recognized in Step 1 as null values, and count the number of non-null points between adjacent points marked as null values. If the number is smaller than a set threshold, the non-null points are regarded as outliers and should be removed. In addition, the points marked as null and points to be removed should be replaced by coordinates of [x; y; na].

Step 3: Further remove the unobvious outliers using the K-nearest neighbor algorithm based on statistics. For a point (P_i) of the point cloud, count the nearest K points (P_{i1}, P_{i2}, ..., P_{iK}) of point P_i and calculate the distance (d_{i1}, d_{i2}, ..., d_{iK}) between point P_i and its K-nearest neighbor points. The average distance of the K-nearest neighbor points of the point P_i is calculated as $\mathrm{dMean}_i = \sum_{j=1}^{K} d_{ij}/K$. Since the structure of the measured point cloud data is ordered, the nearest neighbor points of point P_i are P_1, P_2, P_3, and P_4 (shown in Fig. 6.13), and K is determined as 4. Considering the fact that the nearest neighbor points may contain outliers or missing points (shown in Fig. 6.13b), the value of K is usually determined as $K \leq 4$. The average distance of the point cloud is calculated as $\mathrm{dist} = \sum_{i=1}^{\mathrm{Num}} \mathrm{dMean}_i/\mathrm{Num}$, and the mean square error is calculated as $\sigma = \sqrt{\dfrac{1}{\mathrm{Num}} \sum_{i=1}^{\mathrm{Num}} (\mathrm{dist} - \mathrm{dMean}_i)^2}$.

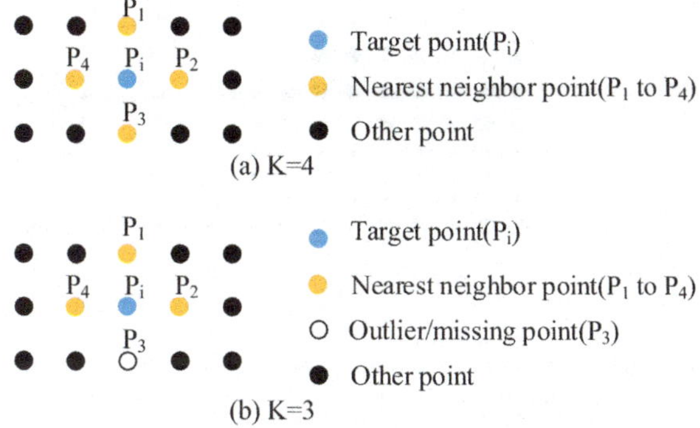

Fig. 6.13 The K-nearest neighbor points of a target point. Reprinted from Ref. [101], copyright 2018, with permission from Elsevier

Here, *Num* is the number of all points of the curved surface point cloud data. Since the average distance (*dMean*) of each point of the curved surface point cloud data is in accordance with the $3\ \sigma$ law of the Gaussian distribution, the outliers do not satisfy the $3\ \sigma$ law of the Gaussian distribution. Set $[\text{dist} - 3\sigma, \text{dist} + 3\sigma]$ as the standard interval and remove the points with $dMean_i$ that does not conform to the standard interval. Coordinate of $[x;\ y;\ na]$ is also used to replace the removed points to keep the ordered structure of the measured point cloud data.

(2)   Boundary recognition algorithm

The measured point cloud data of a workpiece contains usually plane point cloud data and curved surface point cloud data. A curved surface has boundary, which needs to be recognized from the measured point cloud. According to the fact that Z-coordinate values of the points in the plane and the curved surface are quite different, the initial/terminal row and initial/terminal column of the curved surface point cloud data is obtained. The boundary recognition algorithm of identifying the curved surface point cloud data from the measured point cloud data is described as follows.

Step 1: Since the structure of the measured point cloud data is ordered, the rows and columns of the measured point cloud data are marked (shown in Fig. 6.14).

Step 2: Calculate the mean of the Z-coordinate values of each row, determine the serial numbers of the initial and terminal rows according to the change of average Z-coordinate value, and store all the points from the initial row to the terminal row. The stored points are regarded as the selected point cloud data (shown in Fig. 6.15).

Step 3: Calculate the mean of Z-coordinate values of each column of the selected point cloud data, and determine the serial numbers of the initial and terminal columns according to the change of average Z-coordinate values. The location of curved surface point cloud data is shown in Fig. 6.16.

Step 4: Identify the curved surface point cloud data. Although the location of the curved surface point cloud data has been identified, the identified point cloud data in rectangular region (shown in Fig. 6.16) also contains the plane point cloud data. According to the difference of Z-coordinate values of the points in the plane and the curved surface, the points in the plane that are within the initial/terminal row and

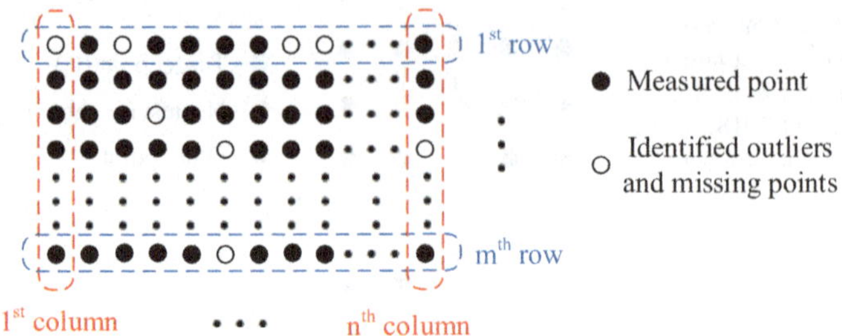

Fig. 6.14 The structure of the measured point cloud. Reprinted from Ref. [101], copyright 2018, with permission from Elsevier

Fig. 6.15 Determination of initial/terminal row of the curved surface point cloud data. Reprinted from Ref. [101], copyright 2018, with permission from Elsevier

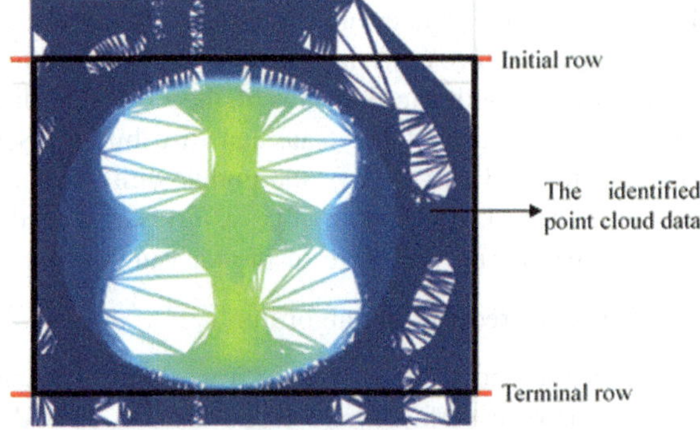

Fig. 6.16 Location of the curved surface point cloud data. Reprinted from Ref. [101], copyright 2018, with permission from Elsevier

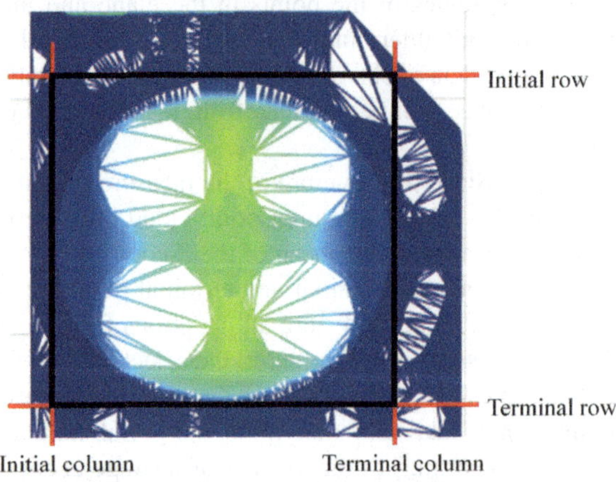

initial/terminal column are selected. In order to ensure that each row/column contains the same number of points, these selected points should be replaced by coordinates of $[x;\ y;\ 0]$ instead of being deleted.

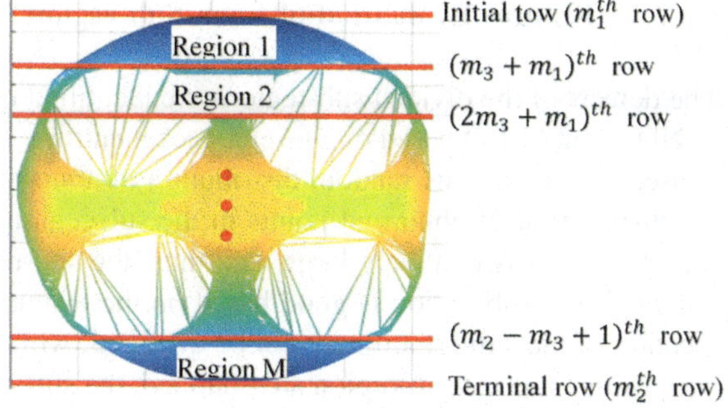

Fig. 6.17 M divided sub-regions of the curved surface point cloud data. Reprinted from Ref. [101], copyright 2018, with permission from Elsevier

(3) Sub-region division

The identified point cloud data is regarded as the curved surface point cloud data since the coordinates of the plane points included in the identified point cloud data are marked as [0 0 0]. In order to monitor the curved surface quality of a workpiece, the curved surface is divided into $M \times N$ (M and N should be integers) sub-regions. Each sub-region contains a small number of points (dozens of points). Assume that the serial number of the initial and terminal rows are m_1 and m_2, and the serial number of the initial and terminal columns are n_1 and n_2. Then the curved surface point cloud data contains $(m_2 - m_1 + 1)$ rows and $(n_2 - n_1 + 1)$ columns.

Step 1: Divide the curved surface into M sub-regions according to the row. Assume that $\text{INT}[(m_2 - m_1 + 1)/M] = m_3$, and $m_4 = (m_2 - m_1 + 1) - M \times m_3$ (the remainder of $(m_2 - m_1 + 1)$ divided by M). Each of the former m_4 sub-regions contains $(m_3 + 1)$ rows, and each of the latter $(M - m_4)$ sub-regions contains m_3 rows. An example is shown in Fig. 6.17.

Step 2: Divide each of the M sub-regions into N sub-regions according to the column. The process of the division is the same as the previous process shown in the previous step. The division of the curved surface is shown in Fig. 6.18.

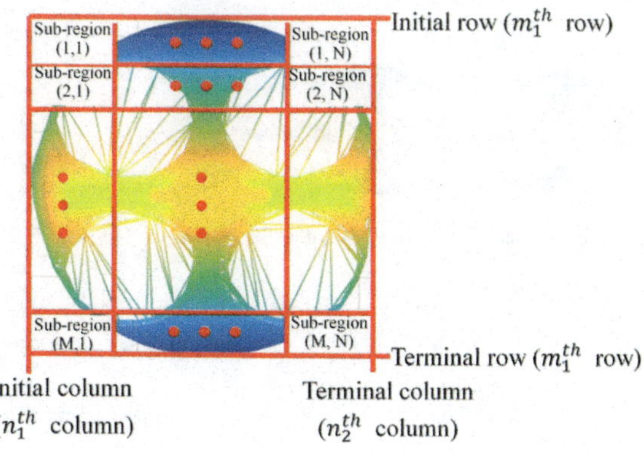

Fig. 6.18 The sub-region division of the curved surface. Reprinted from Ref. [101], copyright 2018, with permission from Elsevier

6.3.2.3 Feature Evaluation of Each Sub-region

The defects of the divided sub-regions are identified by two new evaluation indexes of NDAP and PDD based on wavelet packet entropy and normal vector. The NDAP is used to evaluate the random distribution of abnormal points in the sub-region. If the distribution of abnormal points in the sub-region is not random, the NDAP of the sub-region is relatively large, and thus, the sub-region is considered to be bad. Although the sub-region is good based on the evaluation of the NDAP, the spatial location of the sub-region may not be qualified, which is evaluated by PDD. When both features of the sub-region are qualified, the sub-region is good. The calculation of NDAP and PDD is shown in Fig. 6.19. The NDAP feature is represented by wavelet packet entropy, and the PDD feature is represented by normal vector.

(1) Wavelet packet entropy

Wavelet transform reflects the energy distribution of the data in the time–frequency domain [61, 62]. But wavelet transform only decomposes the low-frequency components of the data (shown in Fig. 6.20a), which ignores the information of high-frequency components. The wavelet packet transform decomposes not only the low-frequency components of the signal but also the high-frequency components (shown in Fig. 6.20b), which is better applied for feature extraction [63].

Entropy is used to describe and measure the uncertainty of the random variable, and can accurately characterize information. Therefore, the combination of wavelet packet and entropy (wavelet packet entropy) is considered to make full use of their advantages in feature evaluation.

For a regular set of data (e.g., a set of single-frequency data), its relative wavelet packet energy is very small since all the energy is only concentrated on a band. The energy of a very complex data (e.g., a random data) will be distributed in each band,

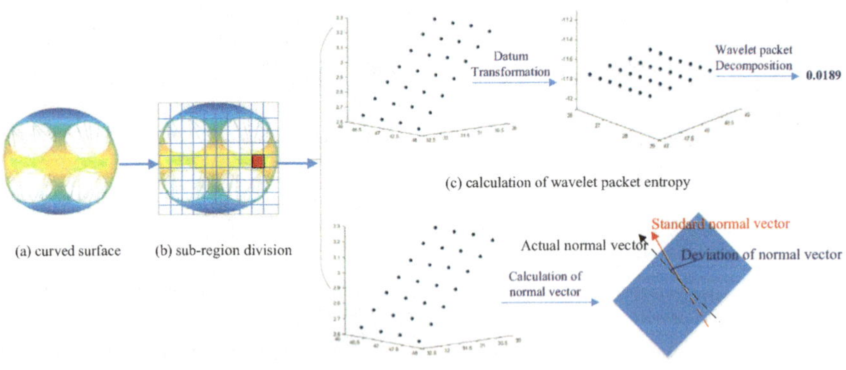

Fig. 6.19 Calculation of NDAP and PDD of a sub-region. Reprinted from Ref. [101], copyright 2018, with permission from Elsevier

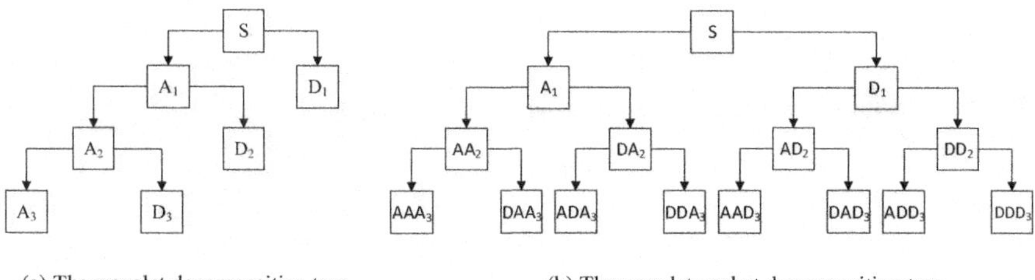

Fig. 6.20 The decomposition trees of wavelet and wavelet packet. Reprinted from Ref. [101], copyright 2018, with permission from Elsevier

and the relative wavelet packet energy will be relatively similar, which makes wavelet packet entropy maximum. Similarly, wavelet packet entropy is used as the measurement of information of the surface morphology, and it represents the randomness and complexity of the surface morphology. If the quality of workpiece surface is good, the height distribution of each point is regular and nonrandom, the energy distributions in the time–frequency domain are relatively concentrated in the low-frequency band (e.g., AAA_3 in Fig. 6.20b), and the entropy is small. If the points are completely equal, the entropy is zero. On the contrary, if the quality of the surface is bad, the surface heights (Z-coordinate values of all points of the surface) distribute randomly and wavelet packet is used for multi-scale analysis of the surface, amplitude and energy of the decomposed components on other scales will increase, and the entropy will be relatively large.

The calculation of wavelet packet entropy is conducted based on wavelet packet transform of the surface. The coefficient matrix of the wavelet packet transform is processed into a probability distribution sequence, and the entropy calculated from it reflects the randomness and complexity of the surface morphology. The procedure of calculating wavelet entropy of each divided sub-region is described as follows.

Step 1: Datum transformation of sub-regions and interpolation. The sub-regions are inclined planes, but current wavelet packet decompositions (1D and 2D wavelet packet transforms) cannot be directly used for feature extraction of the inclined planes. Therefore, the datum of the inclined sub-regions needs to be transformed into flat planes. The transformation is shown in Fig. 6.21 and the specific details can refer to [48]. Since there are sub-regions that only contain few points and the spatial distribution of these points is not a rectangle, rectangular regions should be reselected from these sub-regions and several points are deleted, which would result in information loss of the reselected sub-regions. In order to reduce the information loss, interpolation is applied to add the number of points to be decomposed. The interpolation is triangle-based cubic interpolation and the details can refer to [64].

Step 2: Select input data for wavelet packet decomposition. The distribution of points of each divided sub-region is usually not a regular rectangle, which cannot be directly decomposed by wavelet packet.

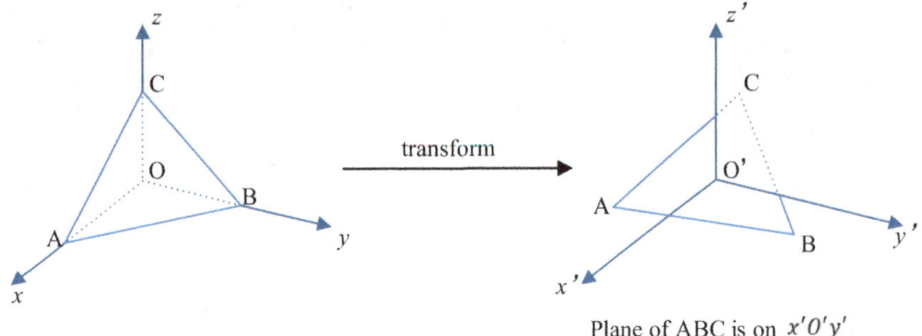

Fig. 6.21 Datum transformation of sub-region plane. Reprinted from Ref. [101], copyright 2018, with permission from Elsevier

The regional point cloud data before and after interpolation is the same in distribution (both are not regular rectangles), so it is necessary to select the rectangular data from the regional point cloud data after interpolation as the input of wavelet packet decomposition. In order to preserve the characteristics of the regional point cloud data, the number of data that the selected rectangle contains should be as large as possible.

Step 2.1: Identify the largest rectangular region that contains valid data. It can be seen from Fig. 6.22 that there are points marked as NaN (null value), which should not be included in the input data. Remove the rows/columns that only contains mark of NaN. The unit of data in Fig. 6.23 is millimeter (mm).

Step 2.2: Count the number of non-null data of each row/column, and delete the row/column that contains the least number of non-null data of all rows and columns until the data contained in each row/column is non-null (shown in Fig. 6.23). The unit of data in Fig. 6.23 is mm. If the number of rows to be deleted is too large, the column that contains the least number of non-null data should be deleted.

NaN	NaN	NaN	NaN	NaN	NaN
NaN	NaN	NaN	NaN	NaN	NaN
NaN	1.7219	NaN	NaN	NaN	NaN
NaN	1.7252	1.6441	NaN	NaN	NaN
NaN	1.7288	1.6479	NaN	NaN	NaN
NaN	1.7331	1.6533	1.5703	NaN	NaN
NaN	1.7388	1.6600	1.5777	NaN	NaN
NaN	1.7442	1.6707	1.5890	NaN	NaN
NaN	1.7466	1.6805	1.6028	NaN	NaN
NaN	1.7477	1.6848	1.6159	NaN	NaN
NaN	1.7486	1.6874	1.6234	NaN	NaN
NaN	NaN	NaN	NaN	NaN	NaN
NaN	NaN	NaN	NaN	NaN	NaN

Fig. 6.22 Selection of largest rectangular region containing valid data. Reprinted from Ref. [101], copyright 2018, with permission from Elsevier

The 1st row should be deleted   The first two rows should be deleted

1.7219	NaN	NaN	1
1.7252	1.6441	NaN	2
1.7288	1.6479	NaN	2
1.7331	1.6533	1.5703	3
1.7388	1.6600	1.5777	3
1.7442	1.6707	1.5890	3
1.7466	1.6805	1.6028	3
1.7477	1.6848	1.6159	3
1.7486	1.6874	1.6234	3
9	8	6	

1.7252	1.6441	NaN	2
1.7288	1.6479	NaN	2
1.7331	1.6533	1.5703	3
1.7388	1.6600	1.5777	3
1.7442	1.6707	1.5890	3
1.7466	1.6805	1.6028	3
1.7477	1.6848	1.6159	3
1.7486	1.6874	1.6234	3
8	7	5	

1.7331	1.6533	1.5703
1.7388	1.6600	1.5777
1.7442	1.6707	1.5890
1.7466	1.6805	1.6028
1.7477	1.6848	1.6159
1.7486	1.6874	1.6234

Fig. 6.23 Selection of rectangular region that only contains valid data. Reprinted from Ref. [101], copyright 2018, with permission from Elsevier

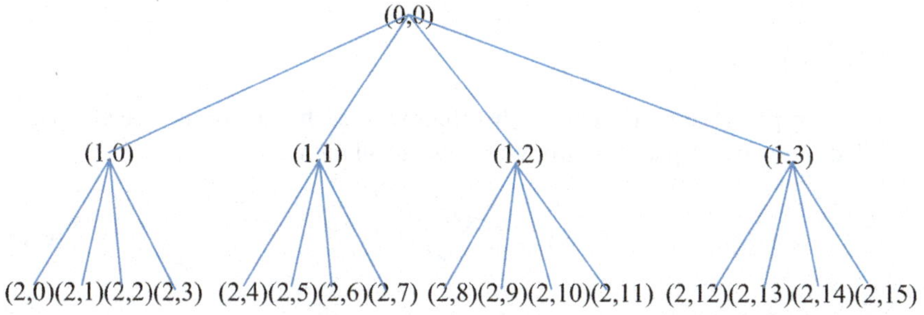

Fig. 6.24 A two-layer decomposition tree of 2D Wavelet packet. Reprinted from Ref. [101], copyright 2018, with permission from Elsevier

Step 3: Wavelet packet decomposition. The wavelet packet decomposition is a multi-level division of the frequency band, which further decomposes the high-frequency components. It can adaptively select the corresponding frequency band to match the signal spectrum according to the characteristics of the analyzed data and improve the time–frequency resolution. 2D Wavelet packet is used to decompose the divided sub-regions. A two-layer decomposition tree of 2D wavelet packet is shown in Fig. 6.24. The decomposition relationship is expressed as $S_{0,0} = S_{2,0} + S_{2,1} + S_{2,2} + S_{2,3} + S_{2,4} + S_{2,5} + S_{2,6} + S_{2,7} + S_{2,8} + S_{2,9} + S_{2,10} + S_{2,11} + S_{2,12} + S_{2,13} + S_{2,14} + S_{2,15}$. The coefficient matrices of all nodes of the last layer are the output of wavelet packet decomposition.

Step 4: Calculation of wavelet packet entropy. The wavelet packet coefficients vector of the ith scale is recorded as

$$S_i = \left(S_{i,1}, S_{i,2}, \cdots, S_{i,n} \right) \tag{6.12}$$

where $S_{i,j}(j = 1, 2, \cdots, n)$ is the wavelet packet coefficient of the ith scale. The coefficient vectors of all scales form a coefficient matrix $\{S_i\}, i = 1, 2, \cdots, m,$

where m is the scale of wavelet packet decomposition. The norm vector of the wavelet packet coefficients matrix to measure the divided sub-region at each scale, that is, the energy of each scale is used to measure the proximity of the decomposed components of each scale. Normalize the energy vector E_i and analyze the structure and complexity of the divided sub-region through distribution of the normalized energy vector. The energy of the ith scale is defined as

$$E_i = \|S_i\|^2 = \sum_{j=1}^{n} |S_{i,j}|^2 \tag{6.13}$$

The distribution of energy vector is defined as the normalized energy at each scale

$$p_i = E_i/E \tag{6.14}$$

and the total energy is $E = \sum_{i=1}^{m} E_i$.

The entropy based on the energy distribution of each scale of the wavelet packet is called the wavelet packet entropy and its calculation is

$$\text{WE} = -\sum_{i=1}^{m} p_i \log_2 p_i \tag{6.15}$$

(2) Normal vector

The normal vector of each divided sub-region is regarded as the feature of the divided sub-regions. The plane parametric equation of $A_x + B_y + C_z + D = 0$ is calculated by fitting the divided regional point cloud data. Then the normal vector (A, B, C) of the divided sub-region is obtained, and the steps are described as follows.

Step 1: Delete the points marked as [0 0 0] in each sub-region. Since the divided sub-regions conclude plane point cloud data, there may be sub-regions that only contain [0 0 0] and these sub-regions should be marked with empty set instead of being deleted in order to ensure that the overall number of the divided sub-regions is not reduced.

Step 2: Count the number of remaining points in each sub-region. For the sub-region (effective sub-region) with the number of points larger than or equal to 3, the random sample consensus (RANSAC) algorithm [65, 66] is used to fit the parametric equation of the effective sub-region and the normal vector is obtained.

6.3.2.4   Quality Parameters Calculation of Curved Surfaces

Before monitoring quality of the curved surface, the quality of the divided multiple sub-regions should be evaluated based on the wavelet packet entropies and normal vector errors. In order to calculate the wavelet packet entropy, the modulus point cloud data is needed. The modulus point cloud data is processed by boundary recognition, sub-region division, and calculation of wavelet packet entropy. The wavelet packet entropies of the modulus point cloud data are regarded as the criteria to determine whether the corresponding sub-regions in the curved surface point cloud data are OOL. If the wavelet packet entropy of the curved surface point cloud data is larger than the wavelet packet entropy of the modulus point cloud data in the same sub-region, the sub-region of the curved surface is considered to be OOL. Similarly, the normal vector error is the angular deviation between the normal vector of the measured curved surface point cloud data and the normal vector of the modulus point cloud data in the same sub-region. If the angular deviation is larger than a threshold, the sub-region of the curved surface is considered to be OOL. The sub-region of the curved surface is determined to be OOL if either of wavelet packet entropy and the angular deviation of normal vector is OOL.

When OOL sub-regions of the curved surface are identified, the quality parameters that represent quality characteristics [59] of the curved surface are defined as (1) the number of OOL sub-regions, (2) the number of OOL sub-regions of the largest cluster, and (3) the number of clusters that contain OOL sub-regions more than a certain value. The last two quality parameters are calculated by a clustering algorithm [67], and the procedure of the clustering algorithm involves the following steps.

Step 1: Initialization and preprocessing. (1) Select the parameter $t \in (0, 1)$ to determine the cutoff distance d_c. Usually, t ranges from 1% to 2%. (2) Calculate the distance $(dd_{i,j})$ between any two points, let $dd_{ji} = dd_{ij}, i < j, i, j \in I_S$. (3) Determine the cutoff distance d_c, which is used to determine the local density of each point in the dataset. Sort the M_1 calculated distances in ascending order, and the order is expressed as $dd_1 \leq dd_2 \leq \cdots \leq dd_{M_1}$. Here, $M_1 = \frac{1}{2}N_1(N_1 - 1)$, and N_1 is the number of all points. Let $d_c = dd_{f(M_1 t)}$, $f(M_1 t)$ represent an integer that is rounded to $M_1 t$. (4) Calculate the local density $(\{\rho_i\}_{i=1}^{N_1})$ of each point according to Eq. (6.16), and generate subscript sequencer $(\{q_i\}_{i=1}^{N_1})$ of the local densities in descending order.

$$\rho_i = \sum_{j \in I_S \setminus \{i\}} \chi(d_{ij} - d_c), \chi(x) = \begin{cases} 1, & x < 0 \\ 0, & x \geq 0 \end{cases} \tag{6.16}$$

(5) Calculate $\{\delta_i\}_{i=1}^{N_1}$ and $\{n_i\}_{i=1}^{N_1}$. δ_i is the distance defined as

$$\delta_{q_i} = \begin{cases} \min_{\substack{q_j \\ j<i}} \{d_{q_i q_j}\} & i \geq 2 \\ \max_{j \geq 2} \{\delta_{q_j}\} & i = 1 \end{cases} \tag{6.17}$$

n_i is the serial number of points closest to x_i in the data points with local densities greater than x_i and is calculated by

$$n_{q_i} = \begin{cases} \arg\min_{\substack{q_j \\ j<i}} \{d_{q_i q_j}\} & i \geq 2 \\ 0 & i = 1 \end{cases} \tag{6.18}$$

Step 2: Determine the clustering center $\{m_j\}_{j=1}^{n_c}$ and initialize the category attribute tag $\left(\{c_i\}_{i=1}^{N}\right)$ of the data point. c_i is determined by

$$c_i = \begin{cases} k, & x_i \text{ is the cluster center and belongs to the } k\text{th cluster} \\ -1, & \text{otherwise} \end{cases} \tag{6.19}$$

Step 3: Classify the non-clustering center points.

Step 4: If $n_c > 1$, further divide the data in each cluster into cluster core and cluster halo. (1) Initialize the tag $h_i = 0, i \in I_S$. (2) Generate an average local density upper bound $\{\rho_i^b\}_{i=1}^{n_c}$ for each cluster. (3) Identify cluster halo.

Step 5: The outputs are clusters and the corresponding clustering centers.

When the clusters of OOL sub-regions have been recognized, the quality parameters that represent quality characteristics of the curved surface are calculated by

$$qc_1 = k_i \tag{6.20}$$

$$qc_2 = \max\left(N_{i,\text{clust}}\right) \tag{6.21}$$

$$qc_3 = \sum_{j=1}^{\text{clust}} S_j \tag{6.22}$$

where k_i is the number of OOL sub-regions of the ith curved surface, clust is the number of clusters found on the curved surface, $N_{i,\text{clust}}$ is the size of the clust[th] cluster on the ith surface, and S_j is a binary variable with a value of one when the jth cluster has a size larger than s_c and zero otherwise.

6.3.2.5 Monitoring the Quality Parameters

The control chart is used in this monitoring phase. The centerline (CL), upper control limit (UCL), and lower control limit (LCL) of the individual control chart are determined by the following equation:

$$\begin{cases} \text{CL} = \mu \\ \text{UCL} = \mu + 3\sigma \\ \text{LCL} = \mu - 3\sigma \end{cases} \tag{6.23}$$

where μ is mean and σ is standard deviation. μ and σ are estimated by

$$\mu = \frac{x_1 + x_2 + \cdots + x_n}{n} \tag{6.24}$$

$$\sigma = \sqrt{\frac{(x_1 - \mu)^2 + (x_2 - \mu)^2 + \cdots + (x_n - \mu)^2}{n}} \tag{6.25}$$

If an OOC condition is found, corrections (e.g., mold maintenance) should be conducted to ensure the quality of the consecutive curved surfaces. Otherwise, the manufacturing process of the curved surfaces should be kept.

6.3.3 Case Study

In this case, the cylinder heads of B12 serial engines with four combustion chambers (see Fig. 6.25) are used to validate the performance of the proposed method on detection and monitoring of defects on 3D curved surface based on high-density point cloud data. Before the measurement, the intake and outtake valves are not put into the combustion chambers.

The point cloud data of combustion chambers of B12 engine cylinder heads is measured by an online HDM measurement machine using laser triangulation metrology [48]. The online measurement machine is shown in Sect. 6.2.2.

(1) Region division of curved surfaces

In this case, the curved surface of only a chamber is considered. In order to remove the outliers generated by ambient light interference, the maximum Z-coordinate value is set as 11 cm (the maximum value is obtained by actual measurement). The threshold for removing outliers caused by the reflective properties of the curved surface of the workpiece is determined as 15 since a group of points less than 15 are considered to be isolated. The result of removing outliers is shown in Fig. 6.26. Boundary extraction of the curved surface is conducted and the result is shown in Fig. 6.27. The point cloud data of the curved surface (an inner surface of the

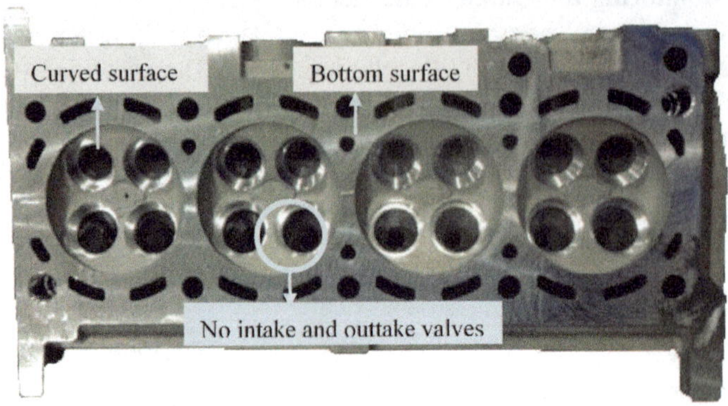

Fig. 6.25 Cylinder head of B12 engine with four combustion chambers. Reprinted from Ref. [101], copyright 2018, with permission from Elsevier

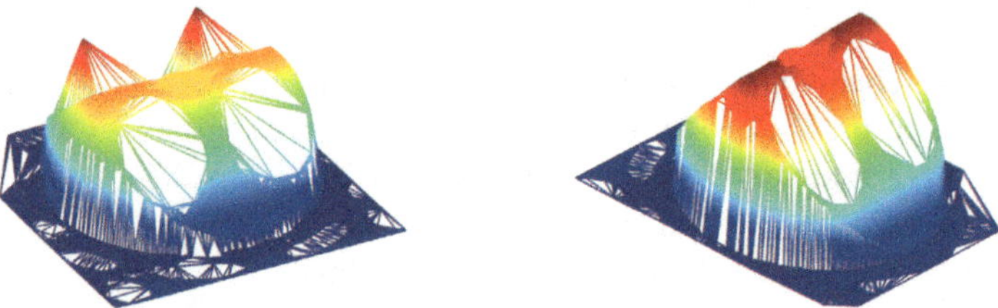

(a) Point cloud data before remove of outliers (b) Point cloud data after remove of outliers

Fig. 6.26 Point cloud data of a curved surface. Reprinted from Ref. [101], copyright 2018, with permission from Elsevier

chamber) is shown in Fig. 6.28. In order to obtain as many small sub-regions as possible, the curved surface is divided into 32×32 sub-regions, and a sub-region of them is shown in Fig. 6.29. It can be seen from Fig. 6.32b that the divided sub-region is similar to a plane. The unit of the axes in Figs. 6.27, 6.28, and 6.29 is mm.

(2) Feature evaluation of each sub-region

After the region division of the curved surface, the effective sub-regions should be selected since there are sub-regions that are in the intake and outtake valve holes of the chamber inner surface. The sub-regions containing less than three points are marked as empty sub-regions, and the distribution of the effective sub-regions is shown in Fig. 6.30. Before calculation of the wavelet packet entropies and normal vectors of the effective sub-regions, datum transformation of each sub-region is conducted and the result is shown in Fig. 6.31. The unit of the axes in Fig. 6.31 is mm.

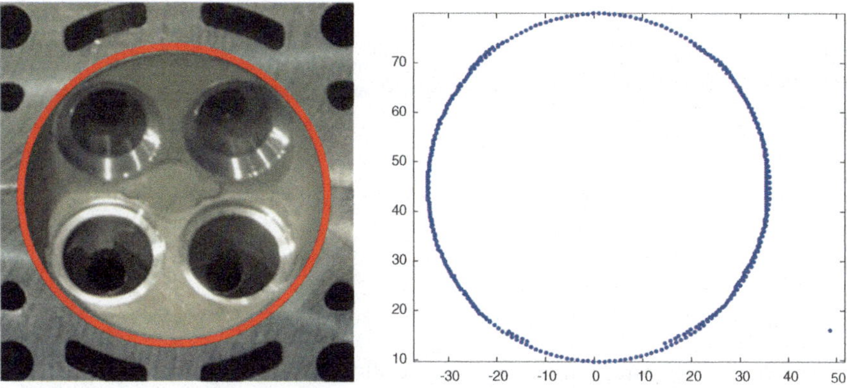

Fig. 6.27 Boundary of the curved surface. Reprinted from Ref. [101], copyright 2018, with permission from Elsevier

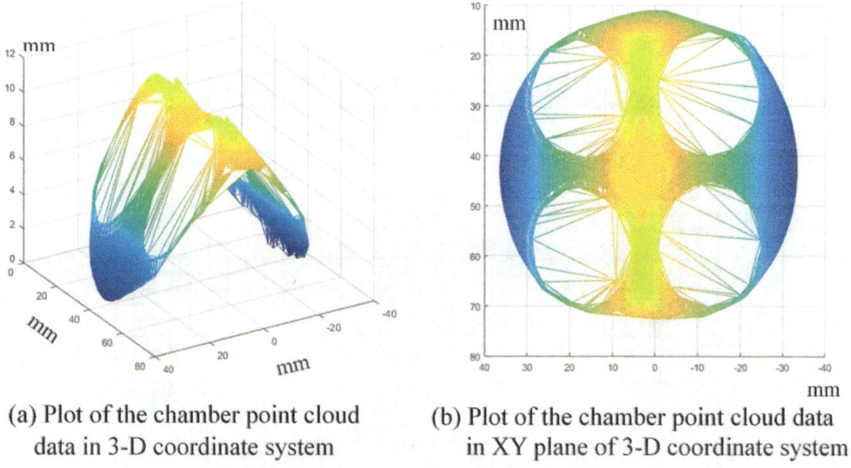

(a) Plot of the chamber point cloud data in 3-D coordinate system

(b) Plot of the chamber point cloud data in XY plane of 3-D coordinate system

Fig. 6.28   Point cloud data of the curved surface after boundary recognition. Reprinted from Ref. [101], copyright 2018, with permission from Elsevier

The wavelet packet entropies and the normal vectors of the sub-regions of the modulus point cloud data are calculated as the criteria to evaluate whether sub-regions of the curved surface point cloud data are qualified. The wavelet packet entropies of some sub-regions are shown in Table 6.2, and the normal vector deviations of the same sub-regions are shown in Table 6.3.

In Table 6.2, the wavelet packet entropy of each measured sub-region is larger than the corresponding modulus sub-region. The reason is that the quality of the curved surface is inferior to the quality of the designed modulus due to the errors in

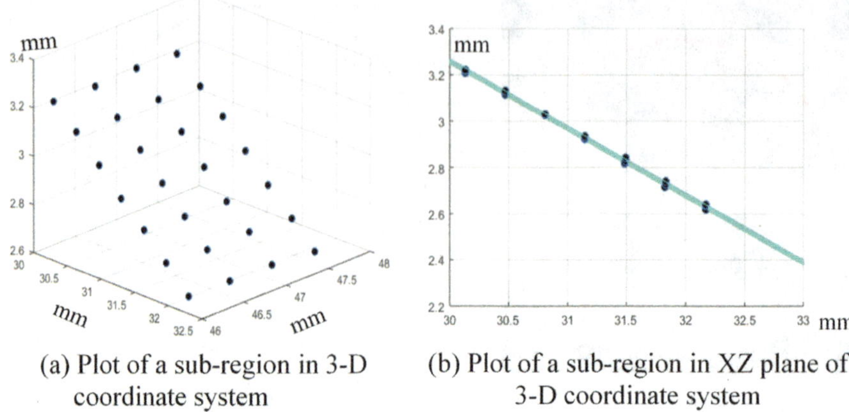

(a) Plot of a sub-region in 3-D
coordinate system

(b) Plot of a sub-region in XZ plane of
3-D coordinate system

Fig. 6.29 Plot of points of a sub-region. Reprinted from Ref. [101], copyright 2018, with permission from Elsevier

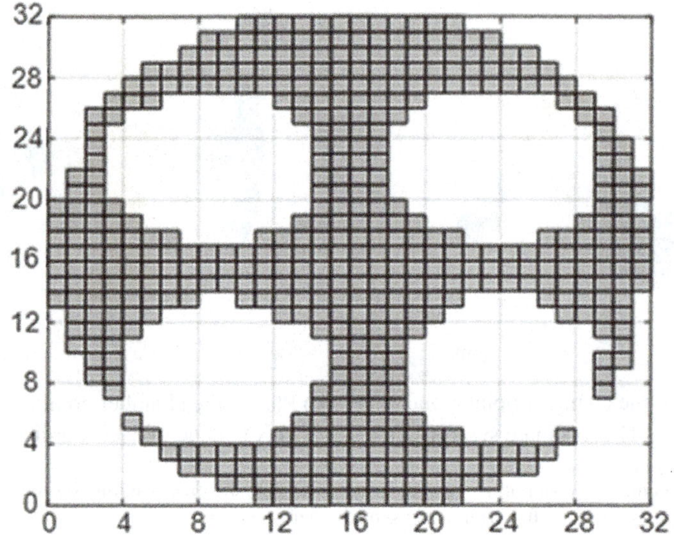

Fig. 6.30 Distribution of effective sub-regions. Reprinted from Ref. [101], copyright 2018, with permission from Elsevier

the manufacturing process. Therefore, it is not reasonable to use the wavelet packet entropies of the modulus sub-regions as the absolute limit to determine whether the sub-regions of the curved surface are OOL. A threshold value should be added to the limit of each sub-region. In this case, the threshold value is set as 0.5. That is, if the wavelet packet entropy difference of the curved surface sub-region and its

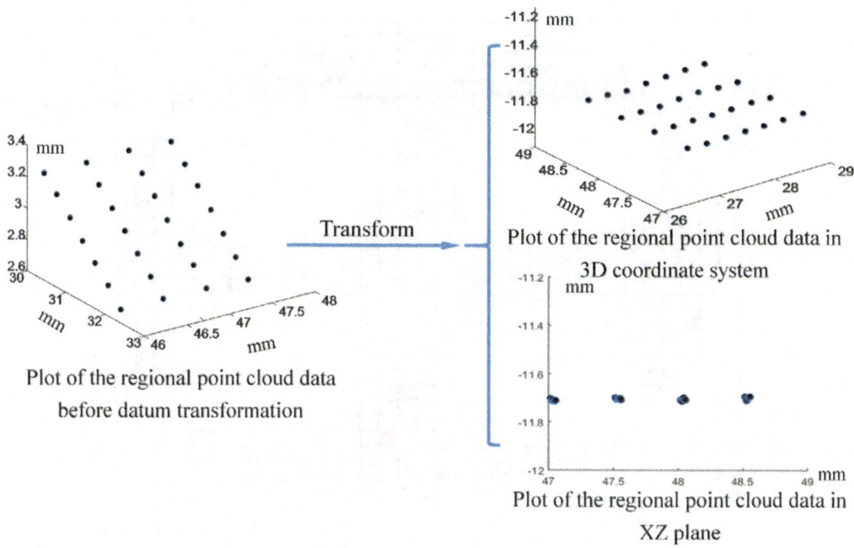

Fig. 6.31 Datum transformation of a sub-region. Reprinted from Ref. [101], copyright 2018, with permission from Elsevier

Table 6.2 Wavelet packet entropies of sub-regions of the curved surface and modulus

Sub-region\Index	1	2	3	4	5	6	7
Measured sub-region	0.0189	0.0169	0.0207	0.0272	0.0234	0.0226	0.0184
Modulus sub-region	0.0177	0.0155	0.0216	0.0240	0.0228	0.0145	0.0135
Sub-region\Index	8	9	10	11	12	13	14
Measured sub-region	0.1085	0.0226	0.0292	0.0254	0.0220	0.0271	0.0229
Modulus sub-region	0.0243	0.0209	0.0256	0.0137	0.0172	0.0135	0.0139

Table 6.3 Normal vector deviations of sub-regions of the curved surface

Index\Sub-region	Measured sub-region	Modulus sub-region	Deviation (degree)
1	(−0.398, −0.004, −1)	(−0.399, −0.003, −1)	0.073
2	(−0.288, −0.008, −1)	(−0.292, −0.011, −1)	0.268
3	(−0.289, −0.007, −1)	(−0.291, −0.007, −1)	0.106
4	(−0.305, −0.002, −1)	(−0.302, −0.003, −1)	0.167
5	(0.004, 0.002, −1)	(0.003, −0.002, −1)	0.237
6	(0.511, −0.021, −1)	(0.509, −0.011, −1)	0.518
7	(0.261, 0.011, −1)	(0.258, 0.009, −1)	0.195
8	(0.341, −0.0003, −1)	(0.342, 0.0007, −1)	0.075
9	(0.272, −0.001, −1)	(0.268, 0.003, −1)	0.308
10	(0.398, −0.014, −1)	(0.369, −0.009, −1)	1.472
11	(−0.280, −0.002, −1)	(−0.283, −0.003, −1)	0.169
12	(−0.302, −0.004, −1)	(−0.298, 0.001, −1)	0.346
13	(−0.013, −0.016, −1)	(−0.010, −0.014, −1)	0.207
14	(0.380, 0.003, −1)	(0.383, 0.005, −1)	0.184

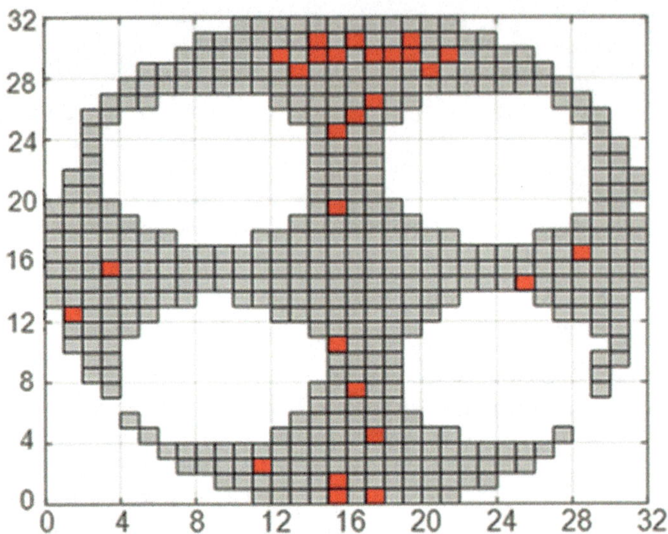

Fig. 6.32 OOL sub-regions in the curved surface. Reprinted from Ref. [101], copyright 2018, with permission from Elsevier

corresponding modulus sub-region is larger than 0.5, the sub-region of the curved surface is considered to be OOL.

In Table 6.2, the 8th sub-region is an OOL sub-region. The limit of deviation degree is set as 0.5 based on engineering experience, and there are two sub-regions (6th and 10th sub-region) are regarded as OOL in Table 6.3. Therefore, the 6th, 8th, and 10th sub-regions are OOL regions of the 14 sub-regions. An example of the curved surface that contains OOL sub-regions is shown in Fig. 6.32 and the OOL sub-regions are marked with red.

(3) Quality parameters calculation of curved surfaces

Once the OOL sub-regions in the curved surface are identified, three quality parameters that represent quality characteristics of the curved surface are quantified by numerical values. For the example shown in Fig. 6.32, the number of OOL sub-regions is 27. Since the values in Figs. 6.32 and 6.33 represent count values, there are no units for the axes in the two figures. Therefore, the first quality parameter is $qc_1 = 27$. The second quality parameter qc_2 is calculated by the clustering algorithm described in Sect. 6.3.2.4, and the clustering result is shown in Fig. 6.33. It can be seen from Fig. 6.33b that the cluster marked with green contains the largest number (the number is 16) of bad sub-regions. Then the second quality parameter is $qc_2 = 16$. The numbers of OOL sub-regions in the four clusters are 16, 7, 2, and 2, respectively. For the curved surface of a chamber, the number of

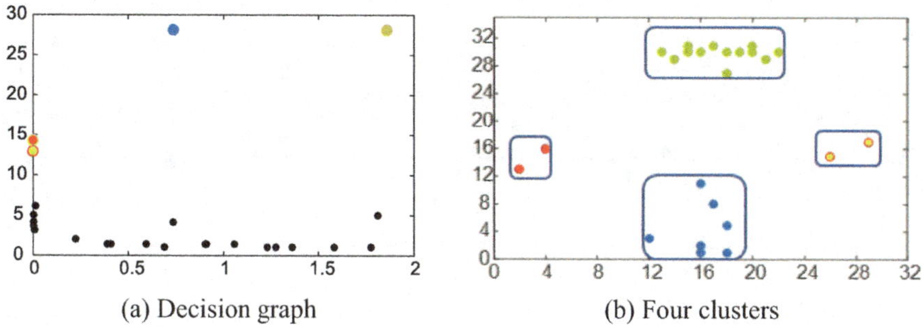

(a) Decision graph   (b) Four clusters

Fig. 6.33 The clustering result of an example of the curved surface. Reprinted from Ref. [101], copyright 2018, with permission from Elsevier

effective sub-regions is about 500 under the region division of 32×32. The threshold s_c is determined to be 3 based on engineering experience, which means that a cluster containing more than 3 OOL sub-regions is considered. In the example, there are two clusters containing more than three OOL sub-regions. Thus, the third quality parameter is $qc_3 = 2$.

(4) Monitoring the quality parameters

In order to monitor the condition of the curved surface of chambers, three individual control charts are conducted to monitor the three quality parameters. In this case, 20 combustion chambers of qualified volumes are randomly sampled with time series and the point cloud data of the 20 curved surfaces is collected. The point cloud data is processed by procedures mentioned in Sects. 6.3.2.2, 6.3.2.3, and 6.3.2.4. The values of the three quality parameters of the 20 curved surfaces are shown in Table 6.4.

The CL, UCL, and LCL of three individual control charts are calculated according to Eqs. (6.26)–(6.28). Then the three individual control charts are conducted and shown in Fig. 6.34.

Table 6.4 The values of three quality parameters of the 20 curved surfaces

Index	1	2	3	4	5	6	7	8	9	10
qc_1	21	19	20	15	18	16	19	21	20	18
qc_2	8	6	11	10	8	9	8	10	5	9
qc_3	2	3	2	1	2	2	1	2	3	2
Index	11	12	13	14	15	16	17	18	19	20
qc_1	14	16	18	21	19	19	15	17	15	14
qc_2	7	8	8	12	6	10	9	8	5	6
qc_3	2	1	2	3	2	1	2	2	3	2

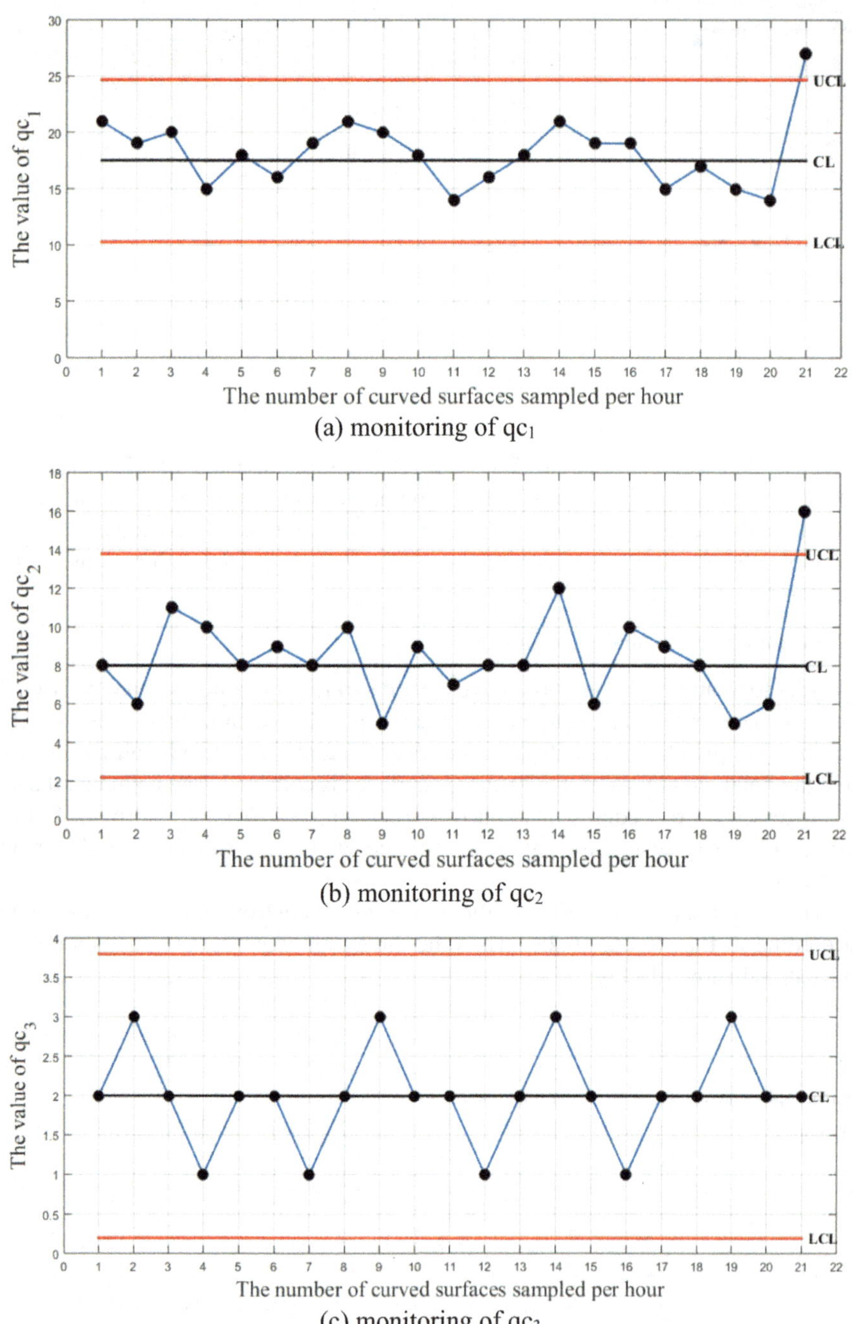

(a) monitoring of qc_1

(b) monitoring of qc_2

(c) monitoring of qc_3

Fig. 6.34 Curved surface defect monitoring. Reprinted from Ref. [101], copyright 2018, with permission from Elsevier

$$\begin{cases} \mathrm{CL}_{qc_1} = \mu_1 = \dfrac{\sum\limits_{i=1}^{20} (qc_1)_i}{20} = 17.5 \\[2em] \mathrm{UCL}_{qc_1} = \mu_1 + 3\sigma_1 = \mu_1 + 3\sqrt{\dfrac{\sum\limits_{i=1}^{20} \left((qc_1)_i - \mu_1\right)^2}{20}} = 17.5 + 3 \times 2.4 = 24.7 \\[2em] \mathrm{LCL}_{qc_1} = \mu_1 - 3\sigma_1 = \mu_1 - 3\sqrt{\dfrac{\sum\limits_{i=1}^{20} \left((qc_1)_i - \mu_1\right)^2}{20}} = 17.5 - 3 \times 2.4 = 10.3 \end{cases} \quad (6.26)$$

$$\begin{cases} \mathrm{CL}_{qc_2} = \mu_2 = \dfrac{\sum\limits_{i=1}^{20} (qc_2)_i}{20} = 8 \\[2em] \mathrm{UCL}_{qc_2} = \mu_2 + 3\sigma_2 = \mu_2 + 3\sqrt{\dfrac{\sum\limits_{i=1}^{20} \left((qc_2)_i - \mu_2\right)^2}{20}} = 8 + 3 \times 1.9 = 13.7 \\[2em] \mathrm{LCL}_{qc_2} = \mu_2 - 3\sigma_2 = \mu_2 - 3\sqrt{\dfrac{\sum\limits_{i=1}^{20} \left((qc_2)_i - \mu_2\right)^2}{20}} = 8 - 3 \times 1.9 = 2.3 \end{cases} \quad (6.27)$$

$$\begin{cases} \mathrm{CL}_{qc_3} = \mu_3 = \dfrac{\sum\limits_{i=1}^{20} (qc_3)_i}{20} = 2 \\[2em] \mathrm{UCL}_{qc_3} = \mu_3 + 3\sigma_3 = \mu_3 + 3\sqrt{\dfrac{\sum\limits_{i=1}^{20} \left((qc_3)_i - \mu_3\right)^2}{20}} = 2 + 3 \times 0.6 = 3.8 \\[2em] \mathrm{LCL}_{qc_3} = \mu_3 - 3\sigma_3 = \mu_3 - 3\sqrt{\dfrac{\sum\limits_{i=1}^{20} \left((qc_3)_i - \mu_3\right)^2}{20}} = 2 - 3 \times 0.6 = 0.2 \end{cases} \quad (6.28)$$

In Fig. 6.34, parameters of the 21st curved surface are $qc_1 = 27$, $qc_2 = 16$, and $qc_3 = 2$. It can be seen that qc_1 and qc_2 are both out of the control range, and qc_3 is in the control range. Therefore, the manufacturing process of the curved surface is OOC. If there appears continuous OOC manufacturing process of curved surfaces, the correction should be conducted. For the bad sub-regions of curved surfaces of combustion chambers in this case, the main reason is that the mold of the chamber is worn and mold maintenance is needed.

In order to evaluate the detection accuracy of the proposed method, ten OOC conditions are recorded, and the entire mold is inspected to find the bad sub-regions. Assume that the number of bad sub-regions of the mold is U_1, the number of the same bad regions of the mold and the curved surface is U_2. Then the detective accuracy is calculated by Eq. (6.29), and the detective accuracy of the proposed method is shown in Table 6.5.

$$\text{Detective accuracy} = \frac{U_1}{U_2} \times 100\% \quad (6.29)$$

According to Table 6.5, the average detective accuracy of the proposed method is 83.95%. The detective accuracy and average detective accuracy can satisfy the

Table 6.5 The detective accuracy of the proposed method

Item\Index	1	2	3	4	5	6	7	8	9	10
U_1	27	30	28	32	30	35	30	31	25	36
U_2	24	25	22	25	24	30	25	28	22	30
Detective accuracy (%)	88.9	83.3	78.6	78.1	80.0	85.7	83.3	90.3	88.0	83.3

requirements of mold maintenance. It is not necessary to inspect the entire mold since the defective sub-regions of the mold can be identified by the OOL sub-regions of the curved surfaces. The locations of OOL sub-regions on the curved surface correspond to the locations of the mold to be inspected. Therefore, the mold maintenance only needs to focus on checking the OOL sub-regions with fixed locations.

(5) Comparison of the profile monitoring method based on Q–Q plot

In order to validate the performance of the proposed method, the profile monitoring technique based on Q–Q plots [15] is used for comparison. For a process where a measured point cloud data is collected as a distribution of points, in-control point cloud data will produce highly linear Q–Q plot while out-of-control point clouds will result in a Q–Q plot deviating from linearity. Wang et al. [14] pointed out that this relationship can be monitored as a linear profile characterized by its y-intercept and slope, which can be monitored by two EWMA control charts. In this case, the data for Q–Q plot is the deviation of Z-coordinate values of corresponding points (coordinate values of (x, y) are the same) between the measured point cloud data and the modulus point cloud data. An example of Q–Q plot for a curved surface in in-control manufacturing process is shown in Fig. 6.35. The monitoring of defects

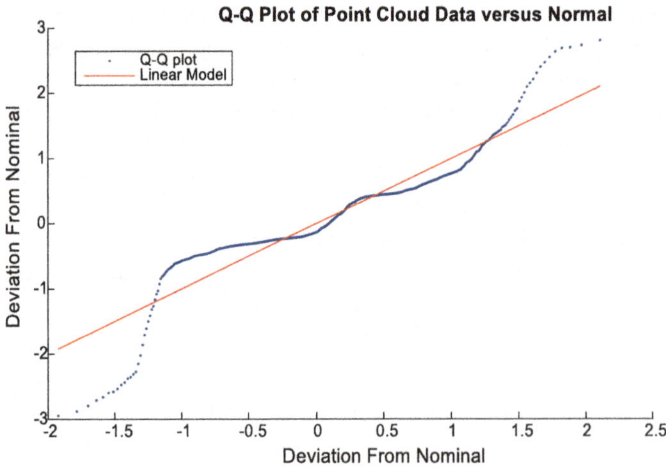

Fig. 6.35 Q–Q plot and fitted linear model for a curved surface in in-control manufacturing process. Reprinted from Ref. [101], copyright 2018, with permission from Elsevier

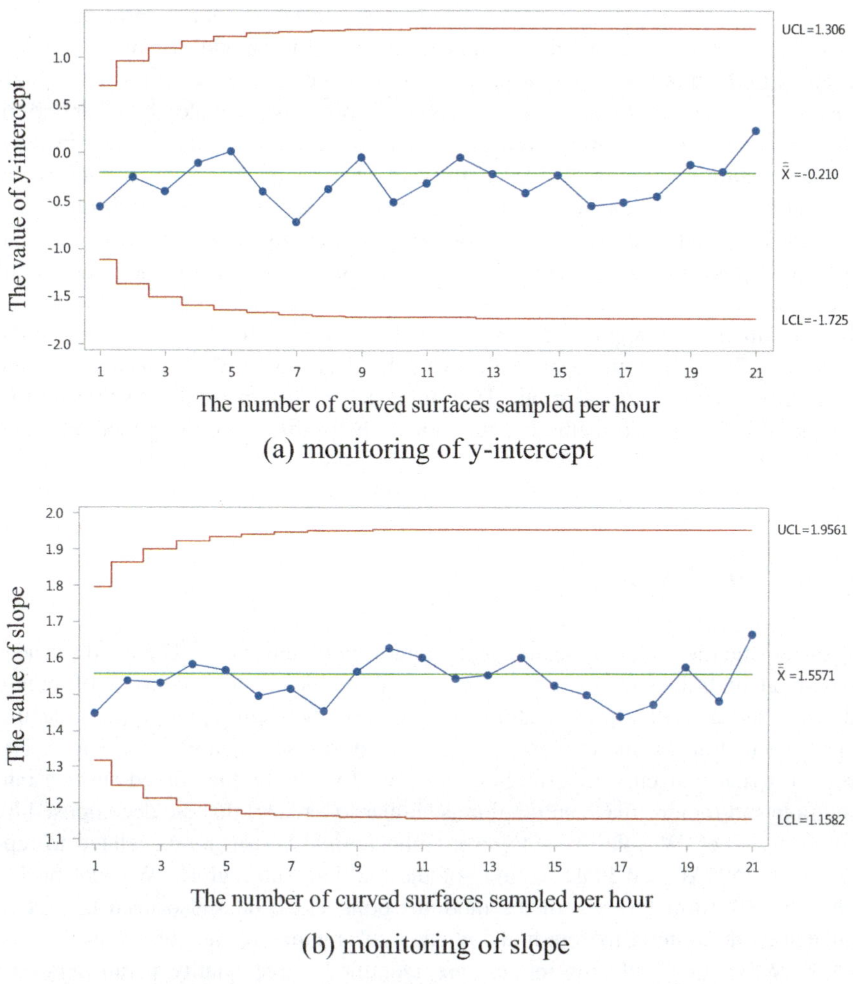

(a) monitoring of y-intercept

(b) monitoring of slope

Fig. 6.36 Two EWMA control charts for monitoring y-intercept and slope. Reprinted from Ref. [101], copyright 2018, with permission from Elsevier

on three-dimensional curved surfaces is transforming into monitoring parameters (y-intercept and slope) of the fitted linear model. Two EWMA control charts for monitoring intercept and slope are shown in Fig. 6.36.

It can be seen from Fig. 6.36 that the two parameters are within the control limits for the point cloud data of the 21 sampled curved surfaces, which indicates that the manufacturing process of the chambers is in control. However, the manufacturing process of the chambers is out of control from the 21st curved surface according to the result of the proposed method. It means that the monitoring method based on Q–Q plot are not able to identify OOC process in time when compared to the proposed method. The reason is that the positions of the measured points (represented by coordinate (x, y)) for different chambers may not be same due to the positioning error in measurement, and the number of the modulus point cloud data is equal to the number of the measured point cloud data. The corresponding points between the measured point cloud data and the modulus point cloud data are not in the same position (coordinate values of (x, y) are different), which results in inaccurate monitoring results. However, the curved areas of the modulus point cloud data and measured point cloud data are the same for sub-region division of the proposed method, and the divided sub-regions of the modulus point cloud data and measured point cloud data are one-to-one correspondence. Thus, the positions of the measured points have no effect on the monitoring result of the proposed method. Besides, there is no appropriate method to detect the location of the defects on the curved surface, since the spatial aspect of the data is lost when transforming the point cloud data into Q–Q plots. Therefore, the monitoring and detecting performance of the proposed method is superior to the profile monitoring method based on Q–Q plot.

6.3.4 Conclusions

A systematic method for detection and monitoring of defects on 3D curved surfaces based on high-density point cloud data is developed, which consists of region division of curved surfaces, feature evaluation of each sub-region, quality parameters calculation of curved surfaces, and monitoring the quality parameters. The region division of curved surfaces is conducted to divide the curved surface into multiple sub-regions (each sub-region is similar to a plane) that are decomposed by wavelet packet. Wavelet packet entropy and normal vector are calculated to represent the NDAP and PDD features of the multiple sub-regions. Wavelet packet entropy and normal vector of the modulus point cloud data also need to be calculated as the criteria to determine whether sub-regions of the curved surface are OOL. When the OOL sub-regions are identified, three quality parameters that represent quality characteristics of the curved surface are calculated based on the clusters of the OOL sub-regions. Then, three individual control charts are proposed to monitor the three quality parameters. As long as any quality parameter is out of the control range, the manufacturing process of the curved surface is determined to be OOC. A case study of curved surfaces of cylinder head combustion chambers sampled in time series is conducted to validate the performance of monitoring and detection of defects on curved surfaces. The results demonstrate that the proposed method can identify the OOC manufacturing process of 3D curved surfaces and accurately detect the locations of the OOL sub-regions form the curved surfaces in OOC manufacturing process. The comparison with other monitoring methods (for example, profile monitoring based on Q–Q plots) for high-density point cloud data of 3D curved surfaces shows that the proposed method is more reliable when positioning error in measurement exists and shows good performance of identifying locations of defects.

6.4 Leakage Monitoring in Static Sealing Interface Based on Three-Dimensional Surface Topography Indicator

6.4.1 Introduction

Leakage is an important issue of concern in the manufacturing industry. When leakage occurs, it is a huge danger which can cause resources waste, product quality decline, equipment damage, and environmental pollution, thereby brings safety accidents and economic losses. Therefore, sealing technology is developed to prevent leakage. A seal is a widely used device for closing a gap and making a joint tight [68]. In many engineering fields, seals play a crucial role to achieve quality and reliability. There are two types of seals: static seal and dynamic seal. Static seal is performed by direct surface to surface contact. Comparing with dynamic seal, static seal is more common in engineering. Direct contact, rubber, and gasket seal are three seal forms corresponding to the seal requirement from low to high, respectively. However, a large number of practices have proved that leakage still exists with the seals in some industrial applications. Therefore, identifying the leakage mechanism and developing the leakage monitoring approach are essential to pre-control the machining process and ensure the product quality in satisfactory condition and reduce the losses.

Numerous studies were conducted in order to understand sealing and especially leakage. Persson et al. [69] had proposed a critical-junction theory of the leak rate of seals, which was based on percolation theory [70] and Persson contact theory [71]. Soon afterward, the critical-junction theory was verified effectively by a series of researches containing theory comparison and experiment investigation [72, 73]. Bottiglione et al. presented a theoretical approach to estimate the fluid leakage mechanism in flat seals by making use of percolation theory and theory of contact mechanics [74, 75]. Marie et al. described an experimental study to characterize fluid leakage through a rough metal contact [76]. From these studies, it is clear that contact pressure, surface topography, and the material property of sealing element are three main factors in static seal. Clearly, contact pressure and sealing element are easy to be controlled, but the full control of surface topography during manufacturing processes is still out of reach. With the limitation of surface characterization and measurement technology, a thorough characterization of surface topography which impacts on the contact efficiency and leakage paths generation is still at the beginning. Therefore, the performance of surface topography on the rate of leakage through seals is the focus of researches among these features.

Due to the existence of roughness and multi-scale on the surface, it is not a simple matter to know the impact of surface topography on leakage in detail. Waviness Motifs based model was proposed by Robbe-Valloire et al. to reveal that both the amplitude and the valley to the peak distribution of surface were influential [77]. Okada et al. investigated the quantitative effect of surface profiles on leakage using surface activated bonding technique [78]. Haruyama et al. developed a metal gasket model with different surface roughness levels to evaluate the sealing performance through simulation study [79]. Based on experimental observations, Marie et al. proved that surface components at the intermediate scale, which is corresponding to waviness, were of major concern for leakage [80]. In their further study, the modal content of surface components was employed to explore the role on leakage of static flat seal [81]. All these studies indicate the direct relationship between surface topography and leakage; thus, surface topography can be considered as an indicator for leakage monitoring.

In general, the vast majority of leakage is detected with a whole part by leak testing machine in the last manufacturing process of production line, and a large number of leakage problems appear during the using of the product. Consequently, it is essential to develop the pre-control measures and leakage monitoring methods. Since the complex leakage mechanism and inconvenient measurement, the leakage monitoring research is less. As a pioneer, Malburg first studied the relationships between the two-dimensional surface topography and the sealing performance. It is one of the important approaches to investigate the sealing problem and leakage monitoring. Malburg adopted the two-dimensional surface waviness profile parameters to monitor the sealing performance based on a morphological-closing filter [82]. Liao et al. pointed the presence of significant middle wavelengths (waviness) would result in leakage, and the tooling marks with the large peak-to-peak variation on the surface were considered as the leakage path [83]. From the data-driven point, Ren et al. first proposed a novel method of modeling and diagnosis of leak areas for surface assembly. A lattice graph model and a color tracking approach were developed to predict potential leak areas and paths in between assembled surfaces [84]. Arghavani et al. proposed a fuzzy logic model to predict sealing performance and leakage rate of gasketed flange joints using inputs including surface roughness and gas pressure [85] Li et al. presented a leakage prediction calculation method for static seal rings in underground equipment [86]. However, these studies neither explain the surface parameters indicator from the leakage mechanism, nor give the threshold when leakage occurs. Nevertheless, surface topography is still the most suitable indicator, which can provide reliable and detectable information for leakage monitoring.

Recently, an advancement of noncontact laser holographic interferometry measurement, called high-definition metrology (HDM), which can generate a surface height map of millions of data points within seconds for 3D inspection of an entire surface has been developed [39]. Figure 6.37 shows an engine block surface measured by HDM, there are about 0.8×10^6 data points generated to cover an area of 320 mm $\times$ 160 mm with 150 µm lateral resolution in x–y-direction and 1 µm accuracy in z-direction. With the precision measurement, a preprocessing method is used to convert the mass data points into a height-encoded and position-maintained gray image to represent the entire surface [41]. The 3D surface topography of the entire surface examined by HDM presents a novel platform, several researches based on HDM such as 3D surface topography evaluation [41, 56], filtering [53, 58, 87], classification [48, 55], forecasting [88] and tool wear monitoring methods [57] have been explored for the pre-control of the manufacturing process.

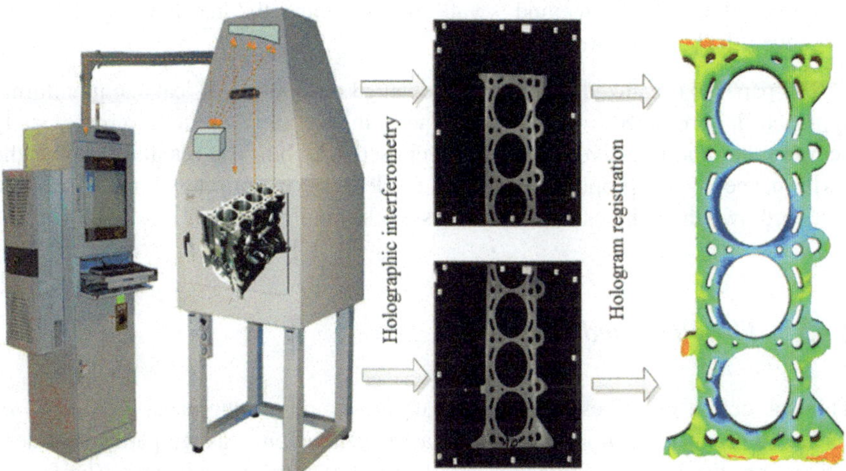

Fig. 6.37 Measurement by HDM. Reprinted from Ref. [102], copyright 2018, with permission from ASME

In the meantime, the full and precision measurement also makes it more possible to monitor leakage condition using 3D surface topography indicator. Malburg pointed out that surface components including form and roughness could be tolerated, but the presence of waviness was highly significant in the static sealing interface. The relationship between surface waviness profile and leakage potential was qualitatively described by several numerical parameters. However, the research only focuses on two-dimensional surface topography and it lacks quantitative analytical information. Meanwhile, the explanation of the relationship between surface parameters and leakage mechanism is absent. Therefore, benefiting from the development of measurement, the research focus is extended from two-dimensional surface topography to three-dimensional surface topography. The main contribution of this paper is to present a new leakage monitoring approach using 3D surface topography indicator based on some conclusions of Malburg's research [82].

- The measured surfaces are first separated into different frequency components including form, waviness, and roughness using spline filter. Then, a virtual gasket is generated by the morphological filtering on the waviness.
- A series of novel surface leakage parameters, which include contact area percentage, void volume, and relative void volume, are defined as indicators for the characteristics of the leakage.
- Meantime, based on the Persson contact mechanics and percolation theory [69–71], the threshold of leakage parameter is found using finite element modeling (FEM).
- Finally, a classical control chart is adopted to monitor the leakage surface of the successive machining process. Results of the engineering case indicate that the

proposed monitoring method is valid to pre-control the machining process and prevent leakage occurring.

The paper is organized as follows: a detailed description of leakage monitoring approach is presented in Sect. 6.2. Then, in the next section, a case study demonstrates the effectiveness of the proposed method. The results illustrate the performance of the proposed approach for leakage monitoring. Finally, the last section draws the conclusions and discusses the future research.

6.4.2 The Proposed Method

This subsection describes an overview of the proposed approach. It consists of surface components extraction, virtual gasket generation, leakage parameters definition, and threshold determination and leakage condition monitoring. The framework is shown in Fig. 6.38. The procedure involves the following steps.

Step 1: Surface components extraction. HDM is employed to measure the engineering surface and generate millions of points. A converted gray image which can represent the entire surface is gained by the points cloud [41]. Then, the high-resolution measured surface is filtered to extract form, waviness, and roughness using spline filter [89], which is the ISO accepted linear filter.

Step 2: Virtual gasket generation. A novel concept called virtual gasket is proposed to simulate the actual situation of contact and deformation. The two-dimensional (2D) and three-dimensional (3D) virtual gasket are generated by 2D and 3D morphological-closing filter based on the waviness profile and surface, respectively.

Step 3: Leakage parameters definition. Three areal leakage parameters CAP (contact area percentage), VV (void volume) and SWvoid (relative void volume) are defined to represent the characteristics of the leakage.

Step 4: Threshold determination and monitoring the leakage condition of the successive machining process. The threshold of leakage parameters is used to determine whether the leakage occurs or not. Based on the Persson contact mechanics and percolation theory, the threshold of leakage parameter is found using FEM analysis. If the value of leakage parameter of a surface region goes beyond the threshold, this surface region is out of limit (OOL), corresponding to leakage area. Subsequently, an individual control chart is used to clearly monitor the leakage conditions of the successive machining process.

6.4.2.1 Surface Components Extraction

Surface topography has profound influences on the part quality as it plays two vital roles: one is to control the manufacturing process and the other is to help functional prediction. In order to clarify the effect mechanism, filtration is mainly employed in

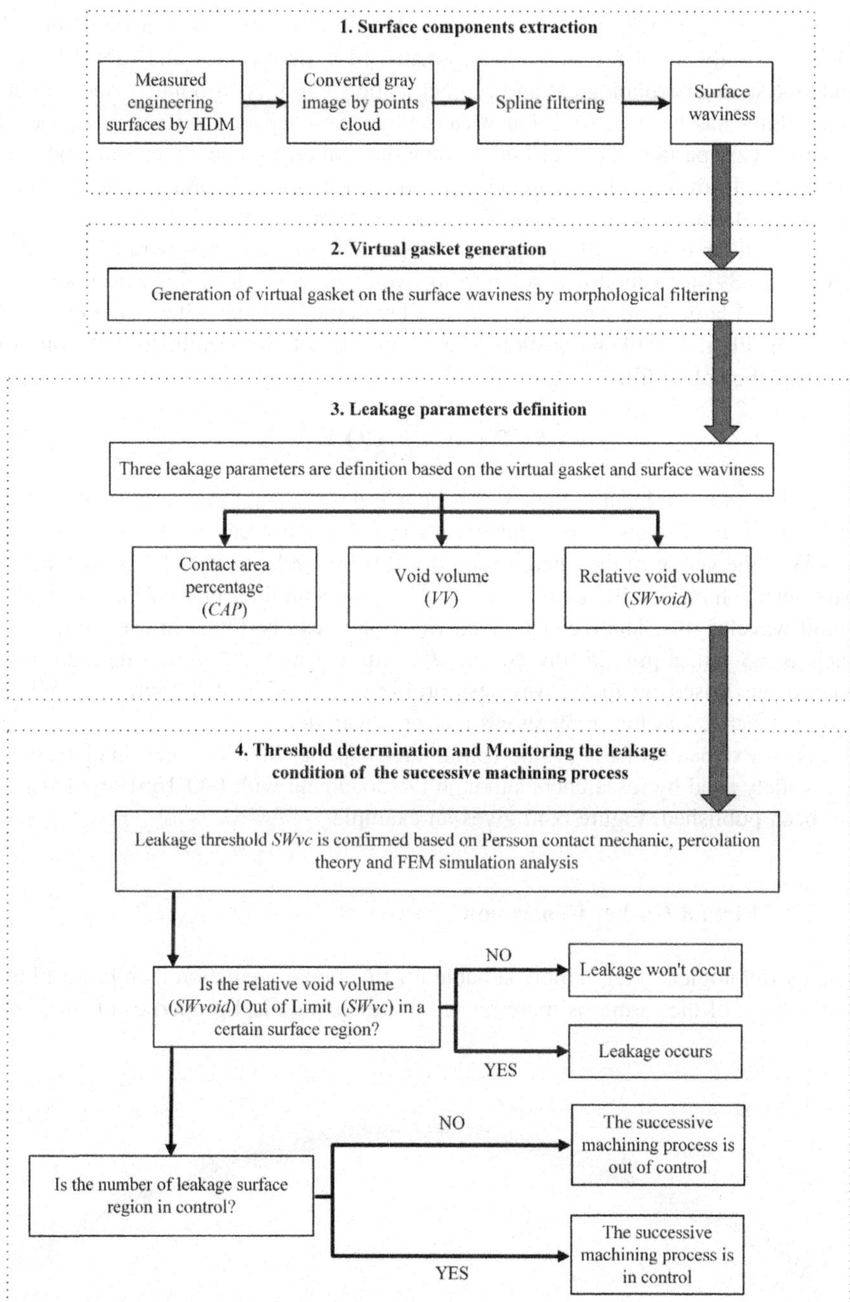

Fig. 6.38 The framework of the proposed approach. Reprinted from Ref. [102], copyright 2018, with permission from ASME

surface metrology, which has been clearly described and specified in ISO 16610-1 [90]. Generally, there are two filters systems: M-system (linear and robust filters) and E-system (morphological and segmentation filters). As a popular linear filter, spline filter has been proved that it can extract and separate surface components exactly. Comparing with classical Gaussian filter, spline filter incorporates improved "form following" capabilities and reduces much boundary effect, which mean that it can be used as a good surface filter for feature extraction.

Spline filter is originally proposed by Krystek [91] and incorporated into ISO 16610-22 [89] as a modified solution to overcome edge distortion and poor performance of large form which are associated with the Gaussian filter. Different from Gaussian filter, the filter equation is used instead of the weighting functions to describe the spline filter:

$$[\mathbf{I} + \beta\alpha^2\mathbf{P} + (1-\beta)\alpha^4\mathbf{Q}]\mathbf{W} = \mathbf{Z} \tag{6.30}$$

where $\mathbf{I}$ is an $n \times n$ identity matrix, $\mathbf{P}$ is an $n \times n$ tridiagonal symmetric matrix, and $\mathbf{Q}$ is an $n \times n$ five-diagonal symmetric matrix. $\mathbf{Z}$ is the vector of the original data, and $\mathbf{W}$ is the vector of the filtered data. $\alpha = 1/(2\sin(\pi\Delta x/\lambda_c))$ and β is the tension parameter which lies between 0 and 1. Δx is the sampling interval and λ_c is the cutoff wavelengths. The recommended value of λ_c can be found in ISO 16610-22, such as 2.5 μm, 8 μm, 25 μm, 80 μm, 250 μm, 0.8 mm, 2.5 mm, 8 mm, 25 mm, and so on. Based on these, waviness profile can be separated from the original profile exactly, and Fig. 6.39 shows a clear illustration.

As an extension of the profile spline, areas spline filter has been implemented and widely used by researchers although the document with ISO 16610-62 has not yet been published. Figure 6.40 gives an example.

6.4.2.2 Virtual Gasket Generation

The morphological filter is more suitable for functional prediction than linear filter, as the logic of the former is more related to the geometrical properties of surfaces.

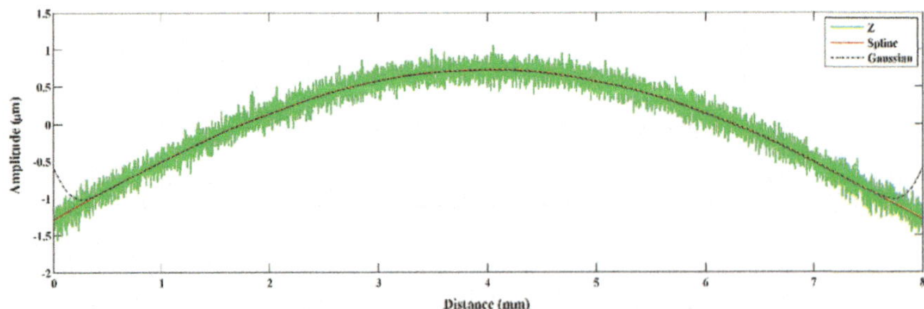

Fig. 6.39 Profile spline filtering. Reprinted from Ref. [102], copyright 2018, with permission from ASME

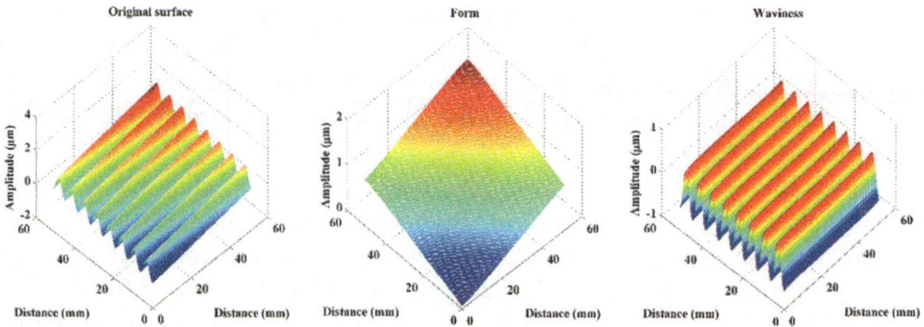

Fig. 6.40   Area spline filtering. Reprinted from Ref. [102], copyright 2018, with permission from ASME

It plays an essential role in understanding static seal interfaces as the various wavelength domains affect the leakage in different ways. Morphological filters were first presented by Magaros and Schafer under the framework of mathematical morphology in 1987 [92], and subsequently they were developed as powerful image processing tools for pattern recognition purposes in various engineering applications. Morphological filters are nonlinear signal transformations that locally modify geometric features of the signal, and they have been accepted in ISO 16610-40 as a part of standard filtration techniques [93].

The natural concept of morphological filter is to perform morphological operations on the surface with structuring elements. There are four basic operations: erosion, dilation, opening, and closing, which are listed as follows:

Erosion:

$$(f \ominus g^s)(x) = \inf_{y \in D \cap G} \{f(y) - g(y - x)\} \tag{6.31}$$

Dilation:

$$(f \oplus g^s)(x) = \sup_{y \in D \cap G} \{f(y) + g(y - x)\} \tag{6.32}$$

Opening:

$$f \circ g(x) = [(f \ominus g^s) \oplus g](x) \tag{6.33}$$

Closing:

$$f \bullet g(x) = [(f \oplus g^s) \ominus g](x) \tag{6.34}$$

where the input surface function $f(x)$ and structuring element $g(x)$ are defined in $D \subset \mathbf{R}^n$ and $G \subset \mathbf{R}^n$, respectively, and $g^s(x) = g(-x)$. The erosion operation decreases the peaks and expands the minima of $f(x)$, while the dilation operation gains the opposite result. The opening and closing operations are both used to smooth $f(x)$, the former one is cutting down its peaks from below, and the other is filling up its valleys form above. According to the monotone increasing and idempotence of opening and closing operations, there are two kinds of morphological filters, namely, opening filter and closing filter in geometrical product specifications (GPS) [93].

The structuring element $g(x)$ is another sticking factor affecting the results of morphological filter. Various kinds of structuring elements have been proposed, whereby circular and horizontal line segments for profile and spherical and horizontal planar segments for surface are recommended by ISO 16610-41 and 16610-85 [94, 95]. According to Malburg's research [82], the circular structuring element is chosen to process the engineering surface profile, which is the better one based on its joint properties on the surface profile component. So in this research, spherical structuring element is employed as an extension for areal surface. Moreover, the radius of the structuring element is equally important to the filter results. Considering the actual part's compressive properties (i.e., its ability to "fill gaps") as well as its bending properties, the radius can be confirmed from the recommended values 1 μm, 2 μm, 5 μm, 10 μm, 20 μm, 50 μm, 100 μm, 200 μm, 500 μm, 1 mm, 2 mm, 5 mm, and so on [94].

As the abovementioned, the gasket is one of the main seal forms to prevent leakage. To some extent, a gasket can conform to the surface when sealing a surface. Given this property of the gasket, it makes more sense to take note of the gaps between the gasket and the surface features than the peak-to-valley height of the surface. However, due to the limits of measurement, it is difficult and inconvenient to get the real gasket topography and the deformation is unknown in practice. Thus, a new concept named virtual gasket is proposed to simulate the actual situation of contact and deformation. The virtual gasket is derived from the result of a morphological-closing filter, which is similar to rolling a disk over the waviness profile or a ball over the waviness surface. As shown in Figs. 6.41 and 6.42, virtual gaskets for profile and surface are generated, respectively, in 2D and 3D view.

6.4.2.3 Leakage Parameters Definition

Once the virtual gasket is determined, the next critical process is to quantitatively describe the difference between the virtual gasket and the underlying waviness, which is related to leakage. Malburg first formulated the surface leakage potential

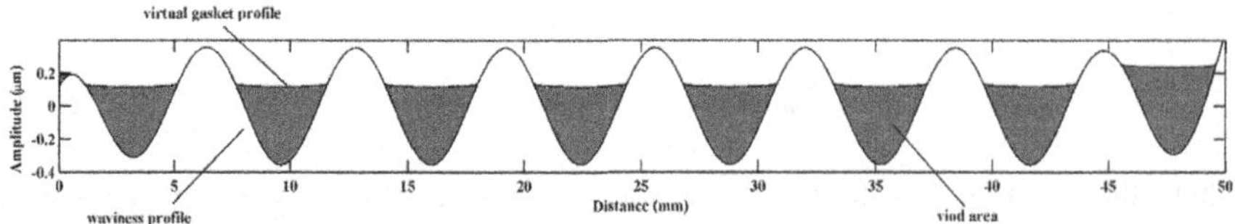

Fig. 6.41 Virtual gasket profile Reprinted from Ref. [102], copyright 2018, with permission from ASME

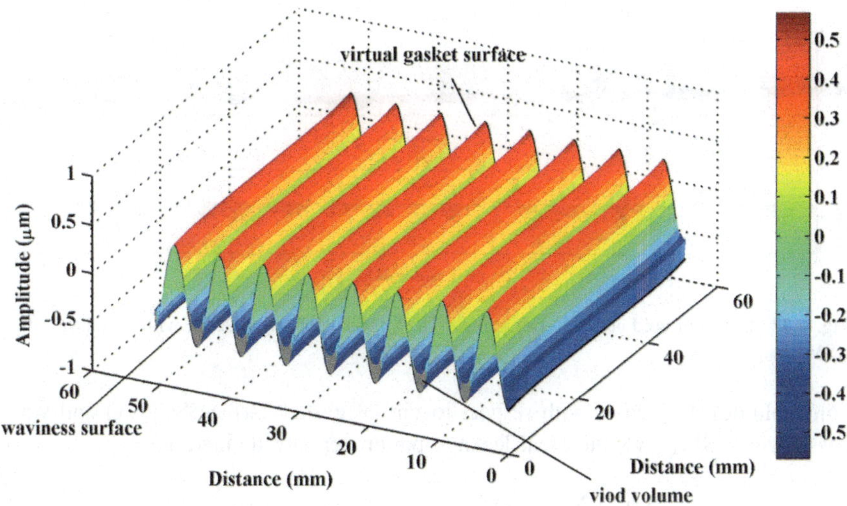

Fig. 6.42 virtual gasket surface. Reprinted from Ref. [102], copyright 2018, with permission from ASME

by using several numerical parameters, such as CLP (contact length percentage) and VA (void area). The contact length percentage is the ratio of the number of contact points between the virtual gasket profile $C(x)$ and waviness profile $W(x)$ to the total number of points, which can be used to assess the load distribution. The void area is the enclosed area between the virtual gasket profile $C(x)$ and waviness profile $W(x)$. It is considered as the related area that may lead to leakage. To be specific, the profile leakage parameters are defined as

$$\text{CLP} = \sum_l [if(C_i(x) = W_i(x))1else0]/l \tag{6.35}$$

$$\text{VA} = \int_0^l (C(x) - W(x))\mathrm{d}x \tag{6.36}$$

where l is the nominal length of the surface profile. It is noted that CLP and VA are relied upon the length l, which are not flexible. So the normalized parameter relative void area Wvoid is preferred to describe a void area per unit length, and it is given by the equation of Wvoid $= \text{VA}/l$.

Benefitting from the development of surface metrology, areal surface parameters have been fully described in ISO 25178-2 [96]. Following the expansion mode of areal surface parameters, an attempt is made to extend the application from 2D surface profile to 3D surface topography, as well as the corresponding areal leakage parameters CAP (contact area percentage), VV (void volume) and SWvoid (relative

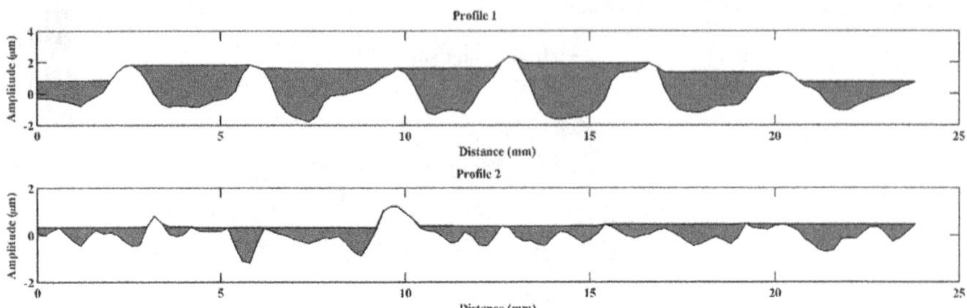

Fig. 6.43 Different void areas. Reprinted from Ref. [102], copyright 2018, with permission from ASME

void volume). Likewise, with respect to virtual gasket surface $SC(x, y)$ and waviness surface $SW(x, y)$, the areal leakage parameters are defined as

$$CAP = \sum_{S} [if(SC_i(x, y) = SW_i(x, y))\, 1\ \text{else}\ 0]/S \qquad (6.37)$$

$$VV = \iint_S (SC(x, y) - SW(x, y))dxdy \qquad (6.38)$$

$$SWvoid = VV/S \qquad (6.39)$$

where S is the nominal area of surface topography. Analogously, CAP and VV are relied upon the nominal area S, which are neither flexible. However, the normalized parameter relative void volume SWvoid is independent of the area of the surface, which is preferred to describe a void area per unit length clearly. Therefore, SWvoid is first selected as the leakage indicator to report the leakage area.

With regard to leakage, traditional surface height parameters [96, 97] like W_a (mean height of waviness profile), W_t (total height of waviness profile), S_a (mean height of waviness surface), and S_z (maximum height of waviness surface) cannot determine whether the leakage occurs or not effectively. The following waviness profiles, as shown in Fig. 6.43, have almost the same W_a and W_t, but the generated virtual gaskets are totally different. Obviously, the leakage parameter Wvoid which reports the leakage area can clearly distinguish them. Analogously, the similar performance of 3D surface by SWvoid is graphically depicted in Fig. 6.44.

6.4.2.4   Threshold Determination and Monitoring the Leakage Condition of the Successive Machining Process

Threshold Determination

Based on the proposed leakage parameter (SWvoid), the key problem of leakage monitoring is to find out the threshold of it, which can be named as SWvc.

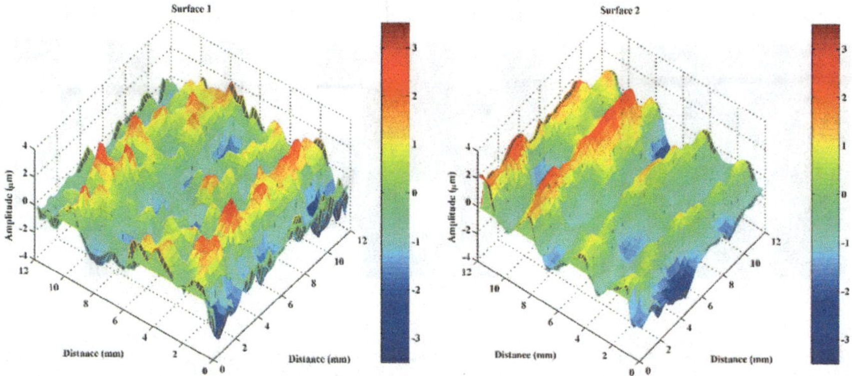

Fig. 6.44 Different void volumes. Reprinted from Ref. [102], copyright 2018, with permission from ASME

The threshold means a limit of leakage; when the value of SWvoid in a surface region goes beyond SWvc, it can be considered as a leakage region. That is to say, this region is out-of-limit (OOL) area on this surface, where the leakage occurs probably. It is noted that the threshold is primarily up to the natural attributes of the part and operating condition, such as material and pressure. As a pioneer on leakage mechanism research, Persson et al. had published a series of research including contact mechanics, leak-rate theory, and the factors affecting leakage. Percolation threshold was first proposed to be regarded as a flag of the occurrence of leakage [71]. Subsequently, the critical-junction theory was extended to present the quantitative relationship [69].

Assume that the nominal contact region between the gasket and the counter surface is rectangular with area $Lx \times Ly$ (see Fig. 6.45, the black color means total contact). The high-pressure fluid and low-pressure fluid are in $x < 0$ region and $x > Lx$ region, respectively. Furthermore, assume that number $N = Ly/Lx$, side $Lx = L$ and area $A_0 = L^2$, N is an integer which does not affect the final results.

In order to understand easily, the study focuses on the contact between the two surfaces within one of the squares as changing the magnification ζ. The magnification ζ is defined as $\zeta = L/\lambda$, where λ is the resolution corresponding to surface scale. As is shown in Fig. 6.46, the apparent contact area $A(\zeta)$ between the two

Fig. 6.45 Contact region. Reprinted from Ref. [102], copyright 2018, with permission from ASME

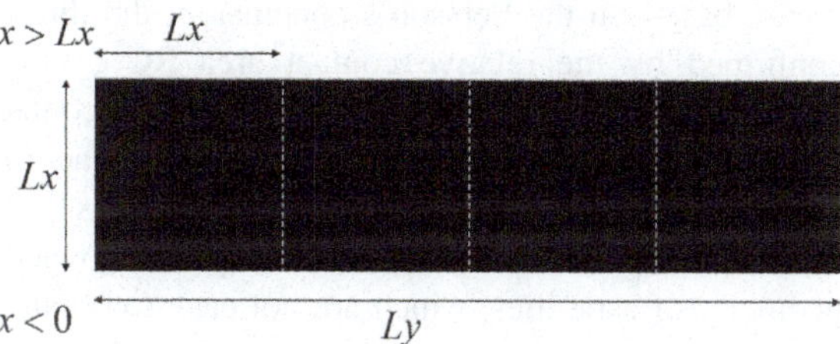

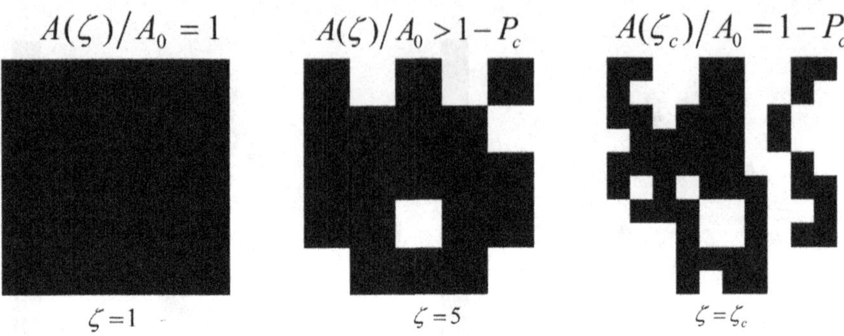

Fig. 6.46 Evolution of the apparent contact area (the black means total contact and the white means no contact). Reprinted from Ref. [102], copyright 2018, with permission from ASME

surfaces is widely divergent with different magnification ζ, and the relative contact area $RC(\zeta) = A(\zeta)/A_0$. Depending on whether the two surfaces contact or not, each square of the lattice which represents the contact area can be black or white (the black means total contact and the white means no contact). When magnification $\zeta = 1$, any surface roughness cannot be observed and the contact appears to be complete, that is $RC(\zeta) = 1$, $A(1) = A_0$. As the magnification ζ increases, some interfacial roughness appears and the apparent contact area $A(\zeta)$ accordingly decreases. Once magnification ζ is high enough, say $\zeta = \zeta_c$, a percolating channel of noncontact area is eventually formed and leakage occurs (see Fig. 6.46c). Instead of determining the critical magnification ζ_c, the relative contact area at this point is given by site percolation theory. Thus, the percolation probability $P(\zeta) = 1 - RC(\zeta) = 1 - A(\zeta)/A_0$ can directly reflect the relative contact area, and the corresponding $P_c = P(\zeta_c) = 1 - RC(\zeta_c) = 1 - A(\zeta_c)/A_0$, where P_c is the so-called percolation threshold [70]. For infinite-sized systems, the percolation threshold P_c is about 0.593 for a square lattice and 0.696 for a hexagonal lattice [70]. For finite-sized systems, the percolation threshold will fluctuate between different realizations of the same physical system. Through molecular dynamics results and experimental verification, Persson et al. [69] pointed that when two elastic solids with randomly rough surfaces were squeezed together, as a function of increasing magnification or decreasing squeezing pressure, a noncontact channel would percolate when the relative contact area $RC(\zeta_c)$ was of the order 0.4, in accordance with percolation theory. That is to say, when it meets the condition of $P_c \approx 0.6$, $RC(\zeta_c) = A(\zeta_c)/A_0 \approx 0.4$, leakage occurs.

So, based on the Persson's conclusion, the threshold SWvc of SWvoid can be confirmed by the relative contact area $RC(\zeta_c) \approx 0.4$ under a certain pressure. Traditionally, the solutions of rough surface contact can be classified into three categories: statistical model, multi-asperity contact model, and deterministic contact model. Many works have shown that there is little difference among the three models, but the former two need more prior knowledge such as the distribution and geometry of asperities, which are not easy to obtain. Fortunately, the deterministic model is based on the directly measured surface data, which reserves the surface geometry as complete as possible. Finite element analysis is a representative and widely used tool in solving deterministic model of rough surface contact [98, 99]. Hence, the relative contact area is obtained by finite element model in case study section, and the threshold SWvc can be confirmed subsequently. But determining the value of RC depends on the results of finite element model which is very time-consuming and is not suitable for engineering practice applications. In order to meet the takt time requirement in the practice applications, SWvoid can be easily calculated and used as the leakage indicator.

Monitoring the Leakage Condition of the Successive Machining Process

Generally, leakage is easily caused by the poorly machined parts, but it can be detected only during later product assembly and the costly and wasteful pressure testing, so that it does not prevent additional parts from being poorly machined. Hence, it is possible to develop monitoring methods of leakage potential prior to product assembly so as to avoid wasted costs, decrease scrap, and enable quick adjustment and control of the machining process before the additional cost of product assembly is incurred.

With a large amount of statistical quality control applications appearing in the industrial field, control chart is a powerful tool for online process monitoring or surveillance. The control chart is a device for describing in a precise manner what is meant by statistical control. That is, sample data is collected and used to construct the control chart, and if the sample values fall within the control limits and do not exhibit any systematic pattern, then the process is in control at the level indicated by the chart. To some extent, a surface which exists leakage is considered as a defective or nonconforming product. Therefore, in monitoring phase, the individual control chart is used to monitor the successive machining process.

As mentioned above, when the leakage parameter SWvoid in a surface region goes beyond the threshold SWvc, this region is out-of-limit (OOL) area on this surface, namely, a leakage region. Assume that n regions are selected in a surface, and the SWvoid of these regions is calculated as $\mathrm{SWvoid}_i (i = 1, \ldots, n)$. Once the parameters SWvoid_i and SWvc are collected, the number of OOL c is defined as

$$c = \sum_{i=1}^{n} k_i \tag{6.40}$$

$$k_i = \begin{cases} 1 & \mathrm{SWvoid}_i \geq \mathrm{SWvc} \\ 0 & \mathrm{SWvoid}_i < \mathrm{SWvc} \end{cases}, \quad i = 1, 2, \ldots, n \tag{6.41}$$

Suppose that m successive machined surfaces are measured and handled by the above procedures, and the numbers of OOL regions of these surface are obtained as $c_1, c_2, \ldots c_m$. The number of OOL c is the control variable. The centerline (CL), the upper control limit (UCL), and the lower control limit (LCL) of the individual chart for leakage can be depicted using the Eq. (6.42):

$$\begin{aligned} \text{UCL} &= \bar{c} + 3\sigma \\ \text{CL} &= \bar{c} \\ \text{LCL} &= \bar{c} - 3\sigma \end{aligned} \tag{6.42}$$

where $\bar{c}$ is the mean value and σ is the standard deviation. They are estimated by the Eqs. (6.43–6.44):

$$\bar{c} = \sum_{j=1}^{m} c_j / m \tag{6.43}$$

$$\sigma = \sqrt{\sum_{j=1}^{m} (c_j - \bar{c})^2 \Big/ m} \tag{6.44}$$

Should these calculations yield a negative value for the LCL, set LCL = 0. If the number of OOL c of a surface is beyond the control line, it can be determined that the successive machining process is out of control and leakage will occur probably in the surface of this machining process. Otherwise, the machining process of the surface should be kept. The detailed control chart can be seen in the case study section.

6.4.3 Case Study

The proposed methodology is applied to a machining process for the top surface of vehicle engine cylinder block. The material of the engine cylinder block is cast iron FC250. This surface is a major sealing surface in automotive power train, and it is manufactured by rough milling, semi-finish milling, and finish milling. The milling process is carried out on an EX-CELL-O machining center using a face milling cutter which has a diameter of 200 mm with 15 cutting inserts intercalated by 3 wiper inserts. Quaker 370 KLG cutting fluid is used. The milling speed is 816.4 m/min, the depth of milling is 0.5 mm, and feed rate is 3360 mm/min. The machining center and cutter are shown in Fig. 6.47.

Leakage is always a serious concern in engine manufacturing, and it may lead to engine overheating, compression loss, and power reduction. Typically, leakage occurs most in the interface between engine cylinder block and head, and the gasket is assembled to prevent it, as seen in Fig. 6.48. As a consequence, the case study on the top surface of engine block cylinder is representative and significant. For example, a top surface of engine block cylinder is measured by HDM, and the converted gray image of the measured result is shown in Fig. 6.49. Thirty typical

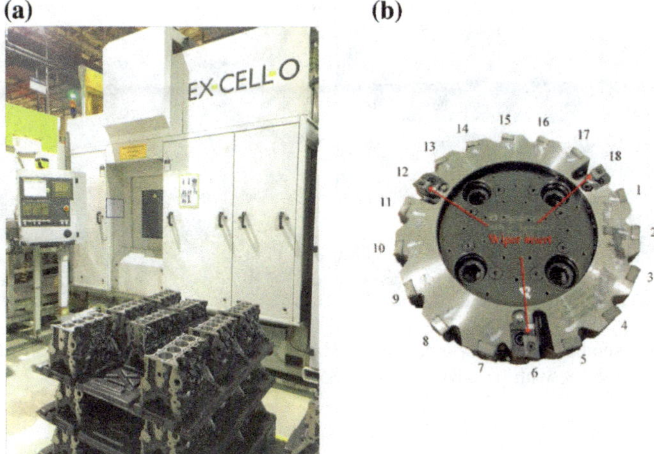

Fig. 6.47 **a** EX-CELL-O machining center **b** The face milling cutter. Reprinted from Ref. [102], copyright 2018, with permission from ASME

surface regions in the same size (6 × 6 mm) are selected to represent the probable leakage areas. The sampling interval is 0.2 mm, and each region has 900 points with 30 × 30 grid.

6.4.3.1 Finite Element Model

In order to determine the threshold of SWvc, an explicit dynamic finite element method is used to simulate the contact process in the subsection. The numerical procedures roughly consist of loading measured surface data, generating the solid, giving the material properties and load conditions, meshing and determining contact deformation.

As in previous researches [100], some well-known results from contact mechanics is used to simplify the contact geometry. If there is no friction or adhesion between two rough surfaces and the surface slope is small, then elastic contact between them can be mapped to contact between a single rough surface and a rigid flat plane. Meanwhile, the contact area has a well-defined thermodynamic limit, that is to say, the percentage of contact area at a given average normal pressure is independent of the system size for a fixed surface.

In this case, the gasket surface and the waviness surface of engine block cylinder both can be considered as elastic solids without friction and adhesion. Thus, a rigid flat surface is used instead of the gasket surface, and the waviness surface of engine block cylinder is considered as a three-dimensional deformable elastic rough surface. One selected 6 × 6 mm size waviness surface with 30 × 30 grid is first loaded into a computer-aided design (CAD) software called Pro/Engineer, boundary hybrid scanning and stretching are adopted to generate the modeling waviness

(a)   **(b)**

Fig. 6.48 **a** Assembled engine cylinder head and block **b** Gasket and block. Reprinted from Ref. [102], copyright 2018, with permission from ASME

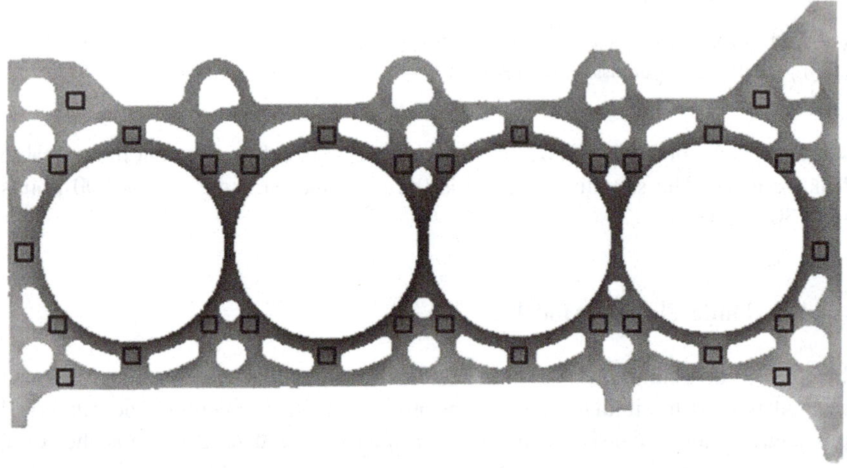

Fig. 6.49 The selected 30 typical regions of the top surface. Reprinted from Ref. [102], copyright 2018, with permission from ASME

surface solid, see Fig. 6.50. Then, the solid is loaded into a computer-aided engineering (CAE) software called ABAQUS for an explicit contact analysis. A same size 3D analytical rigid flat surface is generated by ABAQUS to assemble into the loaded waviness surface solid.

Material properties and load conditions are given according to the real work condition. The material of the waviness surface solid is cast iron FC250, and its Young's modulus, Poisson's ratio, and density are 120GPa, 0.25, and 7.0 g/cm^3, respectively. The boundary condition is that the four sides of the rigid flat surface are fixed, and simultaneously both the four sides and bottom of the waviness surface solid are fixed. The load is applied by tightened force of 10 bolts, and it is equal to about a 2 MPa face pressure. As mentioned above, numerical simulations

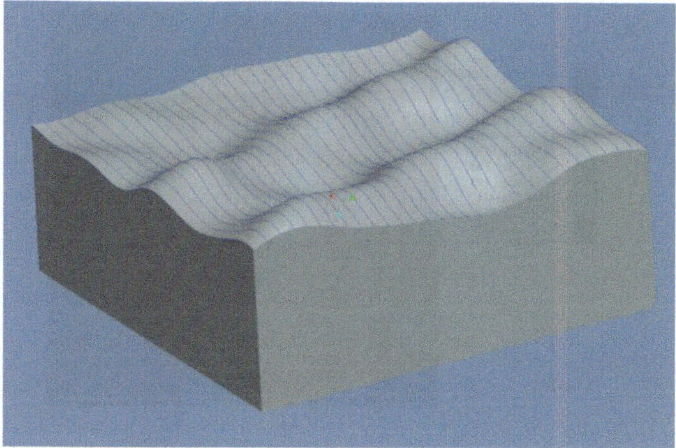

Fig. 6.50 Waviness surface solid. Reprinted from Ref. [102], copyright 2018, with permission from ASME

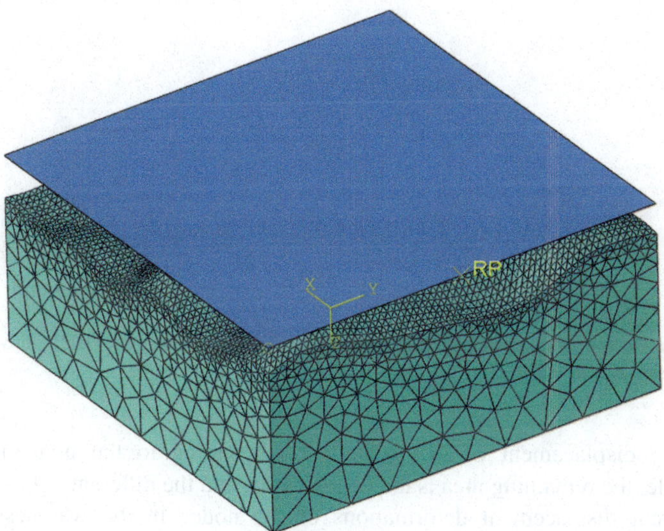

Fig. 6.51 Mesh generation. Reprinted from Ref. [102], copyright 2018, with permission from ASME

are done for an elastic surface in contact with a rigid flat surface. A fine mesh for the elastic waviness surface is illustrated in Fig. 6.51. The mesh is discretized with tetrahedral elements. The mesh for the waviness surface grid contains about 11383 nodes and 54042 elements. By simulation, the results of displacement–deformation and contact area are illustrated in Fig. 6.52. As shown in Fig. 6.52a, the deep red

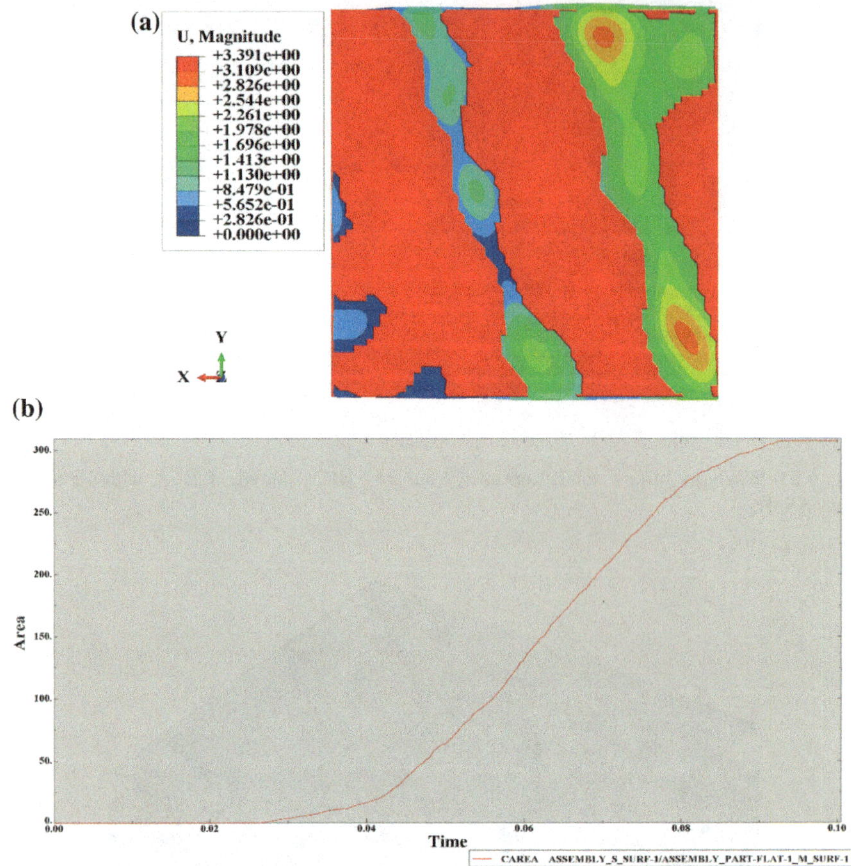

Fig. 6.52 **a** Results of displacement **b** Contact area. Reprinted from Ref. [102], copyright 2018, with permission from ASME

means the displacement of the gasket, corresponding to the noncontact area. Meanwhile, the remaining area is the contact area, and the different colors represent the different displacement–deformations of the nodes in the waviness surface. Considering the face pressure is a progressively load in FEM, Fig. 6.52b displays the changing curves of contact area with time. It is noted that the sampling interval is 0.2 mm and the units are consistent in all directions in FEM. The final contact area is 307.6, that is $307.6 \times 0.2 \times 0.2 = 12.304$ mm, so the relative contact area is $RC = 12.304/36 \approx 0.342$. Obviously, the relative contact area of this surface region is less than 0.4, and this region is a leakage region. It can be seen more clearly in Fig. 6.52a.

Considering the selected 30 surface probable leakage regions in Fig. 6.49, each surface region is analyzed by the same finite element model. At the given load, the contact area is determined for each surface region. The goal is to find out the certain

Table 6.6 Results of surface region

Surface region	RC	CAP	VV	SWvoid
Region 1	**0.401**	0.289	10.76	**0.299**
Region 2	0.452	0.313	11.42	0.317
Region 3	0.440	0.283	10.21	0.284
Region 4	0.496	0.322	9.10	0.253
Region 5	**0.396**	0.298	10.92	**0.303**
Region 6	0.472	0.304	10.30	0.286
Region 7	0.432	0.278	13.01	0.362
Region 8	**0.391**	0.309	11.03	**0.307**
Region 9	0.504	0.343	8.52	0.237
Region 10	0.348	0.300	12.58	0.349
Region 11	0.528	0.341	8.25	0.229
Region 12	0.280	0.311	13.81	0.384
Region 13	0.434	0.331	9.86	0.274
Region 14	**0.409**	0.296	10.56	**0.293**
Region 15	0.449	0.299	10.07	0.280
Region 16	0.360	0.299	13.22	0.367
Region 17	0.457	0.338	9.34	0.260
Region 18	0.380	0.302	11.17	0.310
Region 19	**0.403**	0.330	10.72	**0.298**
Region 20	0.429	0.302	11.30	0.314
Region 21	0.480	0.341	8.90	0.247
Region 22	0.454	0.280	11.58	0.322
Region 23	0.341	0.266	12.72	0.353
Region 24	0.370	0.313	11.63	0.323
Region 25	0.478	0.342	8.89	0.247
Region 26	0.340	0.281	13.54	0.376
Region 27	0.454	0.341	9.40	0.261
Region 28	0.500	0.308	10.16	0.282
Region 29	0.480	0.302	9.76	0.271
Region 30	0.420	0.309	10.38	0.288

surface region when the relative contact area RC is closed to 40%. Here, a small deviation of 5% is tolerated, which means that the deviations of $\pm 2.5\%$ go beyond 0.4 can be considered as 0.4. That is to say, $RC \in [0.390, 0.410]$ is equivalent to $RC \approx 0.4$. Table 6.6 gives the relative contact area of the 30 surface regions and the corresponding CAP, VV and SWvoid.

In order to further clarify the relationships among these parameters, line charts and scatter diagrams are depicted to make correlation analysis and correlation coefficient r is calculated as

Table 6.7 Correlation coefficient

Correlation coefficient	$r(\mathrm{RC}, \mathrm{SWvoid})$	$r(\mathrm{RC}, \mathrm{CAP})$	$r(\mathrm{CAP}, \mathrm{SWvoid})$
r	−0.851	0.493	−0.702

$$r(X, Y) = \frac{\mathrm{Cov}(X, Y)}{\sqrt{\mathrm{Var}(X)\mathrm{Var}(Y)}} \tag{6.45}$$

A line chart is a common chart which displays distribution information as a series of data points, and a scatter diagram indicates the potential relationship between two variables. From the definition, it is clear that VV is in direct proportion to SWvoid. Therefore, the detailed charts and diagrams of RC, CAP, and SWvoid are shown in Fig. 6.53, and the pairwise correlation coefficients are calculated in Table 6.7.

The scatter diagram indicates a strong negative correlation between RC and SWvoid, and the same to CAP and SWvoid. Meantime, RC and CAP exist a medium positive correlation. The results of correlation analysis show that determining the threshold SWvc based on the relative contact area RC is accurate and effective. Therefore, from Table 6.6, the relative contact area of surface region 1, 5, 8, 14, 19 are satisfied with closing to 40%. The threshold SWvc can be approximately calculated as

$$\begin{aligned}
\mathrm{SWvc} &\approx (\mathrm{SWvoid}_1 + \mathrm{SWvoid}_5 + \mathrm{SWvoid}_8 + \mathrm{SWvoid}_{14} + \mathrm{SWvoid}_{19})/5 \\
&= (0.299 + 0.303 + 0.307 + 0.293 + 0.298)/5 \tag{6.46} \\
&= 0.300
\end{aligned}$$

According to contact mechanics, when the parameters such as material, process parameter, and pressure change, the RC and the threshold SWvc will also change. The corresponding value of SWvc can be obtained by the proposed approach. Therefore, the threshold $\mathrm{SWvc} \approx 0.300$ is suitable for the engine cylinder block in the case study condition.

6.4.3.2 Experimental Results

As mentioned in Sect. 2.3, the normalized parameter relative void volume SWvoid is independent of the area of the surface. At the same time, a strong negative correlation between RC and SWvoid is shown by the scatter diagram and the correlation coefficient. Therefore, SWvoid is selected as the leakage indicator to report the leakage area.

From Table 6.6, the SWvoid of the 30 regions are calculated as $\mathrm{SWvoid}_1, \mathrm{SWvoid}_2, \ldots, \mathrm{SWvoid}_{30}$, and the threshold is $\mathrm{SWvc} = 0.3$. The number of OOL c is obtained as

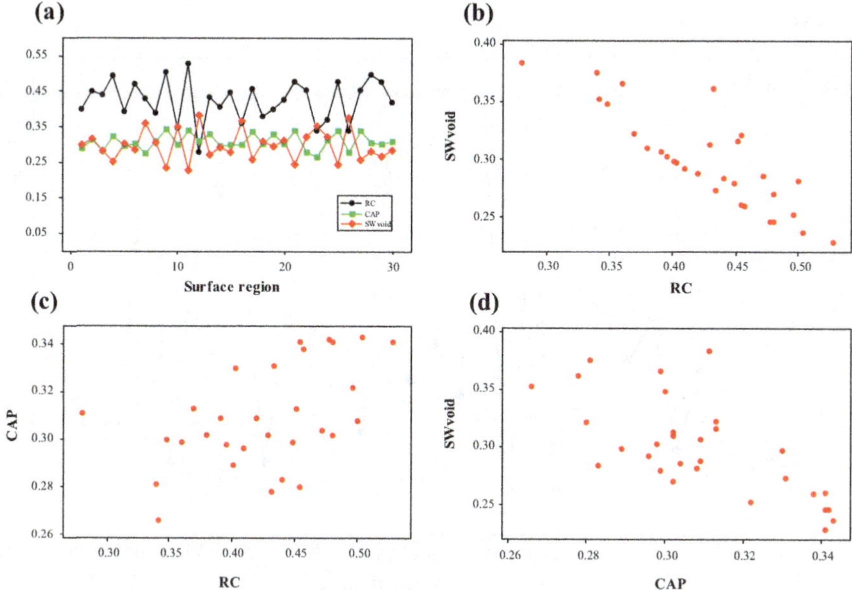

Fig. 6.53 **a** Line charts of RC, CAP and SWvoid **b** Scatter diagram of RC and SWvoid **c** Scatter diagram of RC and CAP **d** Scatter diagram of CAP and SWvoid. Reprinted from Ref. [102], copyright 2018, with permission from ASME

$$c = \sum_{i=1}^{30} k_i \tag{6.47}$$

$$k_i = \begin{cases} 1 & \text{SWvoid}_i \geq 0.3 \\ 0 & \text{SWvoid}_i < 0.3 \end{cases}, \quad i = 1, 2, \ldots, 30 \tag{6.48}$$

The line chart of SWvoid is depicted in Fig. 6.54a; each selected region is determined as a leakage region by the corresponding SWvoid goes beyond 0.3. As shown in Fig. 6.54b, the OOL leakage areas of an engine block top surface are identified in red color. Furthermore, confirming the leakage surface and monitoring the successive machining process are achieved through the abovementioned individual control chart.

Figure 6.55 shows two series of successive machining process of 20 engine block top surfaces, it is clear that the first one is out-of-control process which exists an obvious growing tendency since the 12th block surface, and No. 19 surface is very likely a leakage surface. It indicates that something changes like tool wear or chatting has probably appeared in the machining process which lead to the variation of surface topography. On the contrary, the second machining process is relatively stable and the probability of leakage surfaces is small.

The control chart for leakage monitoring enables leakage susceptibility mensuration long before pressure test, and detects more leak-prone parts which have

(a)

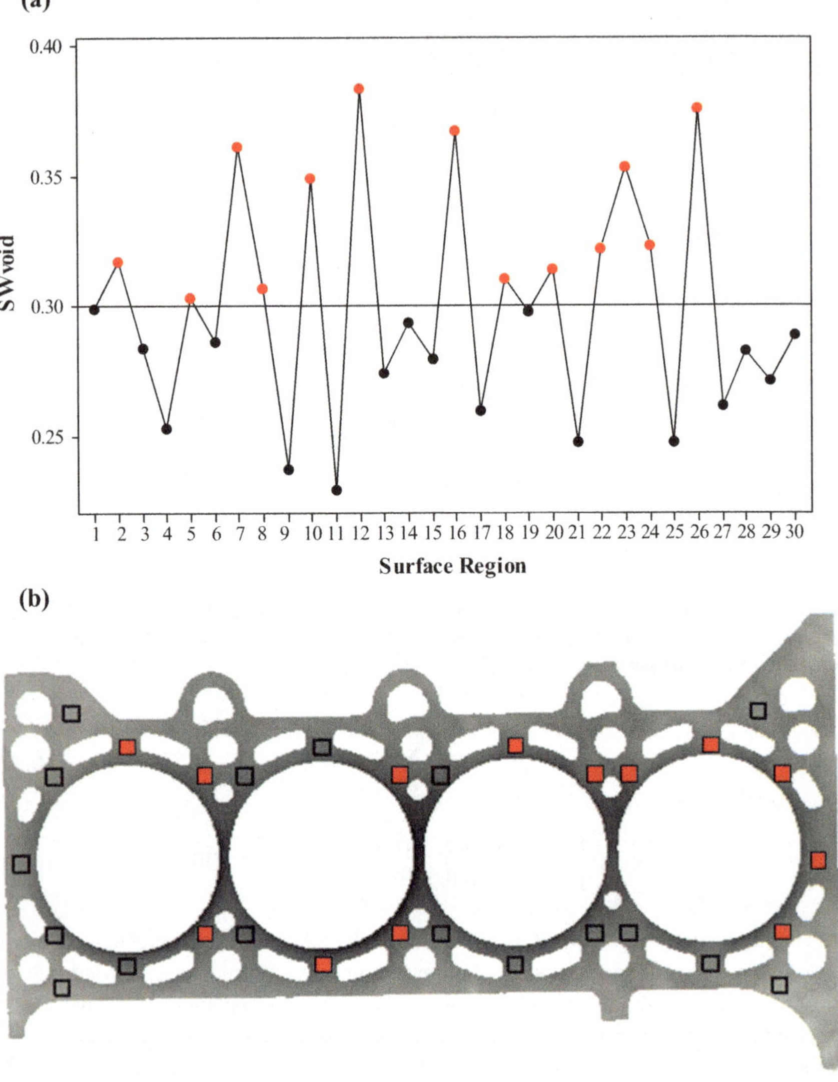

(b)

Fig. 6.54 **a** The line chart of SWvoid **b** OOL leakage regions. Reprinted from Ref. [102], copyright 2018, with permission from ASME

been machined. At the same time, the leakage potential surface can be prevented entering the next costly procedure in time from the control chart. Furthermore, leakage monitoring ensures a higher quality product that will incur lower post-delivery warranty costs and higher customer satisfaction.

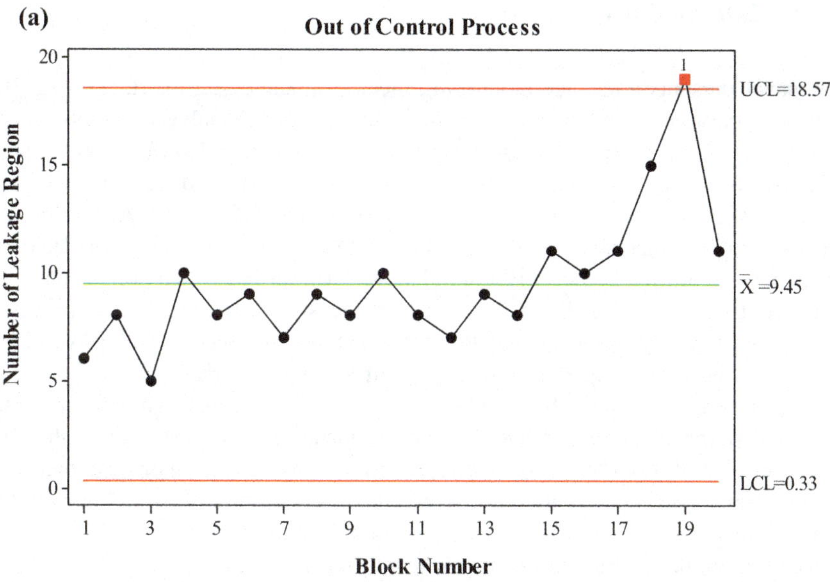

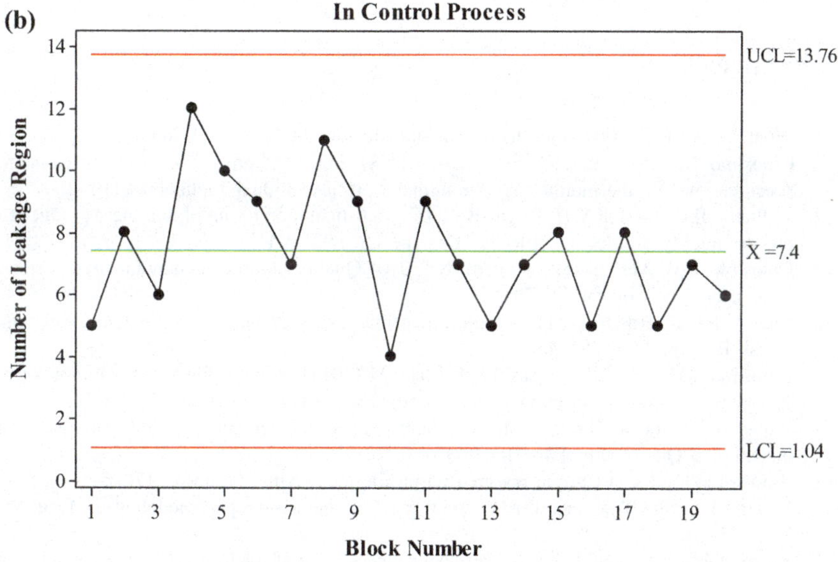

Fig. 6.55 **a** Out-of-control process and **b** In-control process. Reprinted from Ref. [102], copyright 2018, with permission from ASME

6.4.4 Conclusions

This paper presents a leakage monitoring method in static seal interface using 3D surface topography as indicators. To achieve this functional analysis, a combination of spline filter and morphological filter are employed. The 3D surface topography indicators including leakage parameters CAP, VV and SWvoid are calculated by virtual gasket surface $SC(x, y)$ and waviness surface $SW(x, y)$. According to Persson contact mechanics and percolation theory, the threshold of leakage parameter is confirmed using FEM simulation. Then an individual control chart is proposed to monitor the leakage surface of the successive machining process. Experimental results indicate that the proposed monitoring method is valid to pre-control the machining process and prevent leakage occurring.

Furthermore, there have been only a few attempts on leakage monitoring, the proposed approach is a first idea to account for this kind of problems. Thus, there is a large amount of room for important improvements. Moreover, in order to enhance quality control, the quantitative leakage rate experimental investigation will be designed to further reveal and test the potential relationship between leakage and surface topography, which is the next research topic in future.

References

1. Montgomery DC (2007) Introduction to statistical quality control. Wiley
2. Colosimo BM, Semeraro Q, Pacella M (2008) Statistical process control for geometric specifications: on the monitoring of roundness profiles. J Qual Technol 40(1):1–18
3. Williams JD, Woodall WH, Birch JB (2007) Statistical monitoring of nonlinear product and process quality profiles. Qual Reliab Eng Int 23(8):925–941
4. Colosimo BM, Mammarella F, Petrò S (2010) Quality control of manufactured surfaces. Front Stat Qual Control: 55–70
5. Chen S, Nembhard HB (2011) A high-dimensional control chart for profile monitoring. Qual Reliab Eng Int 27(4):451–464
6. Colosimo BM, Cicorella P, Pacella M, Blaco M (2014) From profile to surface monitoring: SPC for cylindrical surfaces via gaussian processes. J Qual Technol 46(2):95–113
7. Wang A, Wang K, Tsung F (2014) Statistical surface monitoring by spatial-structure modeling. J Qual Technol 46(4):359–376
8. Woodall WH (2007) Current research on profile monitoring. Produção 17(3):420–425
9. Etesami F (1988) Tolerance verification through manufactured part modeling. J Manuf Syst 7(3):223–232
10. Xia H, Ding Y, Wang J (2008) Gaussian process method for form error assessment using coordinate measurements. IIE Trans 40(10):931–946
11. Wang H, Suriano S, Zhou L, Hu SJ (2009) High-definition metrology based spatial variation pattern analysis for machining process monitoring and diagnosis. In: ASME 2009 International manufacturing science and engineering conference, pp 471–480
12. Suriano S, Wang H, Hu SJ (2012) Sequential monitoring of surface spatial variation in automotive machining processes based on high definition metrology. J Manuf Syst 31(1):8–14

13. Suriano S, Wang H, Shao C, Hu SJ, Sekhar P (2015) Progressive measurement and monitoring for multi-resolution data in surface manufacturing considering spatial and cross correlations. IIE Trans 47(10):1033–1052
14. Wang K, Tsung F (2010) Using profile monitoring techniques for a data-rich environment with huge sample size. Qual Reliab Eng Int 21(7):677–688
15. Wells LJ, Megahed FM, Niziolek CB, Camelio JA, Woodall WH (2013) Statistical process monitoring approach for high-density point clouds. J Intell Manuf 24(6):1267–1279
16. He K, Zhang M, Zuo L, Alhwiti T, Megahed FM (2017) Enhancing the monitoring of 3D scanned manufactured parts through projections and spatiotemporal control charts. J Intell Manuf 28(4):899–911
17. Roth JT, Djurdjanovic D, Yang X, Mears L, Kurfess T (2010) Quality and inspection of machining operations: tool condition monitoring. J Manuf Sci Eng 132(4):575–590
18. Castejón M, Alegre E, Barreiro J, Hernández LK (2007) On-line tool wear monitoring using geometric descriptors from digital images. Int J Mach Tools Manuf 47(12):1847–1853
19. Jurkovic J, Korosec M, Kopac J (2005) New approach in tool wear measuring technique using CCD vision system. Int J Mach Tools Manuf 45(9):1023–1030
20. Kerr D, Pengilley J, Garwood R (2006) Assessment and visualisation of machine tool wear using computer vision. Int J Adv Manuf Technol 28(7–8):781–791
21. Pfeifer T, Wiegers L (2000) Reliable tool wear monitoring by optimized image and illumination control in machine vision. Measurement 28(3):209–218
22. Shahabi HH, Ratnam MM (2009) In-cycle monitoring of tool nose wear and surface roughness of turned parts using machine vision. Int J Adv Manuf Technol 40(11–12):1148–1157
23. Wang X, Kwon PY (2014) WC/Co tool wear in dry turning of commercially pure aluminium. J Manuf Sci Eng 136(3):031006-1-7
24. Kious M, Ouahabi A, Boudraa M, Serra R, Cheknane A (2010) Detection process approach of tool wear in high speed milling. Measurement 43(10):1439–1446
25. Oraby SE, Al-Modhuf AF, Hayhurst DR (2004) A diagnostic approach for turning tool based on the dynamic force signals. J Manuf Sci Eng 127(3):463–475
26. Kaya B, Oysu C, Ertunc HM (2011) Force-torque based on-line tool wear estimation system for CNC milling of Inconel 718 using neural networks. Adv Eng Softw 42(3):76–84
27. Alonso FJ, Salgado DR (2008) Analysis of the structure of vibration signals for tool wear detection. Mech Syst Signal Process 22(3):735–748
28. Bovic K, Pierre D, Xavier C (2011) Tool wear monitoring by machine learning techniques and singular spectrum analysis. Mech Syst Signal Process 25(1):400–415
29. Salgado DR, Alonso FJ (2007) An approach based on current and sound signals for in-process tool wear monitoring. Int J Mach Tools Manuf 47(14):2140–2152
30. Wang H (2015) Progressive measurement and monitoring for multi-resolution data in surface manufacturing considering spatial and cross correlations. IIE Trans 47(10):1–20
31. Marinescu I, Axinte DA (2008) A critical analysis of effectiveness of acoustic emission signals to detect tool and workpiece malfunctions in milling operations. Int J Mach Tools Manuf 48(10):1148–1160
32. Yen CL, Lu MC, Chen JL (2013) Applying the self-organization feature map (SOM) algorithm to AE-based tool wear monitoring in micro-cutting. Mech Syst Signal Process 34(1–2):353–366
33. Attanasio A, Ceretti E, Giardini C, Cappellini C (2013) Tool wear in cutting operations: experimental analysis and analytical models. J Manuf Sci Eng 135(5):051012-1-11
34. Dutta S, Datta A, Chakladar ND, Pal SK, Mukhopadhyay S, Sen R (2012) Detection of tool condition from the turned surface images using an accurate grey level co-occurrence technique. Precis Eng 36(3):458–466
35. Kassim AA, Mannan MA, Zhu M (2007) Texture analysis methods for tool condition monitoring. Image Vis Comput 25(7):1080–1090
36. Wilkinson P, Reuben RL, Jones JDC, Barton JS, Hand DP, Carolan TA, Kidd SR (1997) Surface finish parameters as diagnostics of tool wear in face milling. Wear 205(1–2):47–54

37. Dutta S, Pal SK, Mukhopadhyay S, Sen R (2013) Application of digital image processing in tool condition monitoring: A review. CIRP J Manufact Sci Technol 6(3):212–232
38. ISO 25178-602:2012 (2010) Geometrical product specifications (GPS)—surface texture: areal-part 602: nominal characteristics of non-contact (confocal chromatic probe) instruments
39. Huang Z, Shih AJ, Ni J (2006) Laser interferometry hologram registration for three-dimensional precision measurements. J Manuf Sci Eng 128(4):887–896
40. Stephenson DA, Ni J (2010) A multifeature approach to tool wear estimation using 3D workpiece surface texture parameters. J Manuf Sci Eng 132(6):1033–1041
41. Wang M, Xi L, Du S (2014) 3D surface form error evaluation using high definition metrology. Precis Eng 38(1):230–236
42. Jr AMDS, Sales WF, Santos SC, Machado AR (2005) Performance of single Si 3N 4 and mixed Si 3N 4 +PCBN wiper cutting tools applied to high speed face milling of cast iron. Int J Mach Tools Manuf 45(3):335–344
43. Astakhov VP (2004) The assessment of cutting tool wear. Int J Mach Tools Manuf 44 (6):637–647
44. Dutta S, Kanwat A, Pal SK, Sen R (2013) Correlation study of tool flank wear with machined surface texture in end milling. Measurement 46(10):4249–4260
45. Al-Kindi Ghassan, Zughaer Hussien (2012) An approach to improved CNC machining using vision-based system. Adv Manuf Process 27(7):765–774
46. Hai TN, Wang H, Hu SJ (2012) Chacterization of cutting force induced surface shape variation using high-definition metrology. J Manuf Sci Eng 135:641–650
47. Haralick RM, Shanmugam K, Dinstein IH (1973) Textural features for image classification. IEEE Trans Syst Man Cybern 3(6):610–621
48. Huang DL, Du SC, Li GL, Wu ZQ (2017) A systemic approach for on-line minimizing volume difference of multiple chambers with casting surfaces in machining processes based on high definition metrology. J Manuf Sci Eng 139(8):081003-1-17
49. Wang K, Wei J, Bo L (2015) A spatial variable selection method for monitoring product surface. Int J Prod Res 54(14):1–21
50. He Z, Zuo L, Zhang M, Megahed FM (2012) An image-based multivariate generalized likelihood ratio control chart for detecting and diagnosing multiple faults in manufactured products. Int J Prod Res 54(6):1771–1784
51. Sullivan JH (2002) Detection of multiple change points from clustering individual observation. J Qual Technol 34(4):374–383
52. Woodall WH, Dan JS, Montgomery DC, Gupta S (2004) Using control charts to monitor process and product quality profiles. J Qual Technol 36(3):309–320
53. Du S, Liu C, Huang D (2015) A shearlet-based separation method of 3D engineering surface using high definition metrology. Precis Eng 40:55–73
54. Du S, Liu C, Xi L (2015) A selective multiclass support vector machine ensemble classifier for engineering surface classification using high definition metrology. J Manuf Sci Eng 137 (1):011003-1-15
55. Du SC, Huang DL, Wang H (2015) An adaptive support vector machine-based workpiece surface classification system using high-definition metrology. IEEE Trans Instrum Meas 64 (10):2590–2604
56. Du S, Fei L (2016) Co-kriging method for form error estimation incorporating condition variable measurements. J Manuf Sci & Eng 138(4):o41003-1-16
57. Wang M, Ken T, Du S, Xi L (2015) Tool wear monitoring of wiper inserts in multi-insert face milling using three-dimensional surface form indicators. J Manuf Sci Eng 137 (3):031006-1-8
58. Wang M, Shao YP, Du SC, Xi LF (2015) A diffusion filter for discontinuous surface measured by high definition metrology. Int J Precis Eng Manuf 16(10):2057–2062
59. Hai N, Wang H, Tai BL, Ren J, Hu SJ, Shih AJ (2016) High-definition metrology enabled surface variation control by cutting load balancing. J Manuf Sci Eng 138(2):021010-1-11

60. Wells, L. J., Shafae, M. S., and Camelio, J. A., 2016, "Automated surface defect detection using high-density data," Journal of Manufacturing Science & Engineering, 138(7), pp. 071001-1-10

61. Chen Q, Yang S, Li Z (1999) Surface roughness evaluation by using wavelets analysis. Precis Eng 23(3):209–212

62. Lu C, Troutman JR, Schmitz TL, Ellis JD, Tarbutton JA (2016) Application of the continuous wavelet transform in periodic error compensation. Precis Eng 44:245–251

63. Xu J, Yamada K, Seikiya K, Tanaka R, Yamane Y (2014) Effect of different features to drill-wear prediction with back propagation neural network. Precis Eng 38(4):791–798

64. Barnhill RE, Pottmann OH (1992) Fat surfaces: a trivariate approach to triangle-based interpolation on surfaces. Comput Aided Geom Des 9(5):365–378

65. Fischler MA, Bolles RC (1981) Random sample consensus: a paradigm for model fitting with applications to image analysis and automated cartography. Commun ACM 24(6):381–395

66. Raguram R, Chum O, Pollefeys M, Matas J, Frahm JM (2013) USAC: a universal framework for random sample consensus. IEEE Trans Pattern Anal Mach Intell 35(8):2022–2038

67. Rodriguez A, Laio A (2014) Clustering by fast search and find of density peaks. Science 344(6191):1492–1496

68. Flitney RK (2011) Seals and sealing handbook. Elsevier

69. Persson BNJ, Yang C (2008) Theory of the leak-rate of seals. J Phys Condens Matter 20

70. Aharony A, Stauffer D (2003) Introduction to percolation theory. Taylor & Francis

71. Persson BNJ, Albohr O, Creton C, Peveri V (2004) Contact area between a viscoelastic solid and a hard, randomly rough, substrate. J Chem Phys 120:8779–8793

72. Lorenz B, Persson BNJ (2009) Leak rate of seals: comparison of theory with experiment. EPL 86

73. Lorenz B, Persson BNJ (2010) Leak rate of seals: effective-medium theory and comparison with experiment. Eur Phys J E 31:159–167

74. Bottiglione F, Carbone G, Mangialardi L, Mantriota G (2009) Leakage mechanism in flat seals. J Appl Phys 106

75. Bottiglione F, Carbone G, Mantriota G (2009) Fluid leakage in seals: an approach based on percolation theory. Tribol Int 42:731–737

76. Marie C, Lasseux D (2007) Experimental leak-rate measurement through a static metal seal. J Fluids Eng 129:799–805

77. Robbe-Valloire F, Prat M (2008) A model for face-turned surface microgeometry. Application to the analysis of metallic static seals. Wear 264:980–989

78. Okada H, Itoh T, Suga T (2008) The influence of surface profiles on leakage in room temperature seal-bonding. Sens Actuators A 144:124–129

79. Haruyama S, Nurhadiyanto D, Choiron MA, Kaminishi K (2013) Influence of surface roughness on leakage of new metal gasket. Int J Press Vessels Pip 111–112:146–154

80. Marie C, Lasseux D, Zahouani H, Sainsot P (2003) An integrated approach to characterize liquid leakage through metal contact seal. Eur J Mech Environ Eng 48:81–86

81. Ledoux Y, Lasseux D, Favreliere H, Samper S, Grandjean J (2011) On the dependence of static flat seal efficiency to surface defects. Int J Press Vessels Pip 88:518–529

82. Malburg MC (2003) Surface profile analysis for conformable interfaces. J Manuf Sci Eng 125:624–627

83. Liao Y, Stephenson DA, Ni J (2012) Multiple-scale wavelet decomposition, 3D surface feature exaction and applications. J Manuf Sci Eng 134

84. Ren J, Park C, Wang H (2018) Stochastic modeling and diagnosis of leak areas for surface assembly. J Manuf Sci Eng 140:041011–10

85. Arghavani J, Derenne M, Marchand L (2002) Prediction of gasket leakage rate and sealing performance through fuzzy logic. Int J Adv Manuf Technol 20:612–620

86. Xin L, Gaoliang P (2016) Research on leakage prediction calculation method for static seal ring in underground equipments. J Mech Sci Technol 30:2635–2641

87. Du S, Liu T, Huang D, Li G (2018) A fast and adaptive bi-dimensional empirical mode decomposition approach for filtering of workpiece surfaces using high definition metrology. J Manuf Syst 46:247–263
88. Shao Y, Du S, Xi L (2017) 3D machined surface topography forecasting with space-time multioutput support vector regression using high definition metrology, V001T02A69
89. ISO 16610-22 (2015) Geometrical product specifications (GPS)-filtration part 22: linear profile filters: spline filter
90. ISO 16610-1 (2015) Geometrical product specifications (GPS)-filtration part 1: overview and basic concepts
91. Krystek M (1996) Form filtering by splines. Measurement 18:9–15
92. Maragos P, Schafer R (1987) Morphological filters–part I: their set-theoretic analysis and relations to linear shift-invariant filters. IEEE Trans Acoust Speech Signal Process 35:1153–1169
93. ISO 16610-40 (2015) Geometrical product specifications (GPS)-filtration part 40: morphological profile filters: basic concepts
94. ISO 16610-41 (2015) Geometrical product specifications (GPS)-filtration part 41: morphological profile filters: disk and horizontal line-segment filters
95. ISO 16610-85 (2015) Geometrical product specifications (GPS)-filtration part 85: morphological areal filters: segmentation
96. ISO 25178-2 (2012) Geometrical product specifications (GPS)-surface texture: areal part 2: terms, definitions and surface texture parameters
97. ISO 4287 (1997) Geometrical product specifications (GPS)-surface texture: profile method: terms, definitions and surface texture parameters.
98. Hyun S, Pel L, Molinari JF, Robbins MO (2004) Finite-element analysis of contact between elastic self-affine surfaces. Phys Rev E-Stat, Nonlinear, Soft Matter Phys 70:026117
99. Megalingam A, Mayuram MM (2012) Comparative contact analysis study of finite element method based deterministic, simplified multi-asperity and modified statistical contact models. J Tribol 134:014503
100. Johnson KL (1985) Contact mechanics. Cambridge University Press, New York
101. Huang D, Du S, Li G et al (2018) Detection and monitoring of defects on three-dimensional curved surfaces based on high-density point cloud data. Precis Eng 53:79–95
102. Shao Y, Yin Y, Du S et al (2018) Leakage monitoring in static sealing interface based on three dimensional surface topography indicator. J Manuf Sci Eng 140(10):101003

Chapter 7
Surface Prediction

7.1 A Brief History of Surface Prediction

Achieving the predefined product quality is the fundamental goal in the manufacturing field. Surface topography characteristics, including flatness, waviness, and roughness, are of great importance for functional behavior of mechanical parts. Due to the complex physical phenomena along with numerous factors in the manufacturing process, it is hard to attain the satisfied surface topography that meets the technical specifications. At this point, prediction of surface topography in advance is essential since it can help optimize process parameters and promote quality control of the manufacturing processes.

In the machining field, a great number of researches on surface topography prediction of machined parts have been published. It is reported that the factors include tool wear, cutting conditions, and the material properties that lead to the formation of surface topography for the parts. These approaches can be roughly classified into two categories: theoretical and experimental approaches.

7.1.1 Theoretical Approach

Theoretical approach is to build the analytical model based on machining theory and cutting kinematics; thereby, it is used for analyzing the processing method and the generation mechanism of surface topography, and studying the effect of the shape of the tool, parameters, and vibration on the surface topography.

The early theoretical model is mainly to consider the difference in processing methods, and the tool and workpiece are regarded as rigid bodies. The influences of tool geometrical shape and feed on the surface residual area are analyzed under different processing methods, and the surface topography is estimated by the height of the residual area. However, this prediction model is relatively simple; it does not

© Springer Nature Singapore Pte Ltd. 2019
S. Du and L. Xi, *High Definition Metrology Based Surface Quality Control and Applications*, https://doi.org/10.1007/978-981-15-0279-8_7

study the formation mechanism of surface topography and the effects of tool deflection, vibration, and so on. In recent years, with the development of computer technology, many researchers have considered the tool hop, feed method, spindle eccentricity, axial turbulence, and machining vibration into the prediction model of surface topography, and the computer simulation is used. The simulated model is closer to the actual manufacturing process, and the prediction results are more accurate. The establishment of the theoretical model is based on the formation mechanism and the influence of various factors on the surface topography. But, due to the complexity and uncertainty of the cutting process, the established model cannot consider all the factors such as cutting parameters, chip type, tool characteristics, and workpiece materials. Therefore, the theoretical model is simplified and the prediction accuracy is low.

7.1.2 Experimental Approach

Unlike the theoretical approaches, the experimental approaches are to explore the mechanism of the influence of each factor on the observed experimental results. These approaches are especially suitable in the cases where there are no analytical relationships between surface topography and the factors such as cutting force and tool vibrations. Based on the principle of data driven, designed experiments and artificial intelligence (AI) are two main kinds of experimental approaches. The designed experiment builds a surface topography prediction model including experimental design, experimental data processing, and analysis. Taguchi's design and regression analysis are the representative approaches that are widely used. Taguchi's design obtains the optimal combination of parameters through the calculation and analysis of experimental data with fewer experiment times. The basic idea of the Taguchi design is to (i) arrange the experimental scheme with orthogonal tables, (ii) simulate the various disturbances that fluctuate the prediction results with the error factors, (iii) find the correlation between the factors by the statistical analysis of various experimental data, (iv) obtain the optimal combination of parameters, and (v) determine the major and minor factors that affect the surface quality. However, the Taguchi method cannot predict surface topography under any processing parameters. Therefore, it is often used in conjunction with other methods in surface prediction. The main idea of regression analysis is to establish a polynomial model. Regression coefficients of the prediction model are obtained by solving the formula by experimental data. The regression analysis method not only reveals the influence of each variable on the dependent variable but also predicts and controls the dependent variable according to the regression equation when the independent variable changes. However, the generalization ability of the regression analysis method is poor, and a large amount of calculation is required when a large number of factors are taken into consideration. The high prediction accuracy depends on the fitting of a large amount of experimental data.

In recent years, AI is developed rapidly and widely used in engineering. AI is implemented in engineering problems by genetic programming, artificial neural network models, fuzzy logic, expert systems, and support vector regression. The surface topography prediction model is built by simulating the way in which human being process information and make decisions. With a series of advantages, AI has gradually become a mainstream method for predicting the surface topography. It is possible to build predict the surface topography by mixing a variety of artificial intelligence methods or developing a new intelligent method, which can reduce the number of cutting experiments and improve prediction accuracy.

7.2 A Space–Time Autoregressive Moving Average Based Predicting Method for 3D Engineering Surface

7.2.1 Introduction

Surface topography characteristics, including flatness, waviness, and roughness, are of great importance for functional performance of mechanical parts. Prediction of three-dimensional surface topography in advance is essential since it can help optimize process parameters and promote quality control of the manufacturing processes. However, the complex physical phenomena along with numerous factors affect the final surface finish, making predicting of surface topography a challenging task.

Various approaches have been employed for the prediction of surface topography. These approaches can be roughly classified into two categories: theoretical approaches and experimental approaches. Theoretical approaches develop analytical models to predict the surface topography based on machining theory and cutting kinematics. Franco et al. proposed a surface roughness model for face milling process considering feed, cutting tool geometry, and radial and axial runout in each cutting tool tooth [1]. Their further research incorporated back cutting to more accurately depict the milled surface [2]. In another research, Tapoglou and Antoniadis [3] simulated surface topography for face milling using tool kinematics in a CAD environment. Buj-Corral et al. [4] established a surface topography prediction model in ball-end milling based on geometric tool–workpiece intersection such as feed, depth of cut, helix angle, and eccentricity. Unlike theoretical approaches, experimental approaches analyze the effects of the related factors through the experimental results. These approaches are especially suitable in the cases where there are no analytical relationships between surface topography and the factors such as cutting force and tool vibrations. Experiments are conducted according to the principles of response surface methodology [5] and Taguchi's design of experiments [6]. Surface roughness is predicted with high precision using various approaches including regression [7], genetic programming [8], artificial neural network models [9], fuzzy logic and expert systems [10], and support vector regression [11].

Considering the abovementioned approaches, the factors that have been used for surface topography prediction can be grouped into four categories. They are (1) machining parameters such as cutting speed, feed rate, and depth of cut; (2) cutting tool properties such as tool shape, nose radius, and runout error; (3) cutting phenomena such as cutting force, vibration, friction, and chip formation; and (4) workpiece properties such as workpiece hardness, workpiece length, and workpiece diameter [12]. Besides all the four groups of factors, numerous factors and their interacting contribute to the formation of surface topography. Although these factors cannot be considered in a single model, they get reflected on the final surface topography. Therefore, the surface topography of manufactured parts should be involved in topography prediction.

Conventional surface topography measurement techniques, such as CMM and roughmeter, only inspect localized area or scattered points and cannot reveal the topography of the entire surface. Thus, those techniques are sample inspection rather than full inspection. A recently developed high-definition metrology (HDM) [13] can measure 3D surface topography of the entire surface with 0.15 mm resolution in x–y direction and 1 μm accuracy in depth direction, which is a full inspection of the whole surface. The HDM measured 3D surface topography of already machined surfaces contains the surface topography evolution information and can be used as input factors to predict the future machined surface. An example of the HDM measured engine block face and the prediction scheme is shown in Fig. 7.1. Once the future machined surface is correctly predicted, surface form error evaluation methods [14], surface classification methods [15–17], and tool wear monitoring methods [18] based on HDM can be adopted for the pre-control of the manufacturing process.

The HDM data measured from successive machined parts is in the form of space–time series. Therefore, an attempt has been made to use HDM data to predict future machined surface through a space–time autoregressive integrated moving average (STARIMA) model. The STARIMA model has been applied to wide

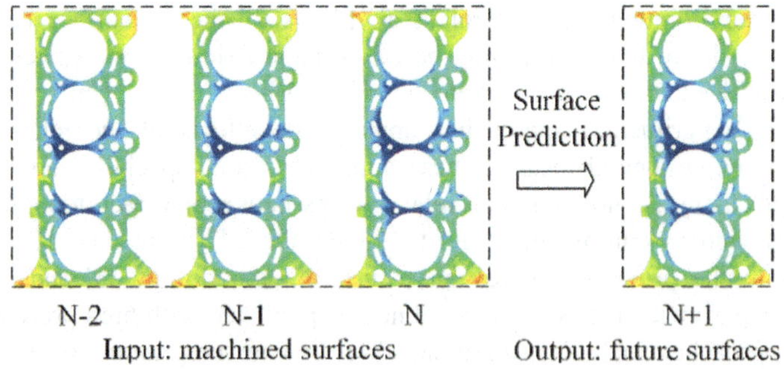

Fig. 7.1 Surface prediction based on HDM measurement

disciplines such as estimating spread of disease, predicting traffic flow, and forecasting household price. STARIMA model is used to describe the spatiotemporal characteristics of three-dimensional surface topography. The spatial characteristics of the surface topography are incorporated through spatial weighting matrix, and the time characteristics are reflected by the autoregressive terms and moving average terms. Compared with the models that can only predict the currently machined surface, the major advantage of the STARIMA model is that it can forecast the future machined surface.

An experiment on the finishing milling process of engine block faces is presented to illustrate the prediction of surface topography using STARIMA model. The experiment shows a good prediction agreement between the predicted surface and the real surface with an average forecast error of 0.181 μm for an area with 80×80 sites. Note that the main purpose is not to optimize the cutting parameters but to forecast surface topography of future machined surface under stable cutting conditions.

7.2.2   STARIMA Model

7.2.2.1   Model Basis

The STAMRA model was first proposed by Pfeifer and Deutsch [19–21] and is briefly reviewed as follows: Assume that $z_i(t)$ is the random variable observed at N fixed sites $(i = 1, 2, \ldots, N)$ at time t. The current observation $z_i(t)$ at time t and site i can be expressed as a weighted linear combination of its past observations of site i and the past observations at its neighboring sites. The spatial dependence among various sites in STARIMA model is expressed by a hierarchical series of $N \times N$ spatial weighting matrices W. Note $W^{(l)}$ as the lth order spatial weighting matrix with elements $w_{ij}^{(l)}$, $(i = 1, 2, \ldots, N, j = 1, 2, \ldots, N)$. The $w_{ij}^{(l)}$ represents the spatial weight to site i by its lth order neighbor site j. The spatial order l represents an order of euclidean distance of all the neighboring sites. Figure 7.2 shows the spatial weighting matrices of one to four orders. The spatial weights have a relation of $\sum_{j=1}^{N} w_{ij}^{(l)} = 1$, which means the total spatial influence to site i by other sites is one.

Given the spatial weight matrix $W^{(l)}$, the STARIMA model is expressed as

$$Z(t) = \sum_{k=1}^{p} \sum_{l=0}^{\lambda_k} \phi_{kl} W^{(l)} Z(t-k) - \sum_{k=1}^{q} \sum_{l=0}^{m_k} \theta_{kl} W^{(l)} \varepsilon(t-k) + \varepsilon(t) \qquad (7.1)$$

where $Z(t)$ is the vector form of $z_i(t)$, ϕ_{kl}, and θ_{kl} are the autoregressive parameters and moving average parameters at temporal lag k and spatial lag l, p, and q are the temporal orders of the autoregressive terms and moving average terms, λ_k and m_k

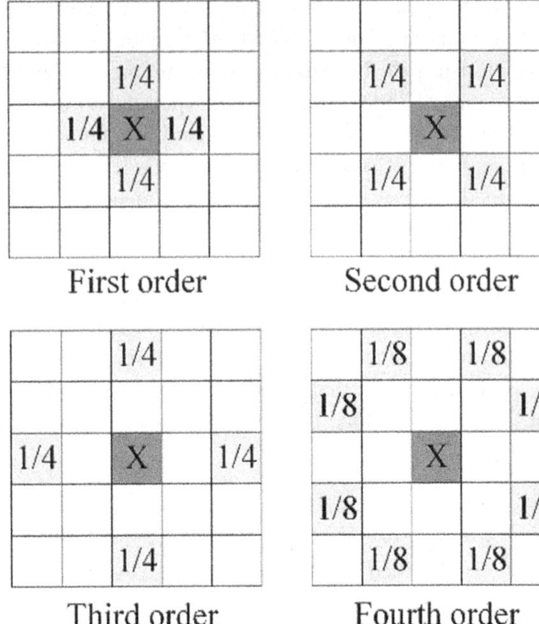

Fig. 7.2 Spatial weighting matrix of one to four orders

are the spatial orders of the kth temporal autoregressive term and the moving average term, and $\varepsilon(t)$ is a random normally distributed error vector with zero mean value and

$$E\left[\varepsilon(t)\varepsilon(t+s)'\right] = \begin{cases} \sigma^2 I_N & s = 0 \\ 0 & \textit{otherwise} \end{cases} \tag{7.2}$$

Equations (7.1) and (7.2) show that only a small number of parameters $(\phi_{kl}, \theta_{kl}, \sigma)$ are needed to be estimated, which improves the efficiency of modeling large spatial scale data.

7.2.2.2 Model Identification

The first stage of building a STARIMA model is to identify the temporal and spatial order of the autoregressive terms and moving average terms. The identification is conducted by calculating the space–time autocorrelation and the partial autocorrelation. The temporal and spatial orders are decided by examining at which order the space–time autocorrelation and partial autocorrelation functions cut off or tail off. The space–time autocorrelation $\rho_{l0}(s)$ between spatial lag l and spatial lag 0 at temporal lag s is calculated as

$$\rho_{l0}(s) = \frac{\gamma_{l0}(s)}{[\gamma_{ll}(0)\gamma_{00}(0)]^{1/2}} \qquad (7.3)$$

where $\gamma_{lk}(s) = E\left\{\left[W^{(l)}Z(t)\right]' \cdot \left[W^{(k)}Z(t+s)\right]\big/N\right\}$. The space–time partial auto-correlation ϕ_{jl} can be calculated from the space–time analog of the Yule–Walker equations

$$\gamma_{h0}(s) = \sum_{j=1}^{k}\sum_{l=0}^{\lambda} \phi_{jl}\gamma_{hl}(s-j) \qquad (7.4)$$

For pure space–time autoregressive (STAR) process, space–time autocorrelation tails off and space–time partial autocorrelation cuts off with space and time. Similarly, for pure space–time moving average (STMA) process, space–time autocorrelation cuts off and space–time partial autocorrelation tails off with space and time. Moreover, for STARIMA process, both the space–time autocorrelation and partial autocorrelation tail off.

7.2.2.3   Model Estimation and Diagnostic Checking

Once the temporal and spatial order is chosen, the second stage is to estimate the model parameters. Nonlinear optimization is used to minimize the error $E(\phi_{kl}, \theta_{kl})$:

$$\min E(\phi_{kl}, \theta_{kl}) = \sum_{i=1}^{N}\sum_{t=1}^{T} \varepsilon_i(t)^2$$

$$s.t. \ \ \varepsilon(t) = Z(t) - \sum_{k=1}^{p}\sum_{l=0}^{\lambda_k} \phi_{kl}W^{(l)}Z(t-k) + \sum_{k=1}^{q}\sum_{l=0}^{m_k} \theta_{kl}W^{(l)}\varepsilon(t-k) \qquad (7.5)$$

$\varepsilon(t)$ is recursively calculated from the observed $Z(t)$. The observations $Z(t)$ and $\varepsilon(t)$ with $t < 1$ are substituted by zero as initial starting values.

After the parameters are estimated, the model should be checked. The residuals from the fitted model should be white noise. One available test is calculating the space–time autocorrelations of the residuals. The variance of the autocorrelations should be zero if the residuals are white noise. Otherwise, the model should be updated to decouple the influence of residuals.

7.2.3 *Experiment*

To illustrate the 3D surface topography prediction procedure based on HDM
measurement using the STARIMA model, an experiment on the finishing milling of
engine block faces is conducted. The milling process uses a CBN milling cutting
tool with a diameter of 200 mm. The cutting speed is 816.4 m/min, the depth of cut
is 0.5 mm, and feed rate is 3360 mm/min. A total of 16 engine blocks are successively machined using the same cutting tool. The first 15 engine blocks are used
to build the STARIMA model, and the 16th engine block is used to evaluate the
model performance.

The total measured site of HDM measured surface is about 0.8 million. Due to
the existence of measurement noise and alignment problem, the raw HDM data is
preprocessed using the method proposed [14]. To reduce the computing difficulty, a
local area of 80 × 80 grid is selected from the measured surface. The position of
the selected area is shown in Fig. 7.3. The area has a total of 3171 nonempty points.
Therefore, the model consists of $N = 3171$ sites and $T = 15$ time periods.

The mean surface heights of the selected area of the first 15 engine blocks are
shown in Fig. 7.4.

The space–time autocorrelation and partial autocorrelation of the differenced
series are calculated according to Eqs. (7.3) and (7.4), and the results are given in
Tables 7.1 and 7.2, respectively. It can be seen that the autocorrelations tail off after
three spatial lags at one temporal lag and the partials seem to tail off after one
temporal lag with zero spatial lag, which indicates the space–time series is a
STARIMA (1, 1, 4) model. Therefore, the model is in the following form:

$$Z(t) = \phi_{10}Z(t-1) - \theta_{10}\varepsilon(t-1) - \theta_{11}W^{(1)}\varepsilon(t-1)$$
$$- \theta_{12}W^{(2)}\varepsilon(t-1) - \theta_{13}W^{(3)}\varepsilon(t-1) + \varepsilon(t) \tag{7.6}$$

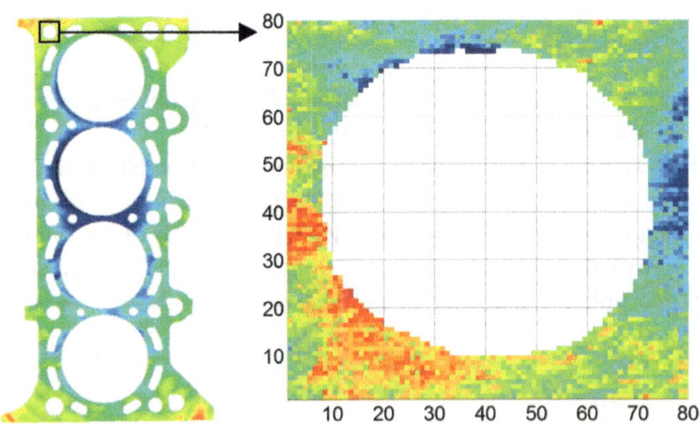

Fig. 7.3 The selected area for model building

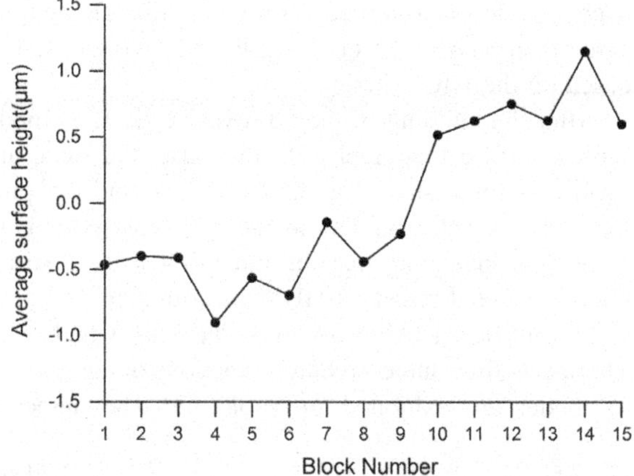

Fig. 7.4 Mean surface height of the selected area

Table 7.1 Space–time autocorrelation of differenced series

Spatial lag(s) Time lag(t)	0	1	2	3
1	−0.498	−0.415	−0.386	−0.367
2	0.034	0.038	0.037	0.037
3	0.037	0.044	0.045	0.045
4	−0.016	−0.019	−0.019	−0.020
5	0.030	0.038	0.038	0.042
6	−0.069	−0.087	−0.088	−0.094
7	0.008	0.017	0.018	0.020
8	−0.046	−0.056	−0.059	−0.057

Table 7.2 Space–time partial autocorrelation of differenced series

Spatial lag(s) Time lag(t)	0	1	2	3
1	−0.498	0.017	0.023	0.021
2	−0.284	0.076	0.059	0.061
3	−0.119	0.179	0.123	0.119
4	−0.055	0.183	0.127	0.114
5	0.018	0.239	0.148	0.154
6	−0.056	0.077	0.044	0.029
7	−0.080	0.052	0.025	0.009
8	−0.145	−0.029	−0.044	−0.006

where $Z(t)$ is the differenced series. The estimation procedure calculated parameter $\phi_{10} = -0.148$, $\theta_{10} = 0.678$, $\theta_{11} = -0.023$, $\theta_{12} = -0.086$, $\theta_{13} = -0.134$, and standard error of $\varepsilon(t)$ is 1.156 μm. By building the STARIMA model, surface height of each site is modeled as a linear combination of the previous measurement at this site plus the previous prediction error at this site and its first three order neighbors plus a random error. STARIMA with higher autoregressive and moving average orders is also tried. However, although the fitting errors of those models are better than STARIMA (1, 1, 4), the STARIMA (1, 1, 4) model has the best results based on the AIC criterion.

After the building of the STARIMA (1, 1, 4) model, it can be used to forecast further surface topography. In this case, the same area of the 16th engine block surface is forecasted. The actual surface and the predicted surface are shown in Fig. 7.5, respectively. The average forecast error is 0.181 μm, and the RMSE is 1.059 μm, indicating that the three-dimensional surface topography can be accurately predicted based on HDM measurement.

Diagnostic checking involves checking whether the residuals are white noise. The space–time autocorrelation functions of the residuals from the STARIMA (1, 1, 4) model are exhibited in Table 7.3. The autocorrelation of the residuals is

Fig. 7.5 The real surface and the predicted surface

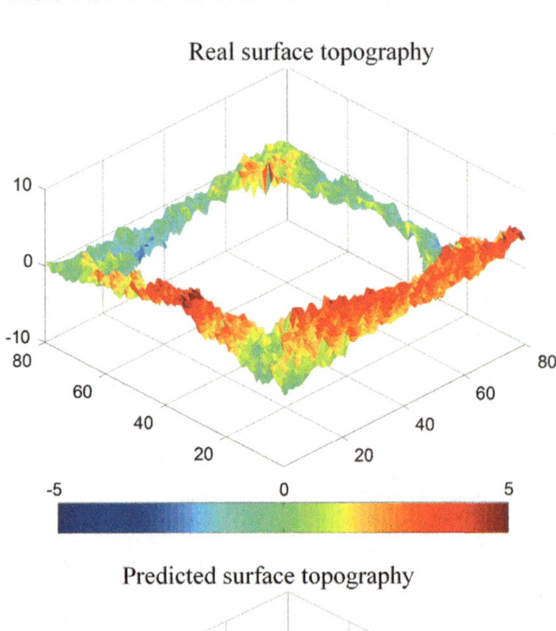

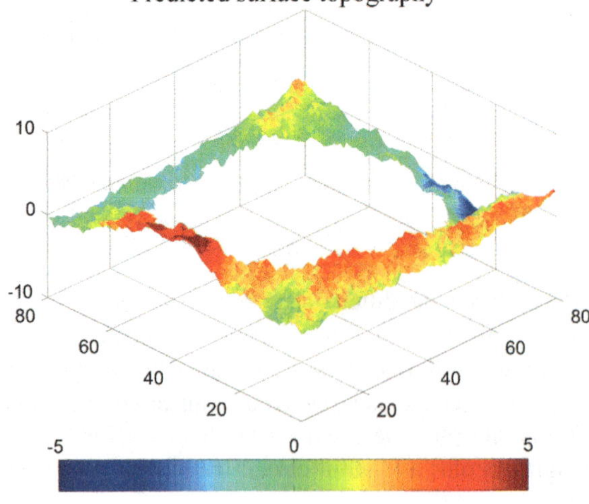

Table 7.3 Space–time autocorrelation of residuals

Spatial lag(s) Time lag(t)	0	1	2	3
1	−0.318	−0.295	−0.274	−0.254
2	−0.006	0.008	0.013	0.016
3	0.028	0.037	0.037	0.035
4	−0.015	−0.020	−0.021	−0.021
5	0.028	0.042	0.044	0.049
6	−0.045	−0.070	−0.072	−0.082
7	−0.016	−0.012	−0.013	−0.010
8	−0.014	−0.022	−0.027	−0.022

relatively large, especially in the first time lag. The reason is that the number of spatial sites ($N = 3171$) is rather large, and not all the sites are strictly stationary. The complicated structure in these sites cannot be described by STARIMA model completely. However, the prediction error is rather small and acceptable compared with the surface flatness tolerance of 50 μm.

7.2.4   Conclusions

A three-dimensional surface topography prediction model based on HDM is presented. HDM provides full inspection of the entire surface topography and the measured HDM data is in the form of space–time series. Therefore, a STARIMA model is adopted to describe the spatiotemporal evolution of the surface topography. An experiment on the finishing milling process of engine block faces is conducted to illustrate the prediction procedure. The experiment shows a satisfactory prediction result with average forecast error 0.181 μm and RMSE 1.059 μm^2. The proposed method gives the opportunity to pre-control the manufacturing process by providing early warning of surface quality deterioration through a single model with a limit number of parameters. For example, one can change the cutting tool based on the predicted surface topography before severe tool wear happens.

7.3   A Space–Time Multi-output Support Vector Regression Based Predicting Method for 3D Engineering Surface

7.3.1   Introduction

Since product quality attracts much attention in manufacturing field, the effects of surface topography on product quality have been a focus. Surface topography is a vital link between a part generated by manufacturing process and the functionality that is expected of it. It is clear that surface topography guarantees functional behavior of the part during manufacturing process. Surface topography can be considered as the linear superposition of flaws, form error, and surface texture that contains waviness and roughness. It is hard and costly to obtain the proper various cutting parameters that influence the final surface topography. At this point, surface topography prediction will be helpful.

In the machining field, a great number of researches on surface topography prediction of machined parts have been published. It is reported that the factors including tool wear, cutting conditions, and the material properties that lead to the formation of surface topography for the parts. To prevent an unexpected surface topography, researchers simulate the conditions during machining so as to build the relationship between different factors and desired part functions. Moreover, with the development of computer-aided technology, various precise prediction models have been proposed. Benardos et al. first reviewed the surface topography prediction methods [12]. The forecasting methods were roughly classified into two types: theoretical method and experimental method. Machining theory and cutting kine-matics were the major theories that conducted to the analytical models. A generic model was proposed to forecast the cutting forces along with three-dimensional surface topography in side milling operation [22]. And soon afterward, simulated models of the formation of machined surfaces were built to visualize surface topography. Tapoglou et al. simulated surface topography for face milling using tool kinematics in a CAD environment [3]. Thasana et al. simulated the shape generation processes to estimate the surface topography considering kinematic motion deviations and the cutting tool geometries [23]. Denkena et al. proposed a combined material removal simulation (MRS) for surface topography prediction, which considered all kinematic properties of the process to predict stochastic influences [24]. The experimental method was to explore the mechanism of the influence of each factor on the observed experimental results. This method was mainly employed in cases without any analytical formulations of the causalities between the various factors such as cutting force and tool vibrations. In this way, surface topography was predicted through various approaches such as spectral analysis [25], regression analysis [26], Taguchi's design of experiment [6, 27], and several artificial intelligent (AI) methods [28, 29].

Conventionally, most of the surfaces are measured by the widely used coordinate measuring machine (CMM). With the low-density sampling, the measurement result of CMM cannot represent the entire surface topography. Simultaneously, it exists a great number of large surfaces in engineering practice, and the precise measurement is required. Therefore, a full inspection and high-density sampling measuring instrument are needed. Recently, a novel measuring instrument called noncontact high-definition metrology (HDM) has been developed [13]. It is a new laser holographic interferometry measurement that can measure surface height of objects which are larger than the field of view. Millions of point clouds are generated within seconds to represent the entire measured surface with 150 μm lateral resolution in x–y direction and 1 μm accuracy in z-direction. The HDM data of 3D machined surfaces contains the surface topography evolution information, which

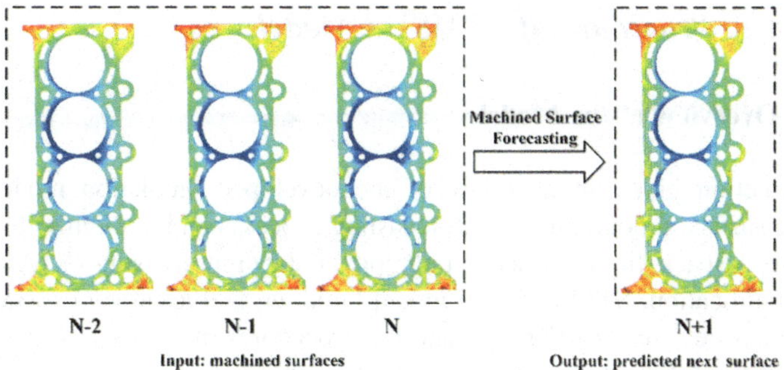

Fig. 7.6 3D surface prediction based on HDM

can be used as input factors to predict the next waiting machined surface. An example of the HDM measured engine block face and the prediction scheme is shown in Fig. 7.6. Once the surface is accurately predicted, 3D surface form error evaluation [30], filtering [17, 31], classification [16, 17], and tool wear monitoring methods [18] based on HDM can be employed for the pre-control of the manufacturing process.

As a typical representative of artificial intelligent methods, support vector machine (SVM) has been widely used to forecast the hidden information in big data. SVM is first proposed by Vladimir N. Vapnik based on Statistical Learning Theory [32]. It has become a recognized tool to solve high-dimensional, nonlinear, and heterogeneous problems efficiently. Support vector regression (SVR) is considered as an extension of SVM to solve nonlinear regression estimation problem. Multi-output support vector regression (MSVR) is developed to overcome the multi-output problems that cannot be disposed by the standard SVR [33, 34]. In order to implement spatiotemporal forecasting, a well-known method called space–time autoregressive moving average (STARMA) [20] is adopted to identify the spatial lag and temporal lag which should be added into the input and output. Notwithstanding, the analysis of space–time series with MSVR in manufacturing process is few. The HDM data measured from these successive machined surface is a form of space–time series that contain the evolution information of machining process. Hence, an attempt has been made to achieve the 3D machined surface topography forecasting with space–time multi-output support vector regression (STMSVR) model using the HDM data. The STMSVR model is the first time to be developed to describe the spatiotemporal of machined surface topography, which can reflect the spatial and temporal characteristics through the training. Furthermore, it is time-saving and stable without overfitting.

7.3.2   The Procedure of STMSVR Model

7.3.2.1   Overview of the Model

This subsection gives an overview of the developed prediction model for 3D machined surface topography. The forecasting process can be roughly divided into four stages: Space–time surface topography data preprocessing, STARMA and model identification, STMSVR and model building, and accuracy evaluation. The framework is described in Fig. 7.7, and the procedures of stages 1–4 are described in detail in the following subsections, respectively.

7.3.2.2   Space–Time Surface Topography Data Preprocessing

Space–time series are the specific time series that contain spatial correlation. With regard to the machined surface under successively stable cutting conditions, the HDM data measured from these surface topographies is the typical kind of space–time series. It is significant for the future machining to extract the implicit facts and spatiotemporal relationships that are contained in the space–time series of successive machined surface topographies. The successively machined surfaces on production line are conveyed to be measured by HDM. Before further image processing, a preprocessing approach is adopted to convert the point clouds into a position-maintained and height-encoded gray image [30]. Therefore, a series of machined surfaces $S_i(t)$ are generated to constitute the dataset, and M is the number of nonempty sites for each machined surface and N is the total of machined surface. To avoid numerical singularity during the machine learning, zero-mean normalization is adopted to make the treated dataset $S_i^*(t)$ obey standard normal distribution. The formulas of the transfer function are followed:

$$x^* = \frac{x - \mu_S}{\sigma_S} \tag{7.7}$$

where μ_S and σ_S are, respectively, average and standard deviation of the dataset.

7.3.2.3   STARMA and Model Identification

Due to the existing of spatiotemporal characteristics, the successively machined surfaces are highly correlated both in space and time domains, so it is necessary to identify the temporal lag s and spatial lag l of surface space–time series $S_i^*(t)$. STARMA is adopted to find out the spatial and temporal lag. The current observation $S_i^*(t)$ at site i and time t can be expressed as a weighted linear superposition of its past observations of site i and the past observations at its neighboring sites. The spatial dependence among various sites in STARMA model is expressed by a

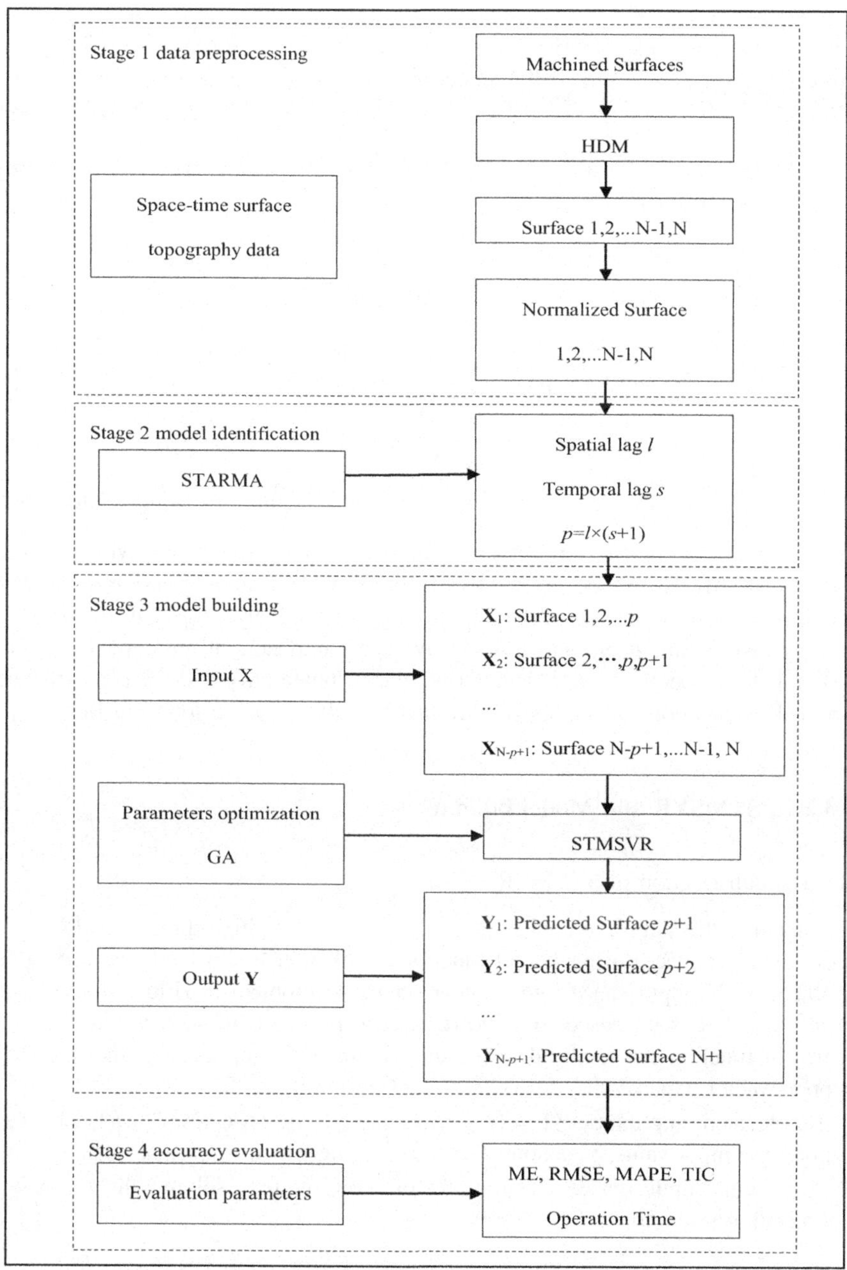

Fig. 7.7 Prediction framework

hierarchical series of $M \times M$ spatial weighting matrix $\mathbf{W}_S$ that is adopted to describe the spatial characteristic of each surface. Thus, the space–time series $S_i^*(t)$, autocorrelation $\rho_{l0}(s)$, and partial autocorrelation φ_{jl} are expressed as

$$S_i^*(t) = \sum_{s=1}^{p} \sum_{l=0}^{\lambda_s} \phi_{sl} \mathbf{W}_S^{(l)} S_i^*(t-s) - \sum_{s=1}^{q} \sum_{l=0}^{m_s} \theta_{sl} \mathbf{W}_S^{(l)} \varepsilon_i(t-s) + \varepsilon_i(t) \qquad (7.8)$$

$$\rho_{l0}(s) = \frac{\gamma_{l0}(s)}{[\gamma_{ll}(0)\gamma_{00}(0)]^{1/2}} \qquad (7.9)$$

$$\gamma_{h0}(s) = \sum_{j=1}^{s} \sum_{l=0}^{\lambda} \varphi_{jl} \gamma_{hl}(s-j) \qquad (7.10)$$

where $\gamma_{ls}(s) = E\left\{ \left[\mathbf{W}_S^{(l)} \mathbf{S}(t) \right]' \cdot \left[\mathbf{W}_S^{(s)} \mathbf{S}(t+s) \right] / M \right\}$ is the space–time autocovari-ance, p and q are the temporal orders of autoregressive terms and moving average terms, respectively. λ_s and m_s are the spatial orders of autoregressive terms and moving average terms at sth temporal lag, respectively. ϕ_{sl} and θ_{sl} are the autoregressive parameters and moving average parameters at temporal lag s and spatial lag l, respectively. $\varepsilon_i(t)$ is a normally distributed error and $E[\varepsilon_i(t)] = 0$. The detailed explanation for the above correlation analysis can be found in Ref. [20].

7.3.2.4   STMSVR and Model Building

(1) The construction of STMSVR

By mapping the sampling data in the input space into a high-dimensional feature space (Hilbert space) using kernel functions, SVR converts a nonlinear separable problem in the input space into a linear separable problem in Hilbert space. It is clear that SVR is proposed to solve regression problems in such a way that the regressor function can totally describe the feature of the input space, the so-called support vector expansion.

Let the sampling dataset $\Omega = \{(x_i, y_i)|i = 1, 2, \ldots n\}$ $(x_i \in R^d, y_i \in R)$, where x_i denotes the input value, y_i denotes the output value which is a scalar quantity, and R^d represents input space. The training of SVR requires solving the following optimization problem:

$$\min \frac{1}{2}\|w\|^2 + C \sum_{i=1}^{n} (\xi_i + \xi_i^*) \qquad (7.11)$$

$$\text{subject to} \begin{cases} y_i - w^T \phi(x_i) - b \leq \varepsilon + \xi_i \\ w^T \phi(x_i) + b - y_i \leq \varepsilon + \xi_i^* \\ \xi_i, \xi_i^* \geq 0, \forall i = 1, \ldots .n \end{cases} \tag{7.12}$$

where $\phi(\cdot)$ is a nonlinear map function to map the input into a higher feature space, w is the weight vector, and b is the bias. The penalization parameter $C > 0$ is a trade-off constant, and $\varepsilon > 0$ is a small positive number which comes from Vapnik ε-insensitive loss function. ξ_i and ξ_i^* are the slack variables.

With the limitation of dimension, the standard SVR cannot solve multi-output problems. The multidimensional regression estimation problem appears when the output variable $\mathbf{y}_i$ becomes a vector for each input vector variable $\mathbf{x}_i$ $(i = 1, \ldots, n)$. When the output is a vector with Q variables, i.e., $\mathbf{y} \in R^Q$, the regressor parameters w^j and $b^j (j = 1, \ldots, Q)$ for each output are needed to be found in the multidimensional regression estimation problem. The developed STMSVR is that adding the spatial lag and temporal lag into MSVR and getting the more suitable input and output which contain the spatiotemporal characteristics. Analogously, the solving method for one-dimensional SVR can be generalized directly to the multidimensional condition, resulting in the minimization of

$$L_P(\mathbf{W}, \mathbf{b}) = \frac{1}{2} \sum_{j=1}^{Q} \left\| w^j \right\|^2 + C \sum_{i=1}^{n} L(u_i) \tag{7.13}$$

where $u_i = \|\mathbf{e}_i\| = \sqrt{\mathbf{e}_i^T \mathbf{e}_i}$, $\mathbf{e}_i^T = \mathbf{y}_i^T - \phi(\mathbf{x}_i)^T \mathbf{W} - \mathbf{b}^T$, $\mathbf{W} = [w^1, \ldots, w^Q]$, $\mathbf{b} = [b^1, \ldots, b^Q]^T$. The Vapnik ε-insensitive loss function can be extended to multiple dimensions, and all dimensions can be considered into a unique constraint yielding a single support vector. The loss function is followed as

$$L(u) = \begin{cases} 0 & u < \varepsilon \\ u^2 - 2u\varepsilon + \varepsilon^2 & u \geq \varepsilon \end{cases} \tag{7.14}$$

which is differentiable [33].

It is noted that the STMSVR returns a solution that is sparse and multidimensional, thus accounting for output relationships. To obtain the solution of the STMSVR, a tool called iterative reweighted least square (IRWLS) algorithm is applied [35]. First, a first-order Taylor expansion of $L(u)$ is adopted to approximate $L_P(\mathbf{W}, \mathbf{b})$ on each iteration k leading to

$$L_P'(\mathbf{W}, \mathbf{b}) = \frac{1}{2} \sum_{j=1}^{Q} \left\| w^j \right\|^2 + C \left(\sum_{i=1}^{n} L(u_i^k) + \frac{dL(u)}{du} \Big|_{u_i^k} \frac{(\mathbf{e}_i^k)^T}{u_i^k} [\mathbf{e}_i - \mathbf{e}_i^k] \right) \tag{7.15}$$

where $u_i^k = \|\mathbf{e}_i^k\| = \sqrt{(\mathbf{e}_i^k)^T \mathbf{e}_i^k}$, $(\mathbf{e}_i^k)^T = \mathbf{y}_i^T - \phi(\mathbf{x}_i)^T \mathbf{W}^k - (\mathbf{b}^k)^T$, $\mathbf{W} = \mathbf{W}^k$, and $\mathbf{b} = \mathbf{b}^k$. Then, a quadratic derivative from $L_P'(\mathbf{W}, \mathbf{b})$ is as follows:

$$L_P'' = \frac{1}{2}\sum_{j=1}^{Q}\|w^j\|^2 + C\left(\sum_{i=1}^{n}L(u_i^k) + \frac{dL(u)}{du}\Big|_{u_i^k}\frac{u_i^2 - (u_i^k)^2}{2u_i^k}\right)$$

$$= \frac{1}{2}\sum_{j=1}^{Q}\|w^j\|^2 + \frac{1}{2}\sum_{i=1}^{n}\alpha_i u_i^2 + \tau \tag{7.16}$$

where

$$\alpha_i = \frac{C}{u_i^k}\frac{dL(u)}{du}\Big|_{u_i^k} = \begin{cases} 0 & u_i^k < \varepsilon \\ \frac{2C(u_i^k - \varepsilon)}{u_i^k} & u_i^k \geq \varepsilon \end{cases} \tag{7.17}$$

and τ is a sum of constant terms. Obviously, Eq. (7.17) is a weighted least squares problem. In order to ascertain the optimization of $L_P(\mathbf{W}, \mathbf{b})$, a descending direction is built by using the optimal solution of L_P'' and a line search algorithm is adopted to compute the next step solution. After the above steps, the solution equation can be expressed as

$$\begin{bmatrix} \mathbf{\Phi}^T\mathbf{D}_\alpha\mathbf{\Phi}+\mathbf{I} & \mathbf{\Phi}^T\alpha \\ \alpha^T\mathbf{\Phi} & \mathbf{1}^T\alpha \end{bmatrix}\begin{bmatrix} w^j \\ b^j \end{bmatrix} = \begin{bmatrix} \mathbf{\Phi}^T\mathbf{D}_\alpha\mathbf{y}^j \\ \alpha^T\mathbf{y}^j \end{bmatrix} \tag{7.18}$$

where $\mathbf{\Phi} = [\phi(x_1), \ldots, \phi(x_n)]^T$, $\alpha = [\alpha_1, \ldots, \alpha_n]^T$, $(\mathbf{D}_\alpha)_{ij} = \alpha_i\delta(i-j)$, $\mathbf{y}^j = [y_{1j}, \ldots y_{nj}]$, and $\mathbf{1}$ is an all-one column vector for $i = 1, \ldots, n$, $j = 1, \ldots, Q$. Generally, the nonlinear mapping is not needed to be exactly known, and "Kernel Trick" is adopted to implement the nonlinear regression model. A specified kernel function $K(x, x_i)$ which fulfills Mercer's theorem is shown as

$$K(x, x_i) = \langle \phi(x), \phi(x_i) \rangle \tag{7.19}$$

Here, a linear combination of the training samples in the feature space can express the learning problem, i.e., $w^j = \sum_i \phi(x_i)\beta^j = \mathbf{\Phi}^T\beta^j$. Then Eq. (7.12) can be converted into:

$$\begin{bmatrix} \mathbf{K}+\mathbf{D}_\alpha^{-1} & \mathbf{1} \\ \alpha^T\mathbf{K} & \mathbf{1}^T\alpha \end{bmatrix}\begin{bmatrix} \beta^j \\ b^j \end{bmatrix} = \begin{bmatrix} \mathbf{y}^j \\ \alpha^T\mathbf{y}^j \end{bmatrix} \tag{7.20}$$

At present, various kernel functions are adopted to achieve the mapping from the input space into feature space, such as linear, polynomial, radial basis function (RBF), Fourier, wavelet kernel functions, and so on. It cannot be concluded that one kernel outperforms another in prediction accuracy. That is to say, different kernel functions with respective properties are applicable to different problems. Due to the disadvantage of having too simple a kernel formation, the kernel function is expanded from a single kernel to a combination of multiple kernels. Finally, the solution ($\mathbf{W}$ and $\mathbf{b}$) and the predicting output can be obtained exactly.

(2) Model building

Due to the nonlinear spatiotemporal relationships in the space–time series of machined surface topographies, STMSVR is adopted to make an attempt to use HDM data to forecast future surface topography. For simplicity, the process of model building is to ascertain the input and output matrix and model parameters including kernel parameters, penalization parameter C, insensitive loss parameter ε, and so on.

The input and output matrices of STMSVR are modified by the spatial lag and temporal lag that are identified in STARMA. Since the STMSVR model is a static model, a kind of technology called "Time Window" is used to achieve the dynamic forecasting. The surface series are rearranged as "Time Window" required to be the input of STMSVR. For surface topography forecasting, every machined surface can be considered as an input vector $\mathbf{x}_i$ and the next machined surface is the corresponding output vector $\mathbf{y}_i$. Assuming the surface series $S_i^*(t)$ is at l spatial lag and s temporal lag, then the input matrix $\mathbf{X}$ is $X_L \times X_W$ and the output matrix $\mathbf{Y}$ is $Y_L \times Y_W$ ($X_L = Y_L = N - p + 1$, $X_W = M \times p$, $Y_W = M \times l$,   and $p = l \times (s + 1)$). The detailed formulation is described in Table 7.4.

Example 1. Consider five normalized successively machined surfaces $S_1^*, S_2^*, S_3^*, S_4^*, S_5^*$; it exists 1000 nonempty sites for each surface, that is, $N = 5, M = 1000$,   and $l = 1, s = 1, p = 2$; then, the input matrix $\mathbf{X}(4 \times 2000)$ and output matrix $\mathbf{Y}(4 \times 1000)$ can be given by:

$$
\mathbf{X} = \begin{bmatrix} \mathbf{X}_1 \\ \mathbf{X}_2 \\ \mathbf{X}_3 \\ \mathbf{X}_4 \end{bmatrix} = \begin{bmatrix} S_1^*, S_2^* \\ S_2^*, S_3^* \\ S_3^*, S_4^* \\ S_4^*, S_5^* \end{bmatrix}, \mathbf{Y} = \begin{bmatrix} \mathbf{Y}_1 \\ \mathbf{Y}_2 \\ \mathbf{Y}_3 \\ \mathbf{Y}_4 \end{bmatrix} = \begin{bmatrix} S_3' \\ S_4' \\ S_5' \\ S_6' \end{bmatrix} \tag{7.21}
$$

Table 7.4   The input and output formulation

Normalized surface 1, 2,…N−1, N	
Input X	Output Y
X_1: Surface 1, 2,…p	Y_1: Predicted surface $p + 1$
X_2: Surface 2,…, p, $p + 1$	Y_2: Predicted surface $p + 2$
…	…
$X_{N\text{-}p+1}$: Surface $N - p + 1,…N - 1, N$	Y_{N-p+1}: Predicted surface $N + 1$

Second, many researches have shown that setting the optimal parameters can immensely improve the prediction accuracy. Genetic algorithm (GA) is adopted to seek out best set of parameters (C and ε) in STMSVR. It is an algorithm of nonlinear global optimization that inspired by the biological evolution mechanism (survival of the fittest, crossover, mutation, etc.).

Finally, the selection of kernel function is also significant to the results forecasting. In accordance with the kernel theory, a convolution kernel is a sort of construction method of kernel function. The operation of the construction can be considered as an expansion of a standard kernel function. A space–time kernel function $K_{ST}(x, y)$ [36] is proposed to satisfy the space–time correlation in STMSVR. Its form is:

$$K_{ST}(x, y) = \sum_{i=1}^{\lambda} (K_S(x, y) \cdot K_T(x, y)) \tag{7.22}$$

The space–time kernel $K_{ST}(x, y)$ is a form of space–time convolution that consists of the spatial kernel $K_S(x, y)$ and the temporal kernel $K_T(x, y)$. With a slightly larger value of the order λ, the learning capacity can promote dramatically. Conventionally, in machine learning, RBF has been adopted to tackle local spatial heterogeneous characteristics without regard for the dimensionality of the sample data. Furthermore, it should be aware that Fourier kernel matches well for modeling periodic series and polynomial kernel is more suitable to sequences without periodicity due to its stronger generalizing ability. Therefore, RBF can be used in the spatial kernel $K_S(x, y)$, and a combination of Fourier kernel and polynomial kernel can be used in the temporal kernel.

7.3.2.5 Accuracy Evaluation

In order to examine the performance of the developed model, four common statistical evaluation parameters that contain mean error (ME), mean absolute percent error (MAPE), root-mean-squared error (RMSE), and Theil inequality coefficient (TIC) are chosen to measure the prediction accuracy. The formulas of the above parameters are listed as follows.

$$ME = \frac{1}{n} \sum_{i=1}^{n} (x_i - x_i') \tag{7.23}$$

$$RMSE = \sqrt{\frac{1}{n} \sum_{i=1}^{n} (x_i - x_i')^2} \tag{7.24}$$

$$MAPE = \frac{1}{n}\sum_{i=1}^{n}\left|\frac{x_i - x_i'}{x_i}\right| \tag{7.25}$$

$$TIC = \frac{\sqrt{\frac{1}{n}\sum_{i=1}^{n}(x_i - x_i')^2}}{\sqrt{\frac{1}{n}\sum_{i=1}^{n}x_i^2} + \sqrt{\frac{1}{n}\sum_{i=1}^{n}(x_i')^2}} \tag{7.26}$$

Furthermore, operation time is also a crucial criterion to assess the efficiency of the model. Integrating these statistical criterions, the predictive effect of the developed model can be evaluated properly by taking the above parameters into an integrated consideration.

7.3.3   Case Study

To illustrate the validity of 3D machined surface topography forecasting with STMSVR based on HDM data, a case study on the finishing milling of top surfaces of engine cylinder blocks is conducted. The top surface is a major sealing surface in automotive power train, and a precise face milling cutter with 15 cutting inserts and 3 wiper inserts was used to implement the milling process. The whole process was completed on an EX-CELL-O machining center, and the diameter of the cutter is 200 mm. The machining center and face milling cutter are shown in Fig. 7.8a and b, respectively. Quaker 370 KLG cutting fluid was adopted. The feed rate is 3360 mm/min, the depth of milling is 0.5 mm, and the milling speed is 816.4 m/min. A total of 16 engine cylinder blocks are successively machined under the same processing condition. The former 15 block surfaces are used to train and test the STSMVR model, and the 16th surface is used to evaluate the model performance. To get the real surface information, all the block surfaces are measured by HDM which provides a great platform to analyze the surface topography. The measured top surface is shown in Fig. 7.9a, and the converted gray image is shown in Fig. 7.9b.

There are about 0.8 million sites for each HDM measured the entire top surface. To improve the computing efficiency, four typical local areas are selected from the measured surface. The position of the selected areas is shown in Fig. 7.10. STMSVR is used to predict the four regions, respectively. Region 1 has 8110 nonempty points with 120×120 grid. Regions 2 and 3 have 4525 and 3171 nonempty points, respectively, with the same 80×80 grid. Region 4 has 8661 nonempty points with 150×150 grid. These nonempty points constitute the datasets for forecasting model. First, the spatial lag l and temporal lag s are identified as $l = 1, s = 2$ through STARMA. Then, GA is adopted to adjust and choose the appropriate parameters ($C = 0.5, \varepsilon = 0.001$) to obtain the best results.

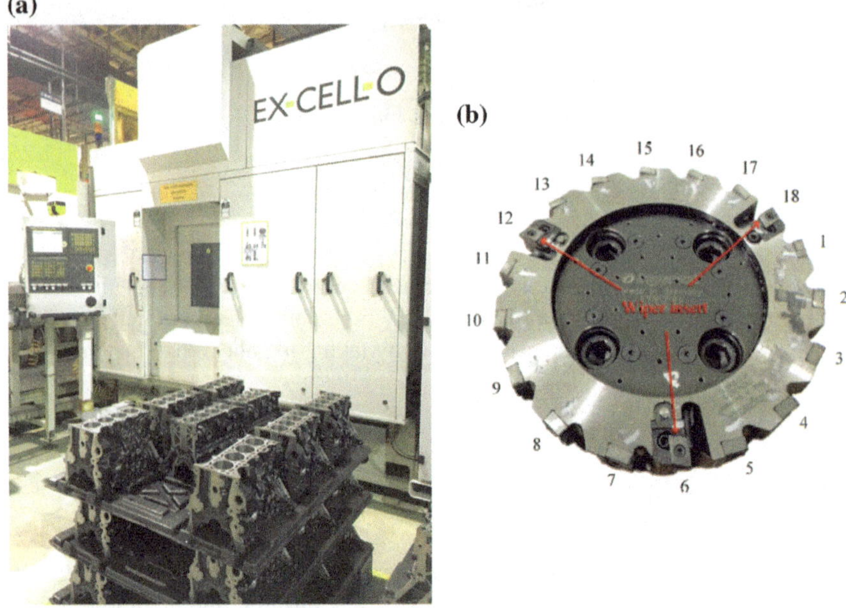

Fig. 7.8 **a** EX-CELL-O machining center **b** The face milling cutter

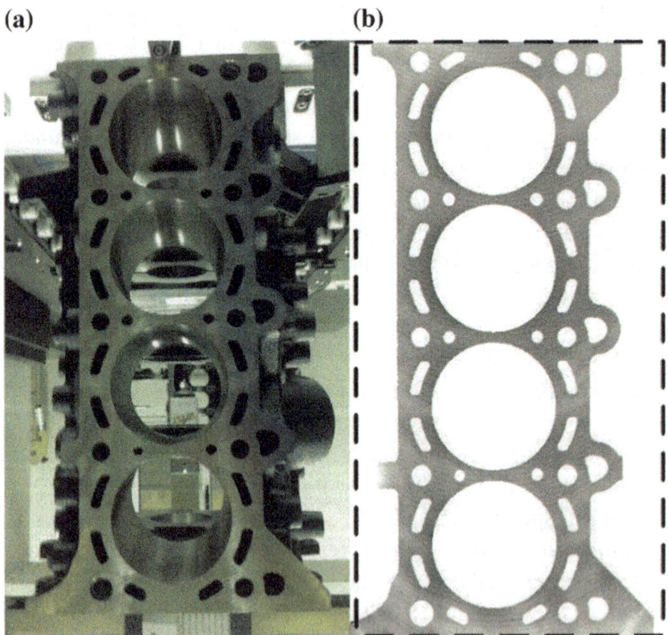

Fig. 7.9 **a** Top surface of engine cylinder block **b** The converted gray image

Next, a space–time kernel function $K_{ST}(x,y) = \left\{ exp(-\|x-y\|^2 / \sigma^2) \bullet \right.$ $\left. ((x \bullet y)+1)^{par} | \sigma = 1, par = 3 \right\}$ which is a combination of RBF and polynomial function is applied to achieve the forecast process. The order of kernel function λ depends on the input, and the recommended value is equal to the temporal lag, such as $\lambda = s = 2$. Finally, the most ordinary testing standard named one-step-ahead prediction is adopted in this case study. All procedures are following the framework in Fig. 7.7. So the input and output in training and testing phase are listed in Eqs. (7.21) and (7.22).

Input and output in training phase:

$$\mathbf{X} = \begin{bmatrix} \mathbf{X}_1 \\ \mathbf{X}_2 \\ \cdots \\ \mathbf{X}_{12} \end{bmatrix} = \begin{bmatrix} \mathbf{S}_1^*, \mathbf{S}_2^*, \mathbf{S}_3^* \\ \mathbf{S}_2^*, \mathbf{S}_3^*, \mathbf{S}_4^* \\ \cdots \cdots \\ \mathbf{S}_{12}^*, \mathbf{S}_{13}^*, \mathbf{S}_{14}^* \end{bmatrix}, \mathbf{Y} = \begin{bmatrix} \mathbf{Y}_1 \\ \mathbf{Y}_2 \\ \cdots \\ \mathbf{Y}_{12} \end{bmatrix} = \begin{bmatrix} \mathbf{S}_4' \\ \mathbf{S}_5' \\ \cdots \\ \mathbf{S}_{15}' \end{bmatrix} \quad (7.27)$$

Input and output in testing phase:

$$\mathbf{X} = [\mathbf{X}_{13}] = \left[\mathbf{S}_{13}^*, \mathbf{S}_{14}^*, \mathbf{S}_{15}^* \right], \mathbf{Y} = [\mathbf{Y}_{13}] = \left[\mathbf{S}_{16}' \right] \quad (7.28)$$

Table 7.5 shows the prediction results of the 16th surface. It is obvious to find that less errors are achieved and the results are stable. Comparing with the surface flatness tolerance of 50 µm, the prediction errors containing ME and RMSE are rather small and acceptable. All the MAPEs of STMSVR are less than 15% that means the high accurate prediction. TIC represents the difference between the actual value and the predicted value, which belongs to 0–1. The smaller TIC means the

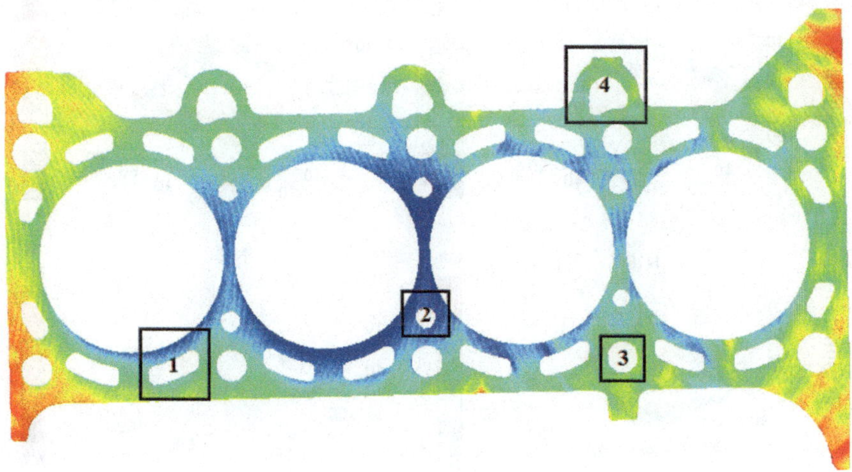

Fig. 7.10 The four selected areas

Table 7.5 Accuracy measures of STMSVR for surface regions

Surface region	Region 1	Region 2	Region 3	Region 4
ME (μm)	0.429	0.222	0.242	-0.283
RMSE (μm^2)	0.919	0.695	1.079	0.997
MAPE	12.41%	11.82%	10.16%	14.61%
TIC	16.26%	25.50%	27.56%	21.86%
Operation time	2.8 s	2.0 s	1.5 s	4.3 s

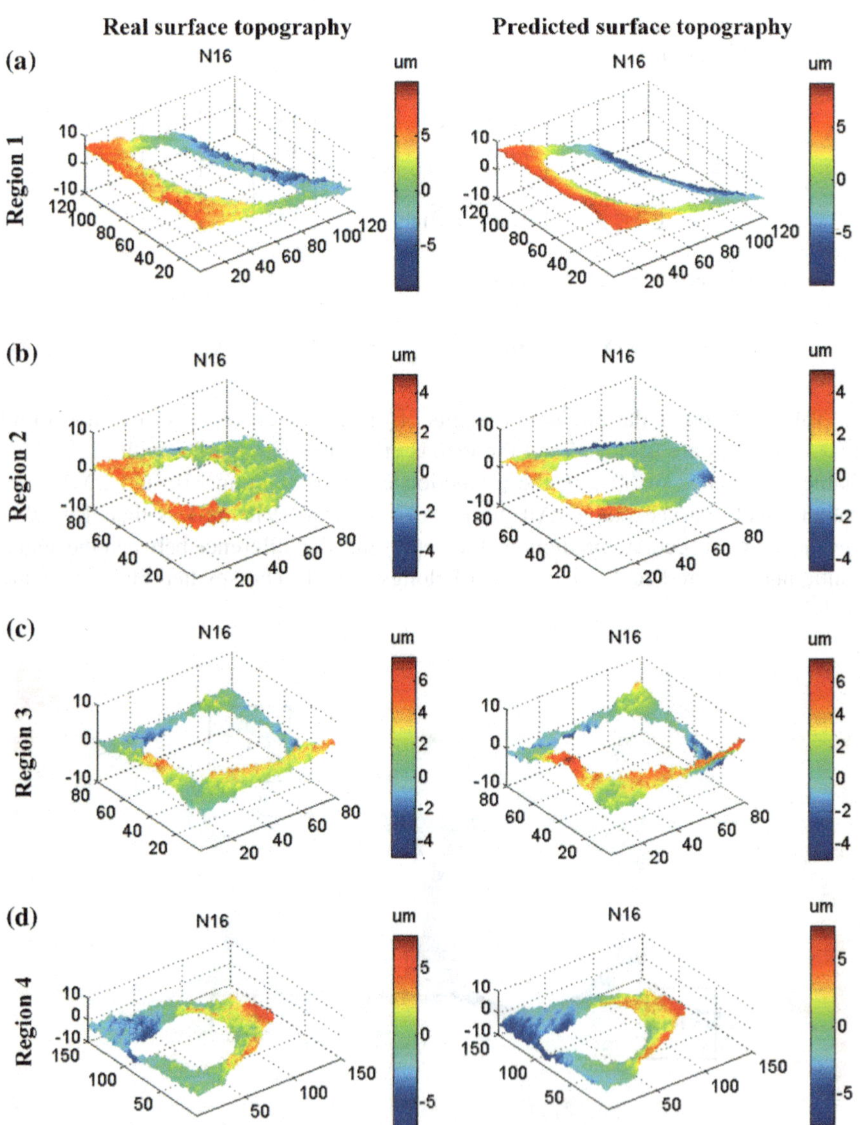

Fig. 7.11 The real surface and the predicted surface by STMSVR

better prediction. Furthermore, operation time shows that the forecasting model is very time-saving. Finally, a clear 3D visualization and comparison of the actual surface and predicted surface is illustrated in Fig. 7.11, indicating that the 3D surface topography can be accurately predicted by STMSVR based on HDM measurement.

7.3.3.1   Conclusions

A space–time multi-output support vector regression model is proposed to achieve the forecasting for 3D machined surface topography. High-definition metrology is adopted to measure the successively machined surfaces and provide the full inspection of the entire surface topographies. The measured HDM data is a form of space–time series that contains the surface topography evolution information. With the real surface as reference, the results show that STMSVR model does well in surface prediction. However, the prediction accuracy of STMSVR is easily affected by the parameters and kernel functions. To ensure the optimal combination of these factors, it requires several repeated attempts. Therefore, a modified STMSVR model will be developed in further studies.

References

1. Franco P, Estrems M, Faura F (2004) Influence of radial and axial runouts on surface roughness in face milling with round insert cutting tools. Int J Mach Tools Manuf 44 (15):1555–1565
2. Franco P, Estrems M, Faura F (2008) A study of back cutting surface finish from tool errors and machine tool deviations during face milling. Int J Mach Tools Manuf 48(1):112–123
3. Tapoglou N, Antoniadis A (2012) 3-Dimensional kinematics simulation of face milling. Measurement 45(6):1396–1405
4. Buj-Corral I, Vivancos-Calvet J, Dominguez-Fernandez A (2012) Surface topography in ball-end milling processes as a function of feed per tooth and radial depth of cut. Int J Mach Tools Manuf 53(1):151–159
5. Hessainia Z, Belbah A, Yallese MA, Mabrouki T, Rigal JF (2013) On the prediction of surface roughness in the hard turning based on cutting parameters and tool vibrations. Measurement 46(5):1671–1681
6. Benardos PG, Vosniakos GC (2002) Prediction of surface roughness in CNC face milling using neural networks and Taguchi's design of experiments. Robot Comput-Integr Manuf 18 (5–6):343–354
7. Upadhyay V, Jain PK, Mehta NK (2013) In-process prediction of surface roughness in turning of Ti-6Al-4 V alloy using cutting parameters and vibration signals. Measurement 46(1):154–160
8. Brezocnik M, Kovacic M, Ficko M (2004) Prediction of surface roughness with genetic programming. J Mater Process Technol 157:28–36
9. Zain AM, Haron H, Sharif S (2010) Prediction of surface roughness in the end milling machining using Artificial Neural Network. Expert Syst Appl 37(2):1755–1768
10. Suksawat B (2011) Development of In-process Surface Roughness Evaluation System for Cast Nylon 6 Turning Operation. Ceis 2011:15

11. Gupta AK (2010) Predictive modelling of turning operations using response surface methodology, artificial neural networks and support vector regression. Int J Prod Res 48 (3):763–778
12. Benardos PG, Vosniakos GC (2003) Predicting surface roughness in machining: a review. Int J Mach Tools Manuf 43(8):833–844
13. Huang Z, Shih AJ, Ni J (2006) Laser Interferometry Hologram Registration for Three-Dimensional Precision Measurements. J Manuf Sci Eng 128(4):1006
14. Wang M, Xi LF, Du SC (2014) 3D surface form error evaluation using high definition metrology. Precis Eng-J Int Soc Precis Eng Nanotechnol 38(1):230–236
15. Du S, Liu C, Huang D (2015) A shearlet-based separation method of 3D engineering surface using high definition metrology. Precis Eng 40:55–73
16. Du S, Liu C, Xi L (2014) A selective multiclass support vector machine ensemble classifier for engineering surface classification using high definition metrology. J Manuf Sci Eng 137 (1):011003
17. Du SC, Huang DL, Wang H (2015) An adaptive support vector machine-based workpiece surface classification system using high-definition metrology. IEEE Trans Instrum Meas 64 (10):2590–2604
18. Wang M, Ken T, Du S, Xi L (2015) Tool wear monitoring of wiper inserts in multi-insert face milling using three-dimensional surface form indicators. J Manuf Sci Eng 137(3):031006
19. Pfeifer PE, Deutrch SJ (1980) A three-stage iterative procedure for space-time modeling phillip. Technometrics 22(1):35–47
20. Pfeifer PE, Deutsch SJ (1980) Identification and interpretation of first order space-time ARMA models. Technometrics 22(3):397–408
21. Pfeifer PE, Deutsch SJ (1980) Independence and sphericity tests for the residuals of space-time arma models: independence and sphericity tests for the residuals. Commun Stat-Simul Comput 9(5):533–549
22. Omar OEEK, El-Wardany T, Ng E, Elbestawi MA (2007) An improved cutting force and surface topography prediction model in end milling. Int J Mach Tools Manuf 47(7–8):1263–1275
23. Thasana W, Sugimura N, Iwamura K, Tanimizu Y (2014) A study on estimation of 3-dimensional surface roughness of boring processes including kinematic motion deviations. J Adv Mech Des Syst Manuf 8(4):0046
24. Denkena B, Böß V, Nespor D, Gilge P, Hohenstein S, Seume J (2015) Prediction of the 3D surface topography after ball end milling and its influence on aerodynamics. Procedia CIRP 31:221–227
25. Alonso FJ, Marcelo A, Cambero I, Salgado DR (2009) Surface roughness prediction based on the correlation between surface roughness and cutting vibrations in dry turning with TiN-coated carbide tools. Proc Inst Mech Eng Part B J Eng Manuf 223(9):1193–1205
26. Vakondios D, Kyratsis P, Yaldiz S, Antoniadis A (2012) Influence of milling strategy on the surface roughness in ball end milling of the aluminum alloy Al7075-T6. Measurement 45 (6):1480–1488
27. Costes J-P (2013) A predictive surface profile model for turning based on spectral analysis. J Mater Process Technol 213(1):94–100
28. Duan C, Hao Q (2014) Surface roughness prediction of end milling process based on IPSO-LSSVM. J Adv Mech Des Syst Manuf 8(3):0024
29. Marani Barzani M, Zalnezhad E, Sarhan AAD, Farahany S, Ramesh S (2015) Fuzzy logic based model for predicting surface roughness of machined Al–Si–Cu–Fe die casting alloy using different additives-turning. Measurement 61:150–161
30. Wang M, Xi L, Du S (2014) 3D surface form error evaluation using high definition metrology. Precis Eng 38(1):230–236
31. Wang M, Shao Y-P, Du S-C, Xi L-F (2015) A diffusion filter for discontinuous surface measured by high definition metrology. Int J Precis Eng Manuf 16(10):2057–2062
32. Vapnik V (1995) The nature of statistical learning theory. Springer, New York

33. Pérez-Cruz F, Camps-Valls G, Soria-Olivas E, Pérez-Ruixo JJ, Figueiras-Vidal AR, Artés-Rodríguez A (2002) Multi-dimensional function approximation and regression estimation. In: Proceedings of International Conference on Artificial Neural Networks, pp 757–762
34. Tuia D, Verrelst J, Alonso L, Perez-Cruz F, Camps-Valls G (2011) Multioutput support vector regression for remote sensing biophysical parameter estimation. IEEE Geosci Remote Sens Lett 8(4):804–808
35. Pérez-Cruz F, Navia-Vázquez A, Alarcón-Diana PL, Artés-Rodríguez A (2000) An IRWLS procedure for SVR. In: Proceedings of EUSIPCO, pp 1–4
36. Wang JQ, Cheng T, Haworth J Space-time kernels. In: Proceedings of ISPRS Archives, pp 57–62

Chapter 8
Online Compensation Manufacturing

8.1 A Brief History of Online Compensation Manufacturing

The manufacturing error is a very important factor that influences the quality of workpieces, which will obviously reduce the manufacturing accuracy of the workpieces. Excessive manufacturing errors may even cause the workpieces to be scrapped and seriously affect the manufacturing efficiency and benefits. There are many causes of manufacturing errors, among which the errors that have a great influence on the quality of workpieces can be divided into the following types: (1) system error of the machine tool, (2) errors caused by the deformation of the machine tools, workpieces and tools, (3) thermal errors generated by machining centers, (4) the principle error of the processing methods, (5) errors caused by tool wear, etc.

Online compensation is a method of real-time monitoring of the machining condition of a machine tool in the machining process, and compensating in real time when deviations occur. Some explorations of online compensation methods are shown as follows. Yang et al. [1] designed a tool deformation compensation system, which was a computer-controlled adaptive system that could measure the cutting force in manufacturing and adjust the tool position in order to compensate for manufacturing errors caused by tool deformation. Johnstone et al. [2] used electromagnetic detectors to monitor the thermal errors in the process in real time and analyzed the detected signals using three-dimensional parameters of finite element. Lee et al. [3] used artificial neural networks based on online training to minimize the errors generated in the manufacturing of rolling surfaces. Watanabe et al. [4] monitored the manufacturing accuracy of the workpieces through an adaptive control system and controlled the errors in the manufacturing by monitoring the bending torque signal on the tool. Liang et al. [5] transplanted the intelligent control system and three-dimensional online compensation software to the CNC system, and displayed the surface errors in real time through the display

© Springer Nature Singapore Pte Ltd. 2019
S. Du and L. Xi, *High Definition Metrology Based Surface Quality Control and Applications*, https://doi.org/10.1007/978-981-15-0279-8_8

window, thereby modified the tool point to compensate for surface manufacturing errors. Yuan [6] installed a laser beam charge coupled detector on the machine tool to monitor the axial misalignment errors during manufacturing and compensated the errors in real time. Chen [7] proposed an online real-time control method for a product manufacturing process to keep the key attributes of the display device within an acceptable specification by adjusting parameters of the production line. However, the above researches are about online compensation for one-dimensional features or two-dimensional flatness surfaces. Online compensation for three-dimensional curved surfaces (for example, volume) is rarely studied.

8.2 A Systematic Method for Online Minimizing Volume Difference of Multiple Chambers in Machining Processes

8.2.1 Introduction

The chamber volumes are very important for some mechanical products. For instance, the volume variations of engine cylinder head combustion chambers (see Fig. 8.1) directly affect the compression ratio of an engine. The interior surfaces of the chambers are usually not being machined after casting processes due to high machining cost. Traditional titration methods are frequently applied offline to evaluate the variations of chamber volumes in machining processes since they are considerably time-consuming (tens of minutes) and also the measurement accuracy is limited to the proficiency level of an operator. It is difficult to online measure and control the volume variation of multiple chambers of a workpiece in machining processes.

With the development of online high-definition measurement (HDM) technologies, great opportunities are provided for online controlling flat surface variation and volume variation of a workpiece. A representative of online HDM for flat surface variation is ShaPix [8], which is based on laser holographic interferometry metrology. Several researches about online controlling flat surface variation based on HDM have been

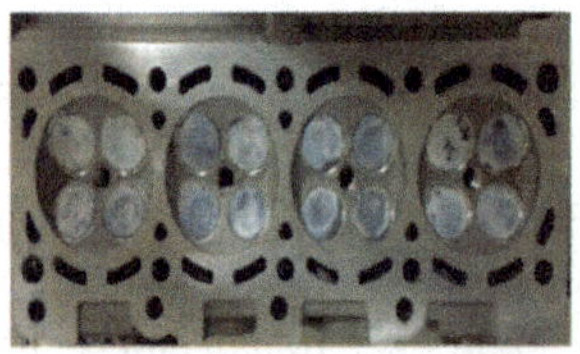

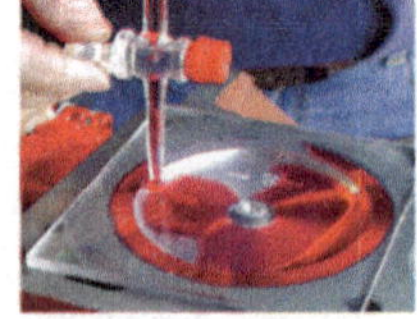

| (a) Cylinder head | (b) A combustion chamber | (c) Titration methods |

Fig. 8.1 A cylinder head with four combustion chambers and offline volume measurement. Reprinted from Ref. [37], copyright 2017, with permission from ASME

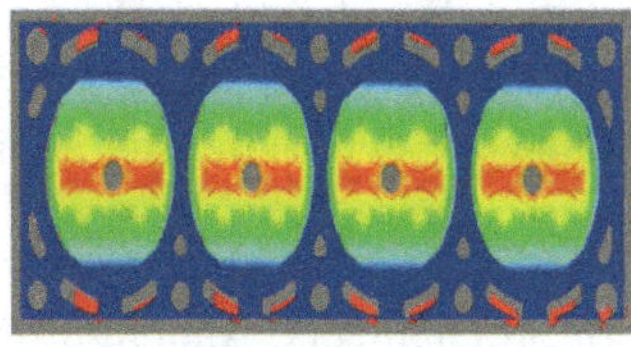

(a) Point cloud of a cylinder head   (b) Point cloud of a combustion chamber

Fig. 8.2 High-density points measured by new HDM technology. Reprinted from Ref. [37], copyright 2017, with permission from ASME

conducted. Du et al. [9–11] proposed a shearlet-based method and support vector machine-based methods to separate and extract different surface components using HDM. Du et al. [12] also presented a co-Kriging method based on multivariate spatial statistics to estimate surface form error using HDM. Wang et al. [13–15] developed a modified gray level co-occurrence matrix to extract features from the images converted from face milled surface measured by HDM. Suriano et al. [16] proposed a new methodology for efficiently measuring and monitoring flat surface variations by fusing in-plant multi-resolution measurements and process information. Nguyen et al. [17] presented a method to reduce flat surface variation in face milling processes based on HDM measurements. However, ShaPix can only be applied for measuring the flat surfaces, but the volume of a chamber. Recently, a new HDM technology using laser triangulation metrology is developed for online measuring the volume variation of multiple chambers. For example, for each interior surface of four chambers of a cylinder head, more than two million measured points (called high-density point clouds) (see Fig. 8.2) can be obtained and the whole measurement time is within 80 s.

Although several methods applicable to the flat surface variation control have been explored based on HDM, they cannot be directly used to control the volume variation of multiple chambers. The control of volume variations based on HDM is confronted with three main difficulties:

(1) Datum transformation. For the online measurement machine, XY plane of the sensor coordinate system is regarded as the datum plane of the obtained point cloud, but the map of partial point cloud is not parallel to the underside of the workpiece. In order to extract the boundary of a chamber from the point cloud, Z-coordinates of the point cloud of underside of the workpiece should be approximately zero. Therefore, it is necessary to transform the datum plane of the point cloud. The conventional methods of plane fitting, such as the least square method (LS) [18, 19] and the characteristic value method [20], involve error analysis, but they cannot eliminate the abnormal points. Therefore, it is difficult to obtain an accurate datum plane of the point cloud.

(2) Accurate volume calculation of multiple chambers. Basically, there are several holes in the interior surfaces of the chambers. For instance, there is a spark hole plugged in the interior surface of an engine head combustion chamber. The hole can cause the missing of the point cloud in the interior surfaces, which adds

difficulty to volume calculation of a chamber. Besides, since the interior surfaces of the chambers are irregular after casting processes, it is difficult to accurately calculate the volume of the chambers based on the three-dimensional coordinates of the point cloud.

(3) Search an optimized machining parameter for depth of chambers to minimize the volume difference of any two ones of all chambers. Although the interior surfaces of the chambers of a workpiece are not being machined due to high machining cost, the underside of the workpiece usually needs to be milled and the milling parameter is closely related to the volumes of all chambers. Therefore, it is desirable to obtain an optimized milling parameter to minimize the volume difference of any two ones of all chambers. Taking a cylinder head as an example, there are n combustion chambers in a cylinder head and V_i is the volume of the ith chamber (see Fig. 8.3). It is unpractical to mill the interior surfaces of each chamber to make the volume of each combustion chamber equal the designed volume due to high machining cost. The combustion chamber volumes can only be controlled by milling the underside of the cylinder head. Therefore, the key to control the volume variation of multiple chambers is to determine the optimal milling parameter (depth h_0). The volume of each combustion chamber is measured by online HDM measurement machine, which consists of the merits of quick speed, high accuracy, and sequential point cloud. After the measurement, the milling parameter h_0 needs to be optimized to minimize the volume difference of any two ones of all combustion chambers.

However, there has been no systematic method for online minimizing volume difference of multiple chambers in machining processes based on HDM. This subsection is intended to contribute to this end.

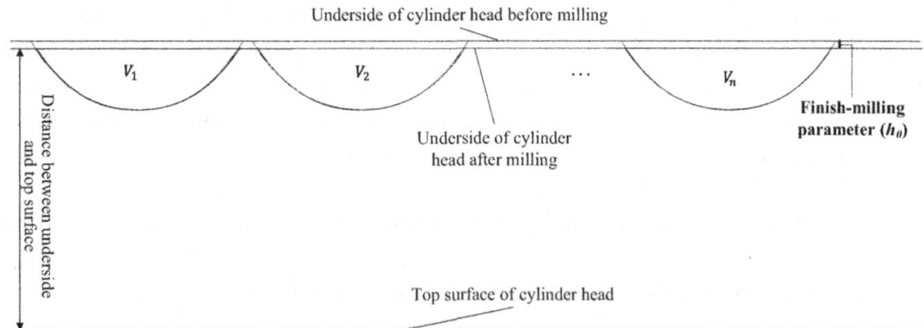

Fig. 8.3 Milling underside of cylinder head with n combustion chambers. Reprinted from Ref. [37], copyright 2017, with permission from ASME

8.2.2   The Proposed Method

The framework of the proposed method is shown in Fig. 8.4, and the main steps of this method are described as follows.

Step 1: Collect three-dimensional and sequential point cloud of multiple chambers of a workpiece using online measurement machine.

Step 2: Develop a method consisting of datum transformation, accurate volume calculation of multiple chambers, and a model obtaining an optimized machining parameter to minimize the volume difference of any two ones of all chambers. First, the datum plane of the high-density points is transformed based on a random sample consensus (RANSAC) algorithm due to its good robustness in fitting; Second, a procedure containing reconstruction of interior curved surface of chambers, boundary extraction, and projection is presented to calculate the accurate volumes of the multiple chambers based on the point cloud. Third, a model is developed to optimize the milling parameter of the underside of a multichamber workpiece to minimize the volume difference of any two ones of all chambers. Finally, multi-objective particle swarm optimization (MOPSO) algorithm is presented to solve the model.

Step 3: Output an optimized milling parameter (depth h_0) from the proposed method and the multichamber workpiece should be milled according to the output parameter.

8.2.2.1   Datum Transformation

In order to transform datum plane of the point cloud from XY plane of sensor coordinate system to the underside of a workpiece, the datum plane parameters of the point cloud based on sensor coordinate system should be calculated. The collected point cloud includes not only all the points on the interior surface but also abnormal points as a result of the influence of measurement environment and

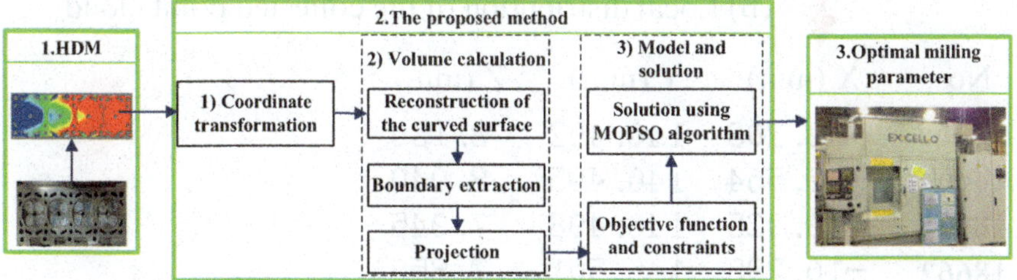

Fig. 8.4   The framework of the proposed method. Reprinted from Ref. [37], copyright 2017, with permission from ASME

reflection of surface finish. Therefore, there appears a problem about how to extract accurate datum plane from the point cloud with many abnormal points. To overcome this problem, a procedure based on RANSAC algorithm to calculate the parameters of the datum plane is explored. The structure of input data collected from a cylinder head by HDM sensor is shown in Fig. 8.5.

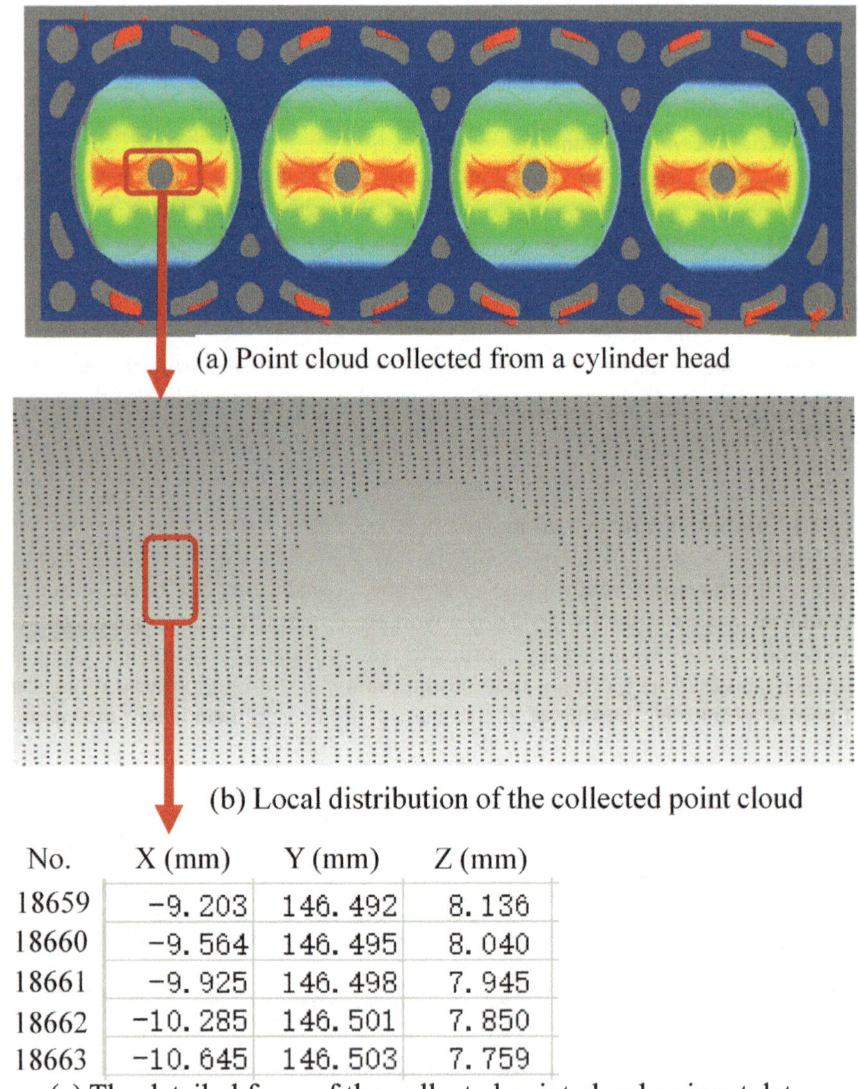

(a) Point cloud collected from a cylinder head

(b) Local distribution of the collected point cloud

No.	X (mm)	Y (mm)	Z (mm)
18659	−9. 203	146. 492	8. 136
18660	−9. 564	146. 495	8. 040
18661	−9. 925	146. 498	7. 945
18662	−10. 285	146. 501	7. 850
18663	−10. 645	146. 503	7. 759

(c) The detailed form of the collected point cloud as input data

Fig. 8.5 The structure of input data collected from a cylinder head. Reprinted from Ref. [37], copyright 2017, with permission from ASME

RANSAC algorithm

Input: U (point cloud), k_{max} (maximum number of iterations), t (threshold)

$k=0$, $n_{max}=0$

while $k<k_{max}$ do

   1. Hypothesis generation

   Randomly select 3 points from the point cloud U

   Estimate plane equation parameters θ_k

   Calculate ds_i (distance from arbitrary point of the point cloud to the estimated plane) by equation (8.1)

   2. Verification

   Calculate n_k (number of inner points, $ds_i \leq t$) and record these inner points as the support set I_k

   If $n_k> n_{max}$ then

   $n_{max}=n_k$, $I^*= I_k$

   end if

   $k=k+1$

end while

estimate θ^* (the parameters of plane equation) based on I^*

Output: θ^*

Fig. 8.6 The description of RANSAC algorithm. Reprinted from Ref. [37], copyright 2017, with permission from ASME

RANSAC algorithm can obtain a robust plane fitting result under the condition of a large number of abnormal points [21–23]. The description of RANSAC algorithm is shown in Fig. 8.6, and the main steps are described as follows.

Step 1: Randomly select three points from the point cloud and calculate the corresponding plane equation of the three points. The plane equation is expressed as $ax + by + cz + d = 0$. Here, a, b, c, and d are the parameters of the plane equation. Then calculate the distance from arbitrary point P_i of the point cloud to the plane and the distance ds_i is calculated by

$$ds_i = \frac{|ax_i + by_i + cz_i + d|}{\sqrt{a^2 + b^2 + c^2}} \tag{8.1}$$

Step 2: Determine the value of threshold t. The strategy to choose the threshold t is referred to [24]. If $d_i \leq t$, point P_i is considered as inner point of the obtained plane. Then count the number (N) of the inner points of the plane.

Step 3: Repeat the above steps for n times and select a plane with the largest number of inner points.

Step 4: The selected plane is refitted with the largest number of inner points according to eigenvalue algorithm, and the final fitted plane equation is calculated.

Assume that the obtained plane equation is $Ax + By + Cz + D = 0$. The normal vector of the plane is (A, B, C), and the plane is through point $(0, 0, D/C)$. Then, the datum plane of the sensor coordinate system is represented as

$$\begin{cases} \overrightarrow{w_1} = (1, 0, -C/A) \\ \overrightarrow{w_3} = (A, B, C) \\ \overrightarrow{w_2} = \overrightarrow{w_3} \times \overrightarrow{w_1} \end{cases} \tag{8.2}$$

The Gram–Schmidt orthonormalization [25] is used to deal with the representation of sensor coordinate system (shown in Fig. 8.7). a_1, a_2, and a_3 are regarded as the workpiece surface coordinate system, and b_1, b_2, and b_3 are regarded as the sensor coordinate system. The datum transformation matrix H transforming datum

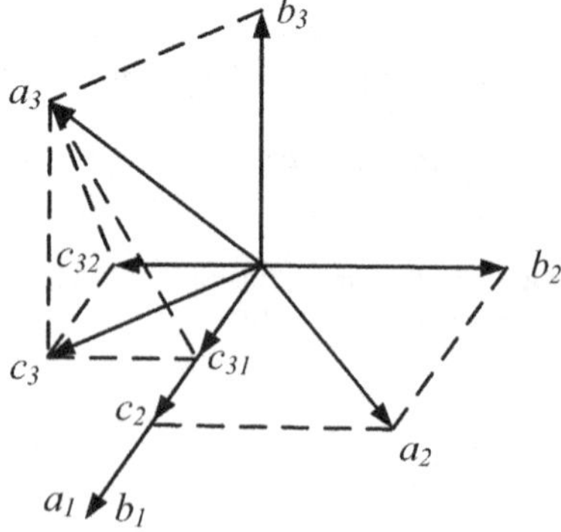

Fig. 8.7 An illustration of Gram–Schmidt transformation. Reprinted from Ref. [37], copyright 2017, with permission from ASME

plane from sensor coordinate system to the workpiece surface coordinate system is obtained by

$$
\begin{cases}
b_1 = a_1 \\
b_2 = a_2 - \dfrac{[b_1,a_2]}{[b_1,b_1]} b_1 \\
b_3 = a_3 - \dfrac{[b_1,a_3]}{[b_1,b_1]} b_1 - \dfrac{[b_2,a_3]}{[b_2,b_2]} b_2 \\
e_1 = \dfrac{1}{\|b_1\|} b_1, e_2 = \dfrac{1}{\|b_2\|} b_2, e_3 = \dfrac{1}{\|b_3\|} b_3 \\
H = [e_1, e_2, e_3]
\end{cases}
\tag{8.3}
$$

For each point (P_i) of the point cloud, the transformation is conducted by

$$
P_i' = inv(H) * P_i
\tag{8.4}
$$

After the point cloud is transformed, the value of Z-coordinate of each point is the distance from the point to the datum plane, and the value of Z-coordinate of each point on the datum plane (workpiece surface) is approximately zero.

8.2.2.2   Accurate Volume Calculation of Multiple Chambers

(1)   Reconstruction of interior curved surface

Due to the interference of the measuring environment and the influence of surface cleanliness of the cylinder head, there exist missing points in the collected point cloud (shown in Fig. 8.8). Besides, there are several holes of spark plug in the cylinder heads, and they cannot be measured by the HDM system. The reflections of these holes are some relatively large blank areas in the point cloud (shown in Figs. 8.8 and 8.9).

The missing points and points that are not collected in the blank area are marked as null and the interpolation is conducted based on the coordinates of the neighborhood points. Ranks of the weighted interpolation are developed to interpolate the points marked as null. The procedure of the interpolation is described as follows.

Step 1: Search each row of the point cloud for the first point P_i that is marked as null, and record the location of the previous point P_{i-1}. The location of a point is the ordinal number of that point in a row. P_{i-1} is an adjacent point of P_i, and it is not a null point.

Step 2: Take point P_i as the starting point and search the same row for the first point P_j that is not marked as null, and record the location of point P_j.

Step 3: For the $(j-i)$ points marked as null from point P_i to point P_j, the interpolation is conducted by

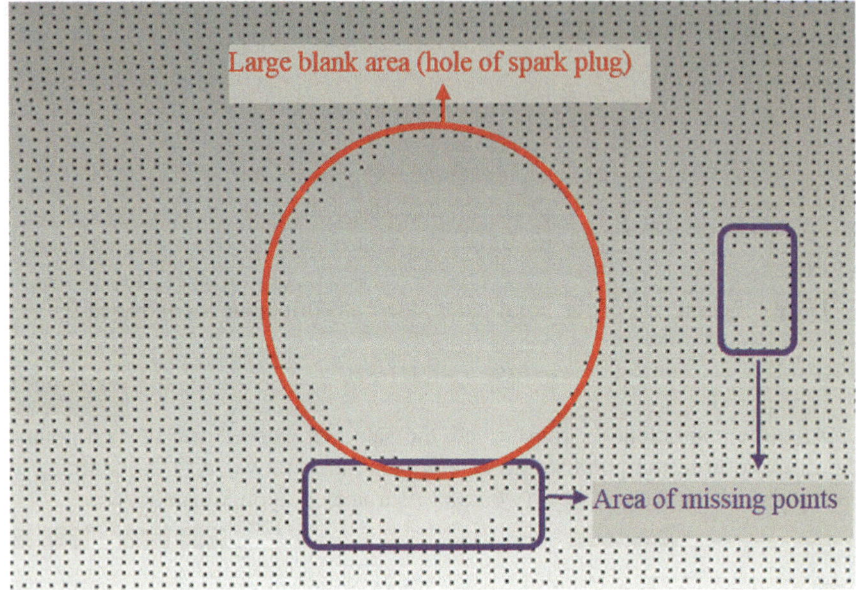

Fig. 8.8 Area of missing points in the collected point cloud. Reprinted from Ref. [37], copyright 2017, with permission from ASME

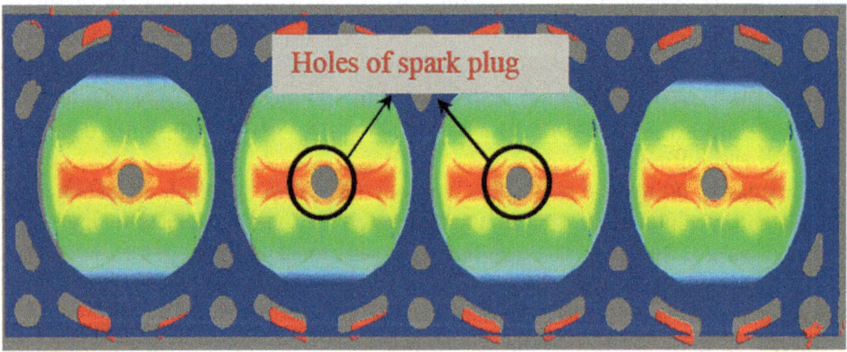

Fig. 8.9 Large blank areas in the collected point cloud. Reprinted from Ref. [37], copyright 2017, with permission from ASME

$$P_k = \frac{k}{j-i}(P_j - P_i) + P_i \quad (i-1 < k < j) \tag{8.5}$$

where P_k is the point marked as null between points P_i and P_j.

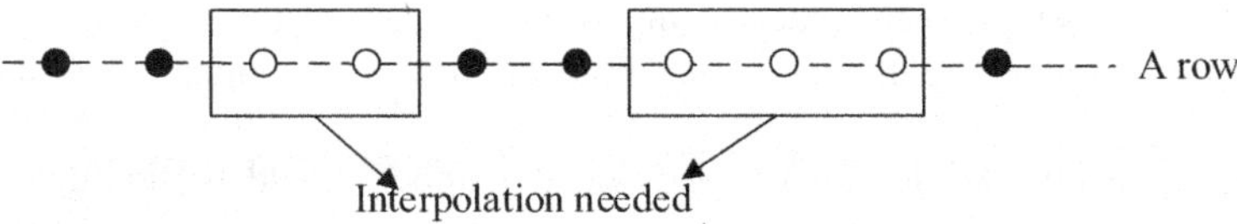

Fig. 8.10 Interpolations needed in a row. Reprinted from Ref. [37], copyright 2017, with permission from ASME

Step 4: Repeat the above steps and complete the interpolation of a row. There may need one more interpolations in a row since the missing points may form one more clusters in that row (shown in Fig. 8.10).

Step 5: Repeat step 1 to step 4 and complete interpolation of all rows.

After the completion of the point cloud interpolation, the point cloud needs to be grid to generate interior curved surface of a chamber. The interpolated point cloud is complete and in order, which means that the numbers of points in each row and each column are the same. There is a corresponding topological relationship in the point cloud, that is, the locations of points in each row are fixed, and each point and its neighborhood can be quickly determined. Then the point cloud mesh is built based on the locations of points in each row and each column of the point cloud.

The reconstruction process of the curved surface of the point cloud is shown in Fig. 8.11. For arbitrary point $P_{i,\,j}$ (point in ith row and jth column) of the sequential point cloud, connect its adjacent points $P_{i\,+\,1,\,j}$ and $P_{i,j\,+\,1}$, and quadrilateral mesh is generated. On the basis of the quadrilateral mesh, connect the points $P_{i,j}$ and $P_{i\,+\,1,\,j\,+\,1}$, and the triangular mesh is also generated. At this point, the reconstruction of the curved surface is completed.

(2)   Boundary extraction

After the completion of reconstruction of the curved surface, the effective area of a chamber on the underside of multichamber workpiece should be determined since it is the area where the reconstructed triangular meshes project. The effective area of a chamber is the boundary of the chamber. Therefore, the boundary of a chamber should be extracted from the point cloud. The point cloud measured from the

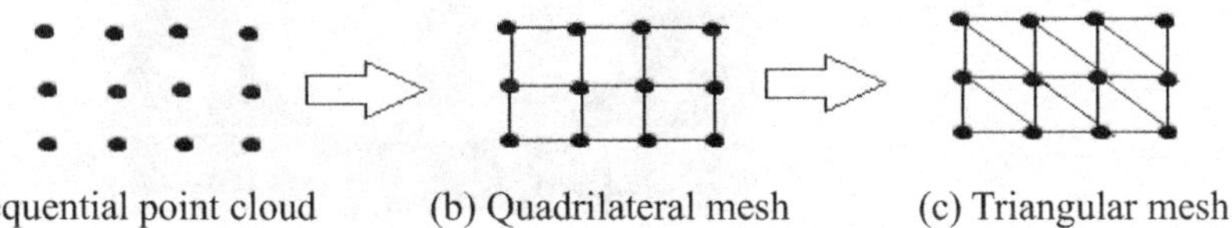

(a) Sequential point cloud   (b) Quadrilateral mesh   (c) Triangular mesh

Fig. 8.11 Schematic diagram of structured point cloud mesh generation. Reprinted from Ref. [37], copyright 2017, with permission from ASME

multiple chambers is sequential and the point numbers of each row in the point cloud are the same. The direction indications of "row" and "column" are shown in Fig. 8.12.

The initial row and terminal row of each chamber should be first determined from the point cloud in the scanning direction of the sensor of the online measurement machine. The Z-coordinate value of a point approximates zero if the point is on the bottom surface where the boundaries of the multiple chambers lie. If not, the point is on the interior surface of the multiple chambers. Similarly, the average Z-coordinate value of points in each row of the point cloud approximates zero if all points in each row are on the surface where the boundaries of the multiple chambers lie. If not, there exist one or more points of a row that are on the interior surface of

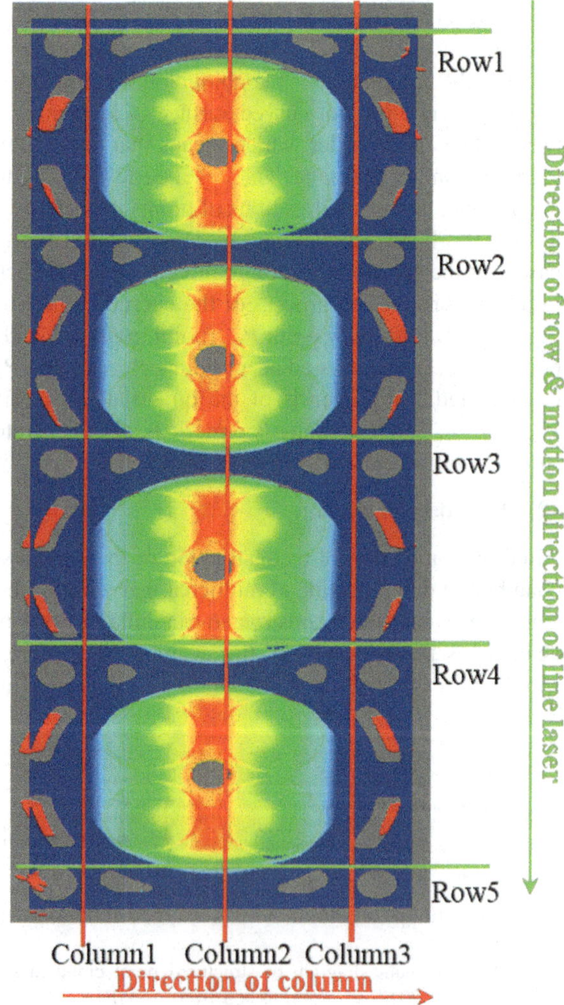

Fig. 8.12 Direction indications of "row" and "column". Reprinted from Ref. [37], copyright 2017, with permission from ASME

the multiple chambers. Then the average Z-coordinate value of each row of the point cloud is calculated and the effect of abnormal points on the accuracy of boundary extraction can be eliminated by a statistical method. The average Z-coordinate value of the kth row of the point cloud is calculated by Eq. (8.6):

$$Z_{kmean} = \left(\sum Z_{km}\right)/n_k \tag{8.6}$$

where Z_{km} is Z-coordinate value of the mth point of the kth row of the point cloud, n_k is the point number of the kth row of the point cloud, and Z_{kmean} is the average Z-coordinate value of the kth row of the point cloud.

The proposed process of boundary extraction of the multiple chambers is described as follows.

Step 1: Calculate the average Z-coordinate value of the points in each row and make a statistical table that consists of the average Z-coordinate value and sequence number of each row.

Step 2: Search the statistical table from middle to both sides. When the average Z-coordinate value is approximately zero for the first time, the corresponding rows are the initial row and terminal row (shown in Fig. 8.13).

Step 3: Search boundary points of each row from the determined initial row to terminal row. Since the Z-coordinate values of points in each of the searched rows obey the same rule of average Z-coordinate value of each row in Step 2, search the boundary points from middle to both sides of each row between the determined initial row to terminal row.

Step 4: Since each column also follows the same rules that are shown in Step 1 to Step 3, the boundary points of each column can be found.

Step 5: Combine the boundary points of each row and each column, and these boundary points form the boundary of each chamber.

(3) Projection

Based on the triangular meshes of the chamber interior surface and the determination of chamber boundary, the volume of a chamber is calculated by projection. The underside of a multichamber workpiece is selected as a projection plane, and the triangular meshes are projected to it, constituting many convex pentahedrons

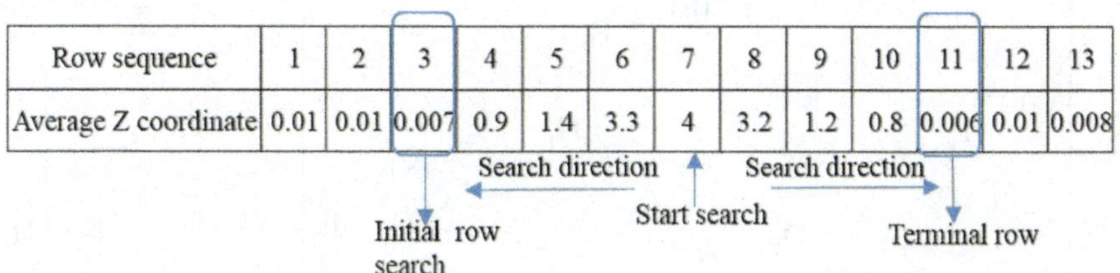

Row sequence	1	2	3	4	5	6	7	8	9	10	11	12	13
Average Z coordinate	0.01	0.01	0.007	0.9	1.4	3.3	4	3.2	1.2	0.8	0.006	0.01	0.008

Fig. 8.13 An illustration of determining initial row and terminal row. Reprinted from Ref. [37], copyright 2017, with permission from ASME

within the extracted boundary. The volume of the chamber is the sum of volumes of all the convex pentahedrons.

In general, the triangular patch is not parallel to the projection plane since the interior surface of the chamber is curved. Therefore, the convex pentahedron is an irregular geometry, and it is difficult to calculate its volume directly. To overcome this problem, the convex pentahedron can be divided into several regular geometric bodies, and then the volume of the convex pentahedron is obtained by calculating the sum of the volumes of the regular geometric bodies.

Due to the similarity between the convex pentahedron and the triangular prism in shape, the convex pentahedron can be divided into a triangular prism and the rest geometric part is a rectangular pyramid (shown in Fig. 8.14a). However, the volume calculation of the rectangular pyramid is also complicated. Thus, the rectangular pyramid also needs to be divided and two triangular pyramids (called tetrahedron) are obtained (shown in Fig. 8.14b). The volume of the tetrahedron can be directly calculated under the circumstances that vertex coordinates of the tetrahedron are known.

The final result is that the convex pentahedron is divided into a triangular prism and two tetrahedrons, and the volume calculation of the convex pentahedron is transformed to the volume calculations of a triangular prism and two tetrahedrons. Figure 8.14c is the complete volume segmentation of the convex pentahedron, and $A_1B_1C_1D_1E_1F_1$ is the original convex pentahedron, $A_1H_1G_1D_1E_1F_1$ is the triangular prism, and $A_1B_1C_1G_1$ and $A_1B_1G_1H_1$ are the tetrahedrons.

For a triangle $D_1E_1F_1$ with known vertex coordinates (see Fig. 8.15), the area of the triangle is calculated by

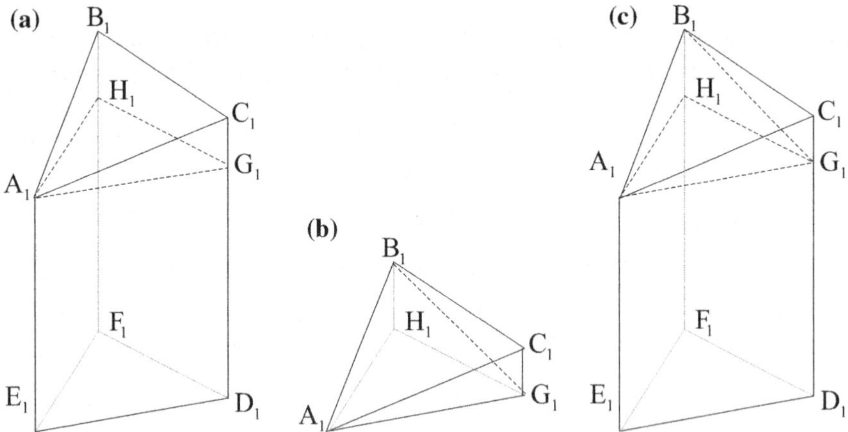

Fig. 8.14 Segmentation of convex pentahedron. Reprinted from Ref. [37], copyright 2017, with permission from ASME

Fig. 8.15 Area calculation of triangle with three known vertex coordinates. Reprinted from Ref. [37], copyright 2017, with permission from ASME

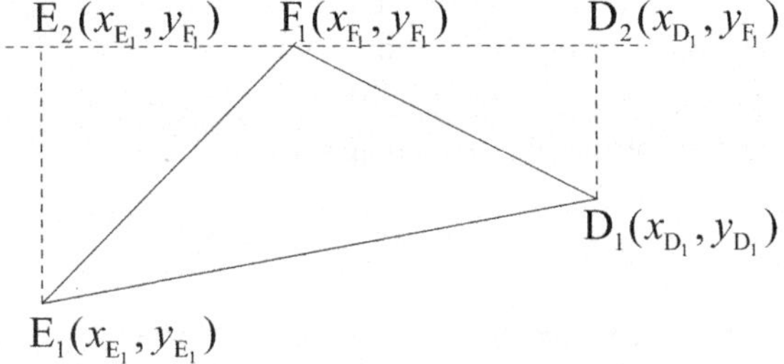

$$S_{\Delta D_1 E_1 F_1} = \frac{\left| x_{F_1} \times y_{E_1} + y_{F_1} \times x_{D_1} + x_{E_1} \times y_{D_1} - x_{F_1} \times y_{D_1} - y_{F_1} \times x_{E_1} - y_{E_1} \times x_{D_1} \right|}{2}$$

(8.7)

The volume of triangular prism $A_1H_1G_1D_1E_1F_1$ is calculated by

$$V_{A_1 H_1 G_1 D_1 E_1 F_1} = S_{\Delta D_1 E_1 F_1} \times h_{A_1 E_1}$$

(8.8)

where $h_{A_1 E_1}$ is the distance between the plane of $E_1F_1D_1$ and the plane of $A_1H_1G_1$.

For the tetrahedron $A_1B_1G_1H_1$ with four known vertex coordinates $(x_{A_1}, y_{A_1}, z_{A_1})$, $(x_{B_1}, y_{B_1}, z_{B_1})$, $(x_{G_1}, y_{G_1}, z_{G_1})$, and $(x_{H_1}, y_{H_1}, z_{H_1})$, the volume is calculated by

$$\begin{aligned}
V_{A_1 B_1 G_1 H_1} &= \frac{1}{6} \times \begin{vmatrix} 1 & 1 & 1 & 1 \\ x_{A_1} & x_{B_1} & x_{G_1} & x_{H_1} \\ y_{A_1} & y_{B_1} & y_{G_1} & y_{H_1} \\ z_{A_1} & z_{B_1} & z_{G_1} & z_{H_1} \end{vmatrix} \\
&= \frac{1}{6} \times \begin{vmatrix} x_{B_1} - x_{A_1} & x_{G_1} - x_{A_1} & x_{H_1} - x_{A_1} \\ y_{B_1} - y_{A_1} & y_{G_1} - y_{A_1} & y_{H_1} - y_{A_1} \\ z_{B_1} - z_{A_1} & z_{G_1} - z_{A_1} & z_{H_1} - z_{A_1} \end{vmatrix}
\end{aligned}$$

(8.9)

8.2.2.3   The Model for Obtaining an Optimized Machining Parameter

(1)   Objective function and constraints

In order to determine the optimal milling parameter to minimize the volume difference of any two ones of all chambers, a model based on multi-objective optimization is proposed. The model is developed based on the following three hypotheses: (1) Milling process for the workpiece is face milling, which means that the milling parameter of each chamber is the same. (2) The interior surfaces of the multiple chambers are not machined. (3) The shape of the interior surface of the chamber is similar to a sphere.

The volume variation control of multiple chambers is equivalent to minimizing the volume difference of any two ones of all chambers. Then the objective functions are expressed as Eq. (8.10). For volume variation control of n chambers, the number of the objective functions is $\frac{n(n-1)}{2}$.

$$\min\left|\left(V_i - f_i(h_0)\right) - \left(V_j - f_j(h_0)\right)\right|(i,j=1,2,3,\ldots,n,i<j) \qquad (8.10)$$

where V_i, V_j are the volumes of the ith, jth chambers before milling, h_0 is finial milling parameter output by the model, and $f_i(\cdot)$ is the function to calculate the volume of the milled part of the ith chamber.

The volume of each chamber can be calculated with the point cloud according to Sect. 8.2.2.2 and this volume is signed as V_i. The ideal milling parameter of the ith chamber is calculated by Eq. (8.11) if the underside of workpiece for the ith chamber is milled to make its volume equal the designed volume.

$$h_i = F_i(V_i - V_0) \qquad (8.11)$$

where V_i is the volume of the ith chamber before milling, V_0 is the designed volume of the chamber, h_i is the milling parameter of the chamber when milled to the designed volume (V_0), and F_i is the relationship between h_i and (V_i-V_0).

According to the third hypothesis, the milled part of the ith chamber is a part of a sphere and the cutaway picture of the ith chamber with milled part is shown in Fig. 8.16. h_0 is the milling parameter to be optimized.

The center O_i and the radius R_i of the sphere can be calculated by LS method when the points on the interior surface of the ith chamber are enough. The distance

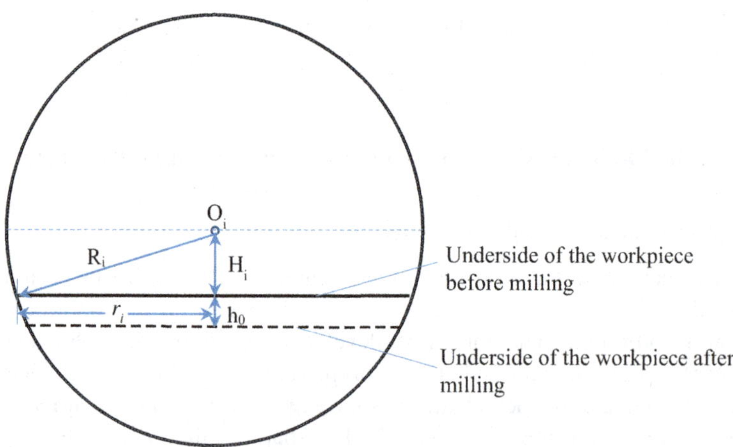

Fig. 8.16 The cutaway picture of the chamber with milled part. Reprinted from Ref. [37], copyright 2017, with permission from ASME

from the center to the underside of workpiece for the ith chamber before milling is calculated as

$$H_i = \sqrt{R_i^2 - r_i^2} \tag{8.12}$$

where H_i is the distance between the center of the fitted sphere of the ith chamber and the underside of workpiece for the ith chamber, R_i is the radius of the fitted sphere of the ith chamber, and r_i is radius of extracted bound from the point cloud of the ith chamber.

The milled part can be regarded as the volume integral of the parallel area (a circle) and the volume of the milled part is calculated by

$$f_i(h_0) = \int_{H_i}^{H_i + h_0} \pi \times \left(\sqrt{R_i^2 - r^2} \right)^2 dr \tag{8.13}$$

where h_0 is finial milling parameter output by the model, $f_i(h_0)$ is the volume of the milled part of the ith chamber.

Then the volume of the ith chamber after milling can be calculated by

$$v_i = V_i - f_i(h_0) \tag{8.14}$$

The volume of the ith chamber milled according to the milling parameter should satisfy the designed volume and tolerance requirement of the chamber, which is shown as

$$v_i \in [V_0 - v_0, V_0 + v_0] \tag{8.15}$$

where v_0 is variable of designed volume of a single chamber.

The range of the milling parameter is given by

$$h_0 \in [\min(h_i), \max(h_i)] \tag{8.16}$$

(2)   Calculation of ideal milling parameter of each chamber

In order to calculate the ideal milling parameter of each chamber, registration is applied to match the measured point cloud and geometric model of the chamber. In the registration, a rotation matrix and a translation vector are obtained. The ideal milling parameter of each chamber is calculated based on the Z-coordinate value of the rotation matrix, and the schematic diagram is shown in Fig. 8.17.

The registration contains two steps of initial registration and high-accuracy registration. Initial registration is aimed to reduce misplacement of the rotation and translation of the point cloud, and thus improves the efficiency and trend of the registration. High-accuracy registration is to minimize the registration error between two point clouds.

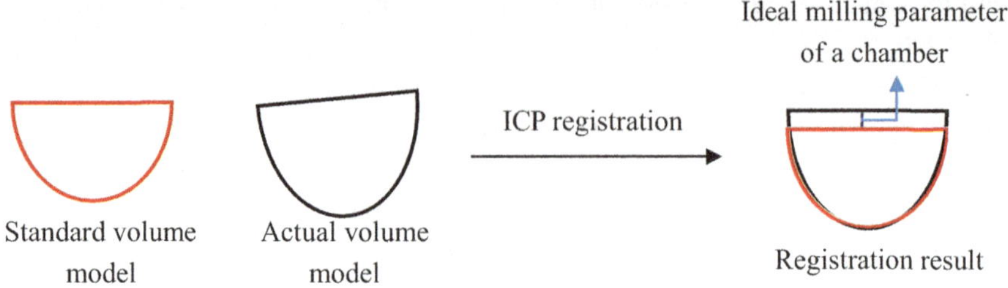

Fig. 8.17 Calculation of the optimal milling parameter of a single combustion chamber. Reprinted from Ref. [37], copyright 2017, with permission from ASME

In initial registration, the main direction of the legitimate is applied to adjust the reference coordinate systems of the measured point cloud and geometric model to be same. The main direction of the point cloud can be fitted by calculating the feature vector of the point cloud. Take the main direction as a coordinate axis, and two directions perpendicular to the main direction are regarded as the other two coordinate axes, then a three-dimensional coordinate system is established. For two point clouds to be registered, establish two such coordinate systems and adjust them to be same, and the initial registration of the point cloud is achieved.

In high-accuracy registration, iterative closest point (ICP) algorithm [26] is one of the most widely used algorithms and the details of ICP algorithm are shown in Appendix A. The procedure of calculating ideal milling depth of each chamber using ICP is included as follows.

Step 1: Determination of initial corresponding point sets. The corresponding point set is determined by directly searching a point of an actual volume model that is closest to a point in standard volume model. The point-to-point search is easy to implement but it cannot satisfy the computational time in terms of huge point cloud. Therefore, k-d tree [27, 28] is applied to accelerate the match of point set.

Step 2: Remove of mismatched point sets. In order to remove the mismatched point sets, distance constraint is used to evaluate whether the corresponding point set is reliable. The distances of all the corresponding point sets are calculated and ordered in descending sequence. Then a number of corresponding point sets in front of the order in descending sequence are regarded as the mismatched point sets and removed from the initial corresponding point sets, which is a substitute for a fixed threshold to remove mismatched point sets.

Step 3: Solution of coordinate transformation. For eventually established corresponding point set, LS method based on unit quaternions [29] is used to iteratively calculate the optimal general coordinate transformation between two point clouds. In the calculation, the translation vector can be ignored and the X-and Y-coordinate values of the rotation matrix can be set as 1 since the purpose of registration in this model is only to obtain the Z-coordinate values of the rotation matrix.

Once the procedure is terminated, a rotation matrix is an output. Then the point cloud of the bottom surface of the cylinder head is rotated with the output matrix. The distance between the rotated bottom surface of the cylinder head and the bottom surface of the reference point cloud is the ideal milling parameter of each chamber.

8.2.2.4   Model Solution

Particle swarm optimization (PSO) algorithm contains the advantages of efficient global search capability, fast search speed, and simple structure [10, 30, 31]. However, the traditional PSO algorithm cannot be directly applied to solve multi-objective optimization (MOO) problem. The following problems should be solved if PSO algorithm is effectively used for solving MOO problem: (1) How to determine which particle is more preferably between two particles? (2) How to select individual extreme value and global extreme value? (3) How to maintain the uniformity of solution distribution of the algorithm?

In order to overcome the above problems, the multi-objective particle swarm optimization (MOPSO) based on Pareto dominance relationship is applied [32–34]. In the algorithm, a total of three sets are used to save the particle swarm, the non-dominated set, and the external set. The relationships of the three sets are shown in Fig. 8.18.

The particle swarm is the main implementation of the search; the non-dominated set and external set are the main sets that save search results. After initialization of the particle swarm and related parameters, select the non-dominated particles from particle swarm and insert them into the non-dominated set. Although non-dominated set is on behalf of the searched optimal particles in current iteration, the optimal particles searched in previous iterations should be isolated. Therefore,

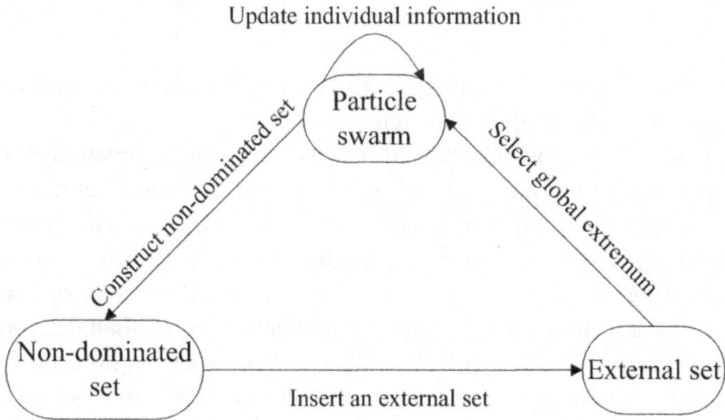

Fig. 8.18 Multi-objective PSO algorithm. Reprinted from Ref. [37], copyright 2017, with permission from ASME

insert the non-dominated particles obtained in each iteration into the external set. The external set can be regarded as a candidate set of global extreme value. The optimization of particle swarm is guided by the global extreme value to constantly search more optimal solution, and then enter the next iteration.

A new procedure of MOPSO algorithm to solve the model for obtaining an optimized milling parameter is shown in Fig. 8.19.

The details of the procedure of MOPSO algorithm are included as follows. Explanation of some useful notations: p_{best} is individual optimal position, g_{best} is global optimal position, x is the current position of a particle and it represents a milling depth, and $F_m(x)$ is the function to calculate the mth volume difference of any two ones of all chamber with the milling depth x.

Step 1: Initialize the position x_t and velocity λ_t of each particle. Set iterative number, population size, and parameters of the algorithm. The position of each particle is randomly generated in uniform distribution but the range of position values is determined by Eqs. (8.11) and (8.16).

Step 2: Calculate the fitness of each particle based on the objective functions of the proposed model. For the first iteration, the position of each particle is regarded as the initial p_{best}. The initial non-dominated set is selected by the dominated relationship among the particles and the initial non-dominated set is saved as initial external set.

Step 3: Update each particle and guide the particle to search the final global optimal position with the information of global optimal position of current iteration and the historical local optimal position. The update of g_{best} by density distance shown is given as follows.

Sort the particles from the initial external set according to the density distance. Randomly select a particle as g_{best} from the first 20% particles with larger density distances according to the order of density distance. The density distance of each particle is calculated as

$$I(x_i) = \sum_{m=1}^{n} \left|[F_m(x_j) - F_m(x_k)]\right|/F_m^{\max} \tag{8.17}$$

where x_j and x_k are the closest particles to x_i, and F_m^{max} is the maximum value of the mth objective function of all the particles.

The update of p_{best} is determined by the dominance relationship between x (current position of the particle) and p_{best} (the best position of current individual particle in its past records). If p_{best} dominates x, then p_{best} remains unchanged. If x dominates p_{best}, x would replace p_{best} as the new p_{best}. If there is no dominance relationship between x and p_{best}, p_{best} remains unchanged or either of x and p_{best} is randomly selected as the new p_{best} with 0.5 probability in traditional algorithm. The selection probability of x and p_{best} is adjusted when there is no dominance relationship between x and p_{best}. For instance, the selection probability of x is set as 0.7 while the selection probability of p_{best} is 0.3. The set of selection probability makes p_{best} in each iteration be different from the last generation, which can prevent the

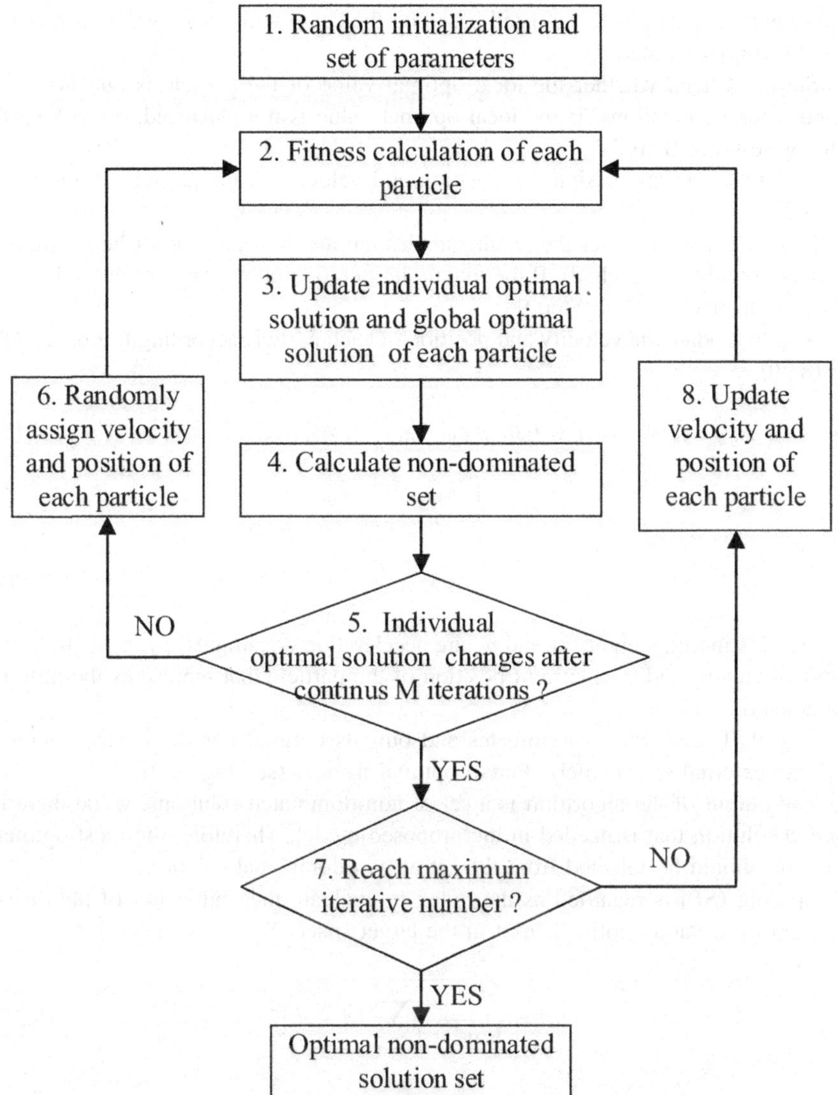

Fig. 8.19 The procedure of MOPSO algorithm. Reprinted from Ref. [37], copyright 2017, with permission from ASME

particle from searching the region once searched under the condition of several continuous iterations of the p_{best} without changing.

Step 4: Calculate the non-dominated set. The particles are sorted based on the non-inferior ordering and the density distance. The number of individuals in non-dominated set can be adjusted based on the density distance, which distributes Pareto frontier evenly.

Step 5: Check whether the local optimal value of the particle is enhanced for continuous M iterations. If the local optimal value is not enhanced, go to Step 6, otherwise go to Step 7.

Step 6: Randomly assign the position and velocity of the particle. Then go to Step 2.

Step 7: Check whether the result satisfies the iteration condition (the maximum iterative number is reached). If the maximum iterative number is not reached, go to Step 8. Otherwise, go to Step 9.

Step 8: Update the velocity and position of each particle according to Eqs. (8.18) to (8.20).

$$\lambda_t = \omega\lambda_{t-1} + \alpha_1\mu_1\left(p_{best} - x_{t-1}\right) + \alpha_2\mu_2\left(g_{best} - x_{t-1}\right) \tag{8.18}$$

$$\lambda_t = \begin{cases} \lambda_{\max}, \lambda_t > \lambda_{\max} \\ -\lambda_{\max}, \lambda_t \leq \lambda_{\max} \end{cases} \tag{8.19}$$

$$x_t = x_{t-1} + \lambda_t \tag{8.20}$$

where ω is inertia weight, α_1 and α_2 are acceleration constraints, $\mu_1, \mu_2 \in [0, 1]$ are random values, and x_t is current position of the particle that represents the milling parameter.

Step 9: The algorithm terminates and outputs optimal non-dominated solution set (the external set), namely, Pareto optimal frontier (see Fig. 8.20).

The output of the algorithm is a set of non-dominated solutions, while there is only a solution that is needed in the proposed model. Therefore, the most optimal solution should be selected from the set of non-dominated solutions.

Spacing (SP) is regarded as the index to evaluate the uniformity of the distribution of the Pareto optimal front in the target space. SP is calculated by

$$SP = \sqrt{\frac{1}{n_0 - 1}\sum_{i=1}^{n_0}\left(\overline{d} - d_i\right)^2} \tag{8.21}$$

$$\overline{d} = \frac{1}{n_0}\sum_{i=1}^{n_0} d_i \tag{8.22}$$

where n_0 is the number of Pareto optimal solutions, d_i represents the minimum value of the Euclidean distance between the target point of the ith Pareto optimal solution and all points of the real Pareto front of the problem.

The number of the objective function is m, and d_i is calculated as

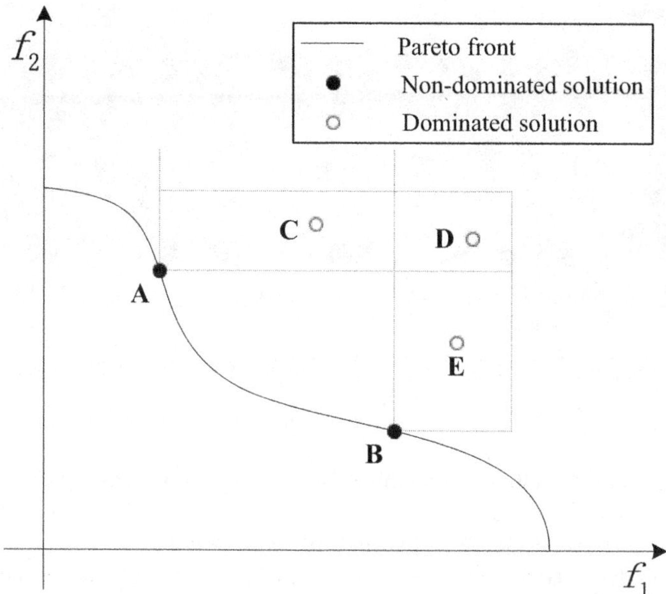

Fig. 8.20 The diagram of Pareto front. Reprinted from Ref. [37], copyright 2017, with permission from ASME

$$d_i = \min_{j}\left(\left(F_1(x_i) - F_1(x_j)\right)^2 + \left(F_2(x_i) - F_2(x_j)\right)^2 + \cdots + \left(F_m(x_i) - F_m(x_j)\right)^2 \right)^{1/2}$$

$$(8.23)$$

where $j \neq i$, $j = 1, 2, \ldots, m$.

If SP = 0, it indicates that the Pareto optimal front is evenly distributed. In order to choose the best solution from the Pareto optimal front, the solution with the least SP value is regarded as the final solution of the proposed model.

8.2.3   Case Study

8.2.3.1   Machining Process Description

In this case, the cylinder heads of B12 serial engine with four combustion chambers (see Fig. 8.21) are used to validate the performance of the proposed method on volume variation control of multiple chambers. Before the measurement, the intake and outtake valves are put into the combustion chambers and the hole of the spark plug has been reamed.

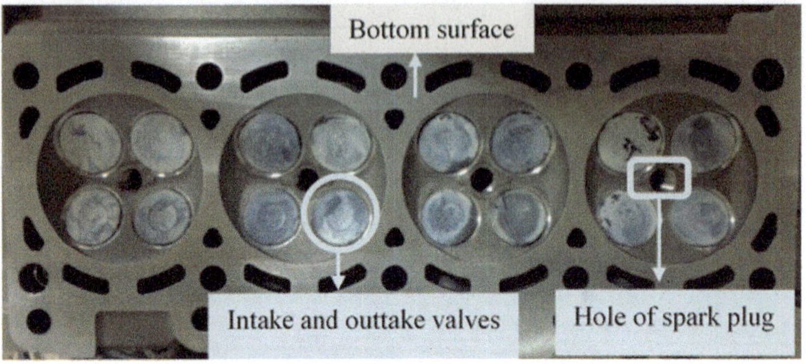

Fig. 8.21 Cylinder head of B12 engine with four combustion chambers. Reprinted from Ref. [37], copyright 2017, with permission from ASME

The volumes of combustion chambers of B12 engine cylinder head are measured by an online HDM measurement machine using laser triangulation metrology [35, 36]. The online HDM measurement machine is shown in Sect. 2.2.2.

The cylinder head milling operations influencing the volumes of combustion chambers are shown in Fig. 8.22. F1000 is the underside of the cylinder head and F2000 is the top surface of the cylinder head, and OP10, OP20, OP80, and OP140 are the operation sequence numbers. Among the four operations, the measurement of the cylinder head is conducted after OP80. The optimized finish milling parameter is input into the computer numerical control (CNC) machine on OP140, and the control of volume variation of the four combustion chambers is implemented on OP140. The CNC machine to mill the cylinder head is EX-CELL-O XS211. Ten cylinder heads are measured and machined in this case.

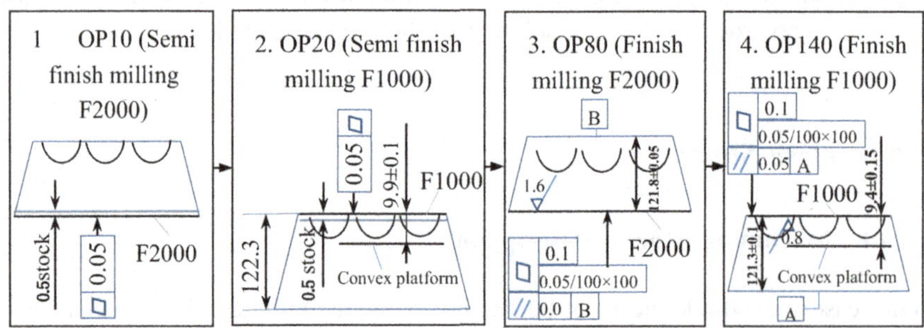

Fig. 8.22 Process of machining combustion chambers of cylinder head. Reprinted from Ref. [37], copyright 2017, with permission from ASME

8.2.3.2   Results and Analysis

(1)   Datum transformation

The point cloud of the cylinder head measured by the online measurement machine is shown in Fig. 8.23a. It can be seen that the underside of the cylinder head mapped by the point cloud is inclined, which means that Z-coordinates of points on the underside of the cylinder head are not approximately zero. Therefore, datum transformation of the measured point cloud is necessary. The point cloud after datum transformation is shown in Fig. 8.23b. The filtering result of the point cloud is shown in Fig. 8.24.

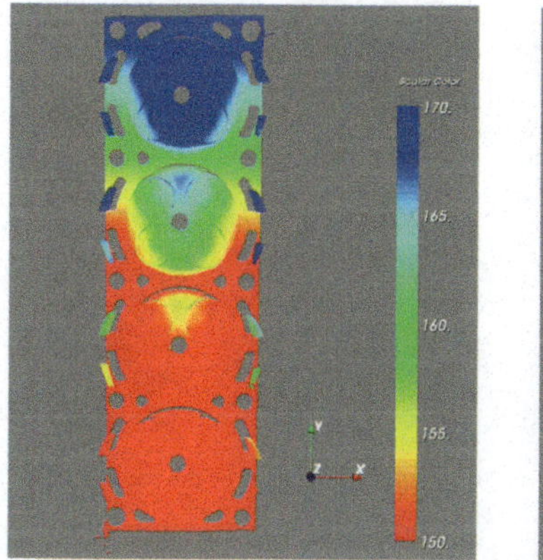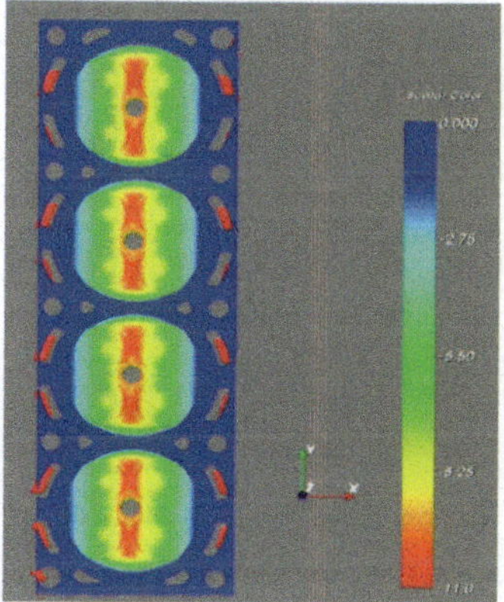

(a) Point cloud measured by HDM   (b) Point cloud transformed by RANSAC

Fig. 8.23 Point cloud before and after datum transformation. Reprinted from Ref. [37], copyright 2017, with permission from ASME

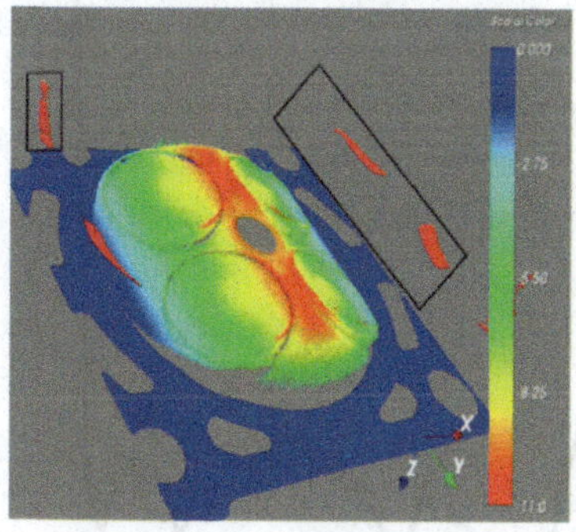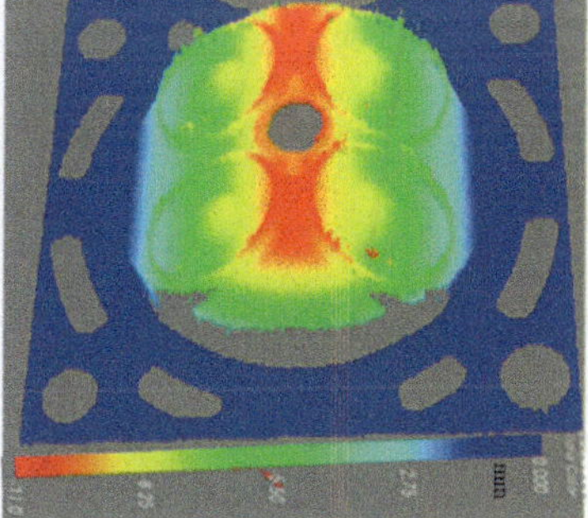

(a) Point cloud before filtering   (b) Point cloud after filtering

Fig. 8.24 Point cloud before and after filtering. Reprinted from Ref. [37], copyright 2017, with permission from ASME

(2)   Volume calculation

The reconstruction result of curved surface of combustion chambers is shown in
Fig. 8.25. It can be seen from Fig. 8.25a that the hole of the spark plug has been
completely interpolated. The meshless algorithm can effectively carry out the triangulation of the point cloud since the local grid of cylinder head combustion chamber
(shown in Fig. 8.25b) is relatively smooth and an obvious grid shape is presented.

The boundary extraction includes two steps. The first step is to determine the
numbers of initial and terminal rows according to the average Z-coordinate value of
each row. An example of a combustion chamber is shown in Fig. 8.26 and the
initial and terminal rows are 17 and 140, respectively.

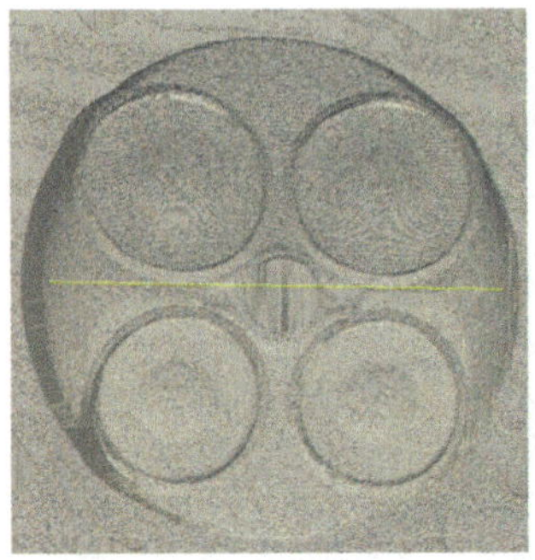

(a) Reconstruction result of a combustion

chamber

(b) Local amplification of the reconstructed

surface of combustion chambers

Fig. 8.25 Cylinder head combustion chamber surfaces after reconstruction. Reprinted from Ref.
[37], copyright 2017, with permission from ASME

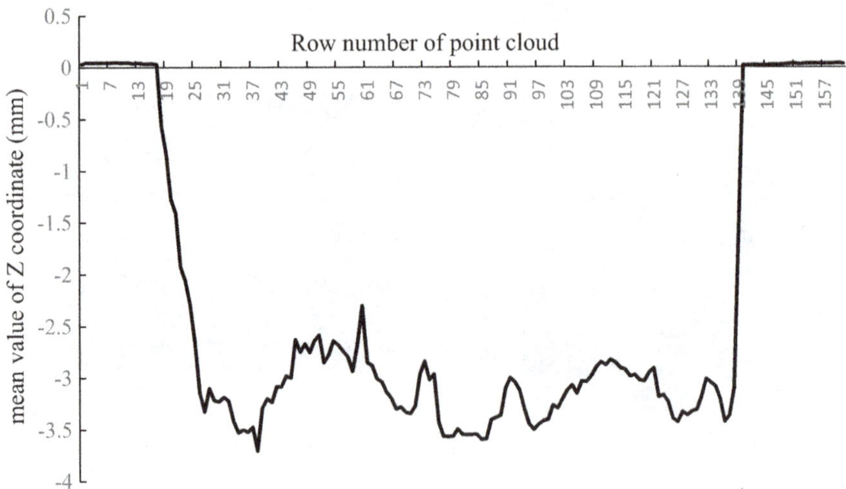

Fig. 8.26 Statistical averages of Z-coordinates of each point cloud row. Reprinted from Ref. [37],
copyright 2017, with permission from ASME

The second step is to determine the numbers of initial and terminal points of each row determined in the first step according to Z-coordinate value of each row. Figure 8.27 shows point numbers of initial and terminal points of a determined row.

Repeat the process of the second step from the 17th row to the 140th row, and the efficient boundary point set of cylinder head combustion chamber can be obtained (see Fig. 8.28). Then the parameters of the boundary can be obtained by the fitting algorithm of LS.

With the projection algorithm calculating the volumes of multiple chambers before milling, the volumes of the four combustion chambers (take a cylinder head for example) are 26.03 ml, 26.26 ml, 26.34 ml, and 26.26 ml, respectively.

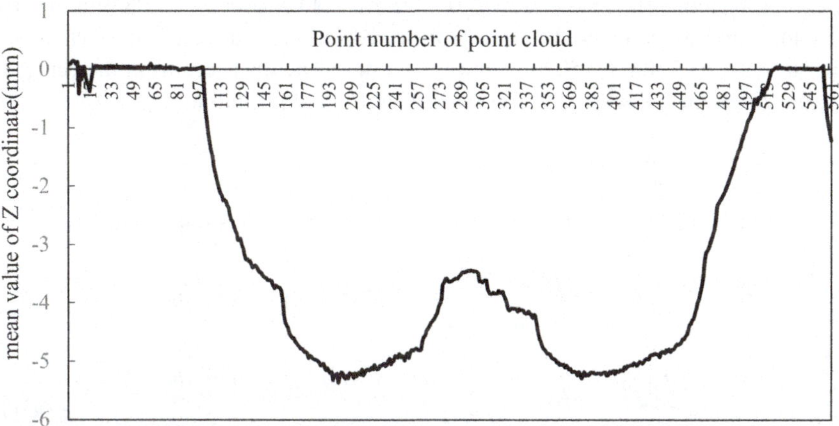

Fig. 8.27 Efficient point cloud boundary extraction. Reprinted from Ref. [37], copyright 2017, with permission from ASME

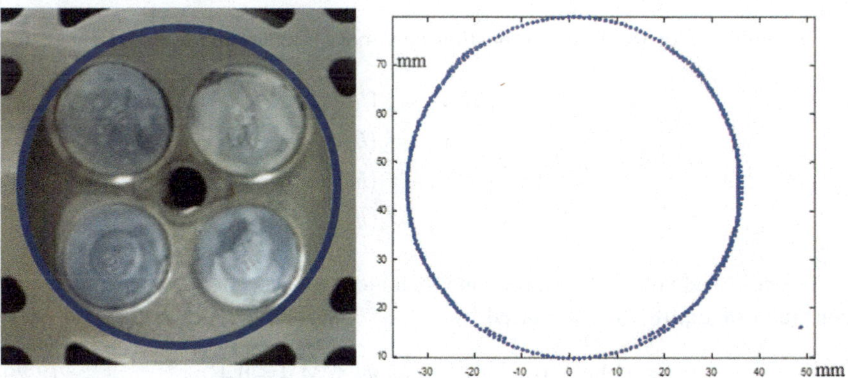

Fig. 8.28 Cylinder head combustion chamber boundary. Reprinted from Ref. [37], copyright 2017, with permission from ASME

(3) The model for obtaining an optimized milling parameter

For the cylinder head with four combustion chambers, the six objective functions are expressed as

$$\text{Min}(|v_1 - v_2|, |v_1 - v_3|, |v_1 - v_4|, |v_2 - v_3|, |v_2 - v_4|, |v_3 - v_4|) \qquad (8.24)$$

The volumes of the four combustion chambers before OP140 workstation are expressed as

$$V_1 = 26.03, V_2 = 26.26, V_3 = 26.34, V_4 = 26.26 \qquad (8.25)$$

According to the boundary extraction shown in Sect. 2.2.2, boundary radii of the four combustion chambers are 35.0378 mm, 34.8576 mm, 35.0829 mm, and 35.0445 mm, respectively. The fitted sphere radii of the four combustion chambers are 60.1581, 61.7508, 60.5026, and 62.8705 mm. The distances from the center to the underside of the workpiece for each chamber before milling are

$$H_1 = 48.9014, H_2 = 50.9716, H_3 = 49.2925, H_4 = 52.1975 \qquad (8.26)$$

Then the volumes of the milled parts of the four combustion chambers are

$$
\begin{aligned}
f_1(h_0) &= \int_{48.9014}^{48.9014 + h_0} \pi \times \left(\sqrt{61.1581^2 - r^2} \right)^2 dr \\
f_2(h_0) &= \int_{50.9716}^{50.9716 + h_0} \pi \times \left(\sqrt{61.7508^2 - r^2} \right)^2 dr \\
f_3(h_0) &= \int_{49.2925}^{49.2925 + h_0} \pi \times \left(\sqrt{60.5026^2 - r^2} \right)^2 dr \\
f_4(h_0) &= \int_{52.1975}^{52.1975 + h_0} \pi \times \left(\sqrt{62.8705^2 - r^2} \right)^2 dr
\end{aligned}
\qquad (8.27)
$$

The volumes of the four combustion chambers after milling are calculated by

$$
\begin{aligned}
v_1 &= V_1 - f_1(h_0) \\
v_2 &= V_2 - f_2(h_0) \\
v_3 &= V_3 - f_3(h_0) \\
v_4 &= V_4 - f_4(h_0)
\end{aligned}
\qquad (8.28)
$$

The designed volume is 24.4 ml and tolerance requirement is $\pm$ 0.4 ml, then the constraint of Eq. (8.16) is replaced by

$$24.4 - 0.4 \leq v1, v2, v3, v4 \leq 24.4 + 0.4 \qquad (8.29)$$

With the ICP algorithm calculating the ideal milling parameters of the four combustion chambers, the four ideal milling parameters are expressed as

$$h1 = 0.426\,\text{mm}, h2 = 0.488\,\text{mm}, h3 = 0.509\,\text{mm}, h4 = 0.488\,\text{mm} \qquad (8.30)$$

The range of the optimal milling parameter is [min (h_1, h_2, h_3, h_4), max (h_1, h_2, h_3, h_4)], which is expressed as

$$0.426 \le h_0 \le 0.509 \qquad (8.31)$$

(4)   Model solution

With MOPSO algorithm solving the model, the optimal milling parameter of the four combustion chambers is 0.468 mm, and the volumes of the four combustion chambers after milling with the optimal parameter are 24.24 ml, 24.44 ml, 24.53 ml, and 24.48 ml, respectively. In this case, other nine cylinder heads are milled with the parameter optimized by the proposed method. The optimal milling parameters (h_0), volumes of the ten cylinder heads, and volume differences of any two ones of all chambers of the ten cylinder heads after milling are shown in Table 8.1. The volumes of the combustion chambers of ten cylinder heads and the volume differences of any two ones of all chambers of the ten cylinder heads before milling are shown in Table 8.2.

For comparison, the volume differences of any two ones of all chambers of the cylinder heads before and after milling are shown in Appendix B. In order to evaluate the volume variation of cylinder head combustion chambers reasonably, reducing the maximum difference of any two ones of all chambers of a cylinder head can be considered as an indicator of improvement by the proposed method. The comparison of the maximum volume difference between two chambers of ten cylinder heads is shown in Fig. 8.29.

Table 8.3 shows the control performance of maximum volume variation of each cylinder head after milling. The ratio of reducing the maximum volume difference should be calculated by Maximum Variation Reduction Ratio (MVRR) = ($V_{\text{before milling}}$ − $V_{\text{after milling}}$)/0.8, where $V_{\text{before milling}}$ is maximum volume variation

Table 8.1 Optimal milling parameters (h_0), volume of each cylinder head chamber, and volume differences of any two ones of all chambers after milling

Volume (ml)	1	2	3	4	5	6	7	8	9	10
h_0 (mm)	0.534	0.539	0.552	0.511	0.538	0.541	0.538	0.539	0.468	0.489
v_1	24.36	24.35	24.24	24.49	24.31	24.28	24.29	24.37	24.24	24.26
v_2	24.45	24.53	24.39	24.65	24.55	24.56	24.43	24.46	24.44	24.46
v_3	24.41	24.31	24.38	24.2	24.35	24.55	24.45	24.43	24.53	24.45
v_4	24.22	24.32	24.17	24.19	24.30	24.20	24.28	24.26	24.48	24.38
$|v_1-v_2|$	0.09	0.18	0.15	0.16	0.24	0.28	0.14	0.09	0.20	0.20
$|v_1-v_3|$	0.05	0.04	0.14	0.29	0.04	0.27	0.16	0.06	0.29	0.19
$|v_1-v_4|$	0.14	0.03	0.07	0.30	0.01	0.08	0.01	0.11	0.24	0.12
$|v_2-v_3|$	0.04	0.22	0.01	0.45	0.2	0.01	0.02	0.03	0.09	0.01
$|v_2-v_4|$	0.23	0.21	0.22	0.46	0.25	0.36	0.15	0.20	0.04	0.08
$|v_3-v_4|$	0.19	0.01	0.21	0.01	0.05	0.35	0.17	0.17	0.05	0.07

Table 8.2 Volumes of each cylinder head and volume differences of any two ones of all chambers of each cylinder head before milling

Volume (ml)	1	2	3	4	5	6	7	8	9	10
V_1	26.48	26.39	26.37	26.44	26.42	26.36	26.39	26.43	26.03	26.16
V_2	26.55	26.59	26.56	26.6	26.65	26.64	26.53	26.54	26.26	26.37
V_3	26.49	26.38	26.56	26.13	26.41	26.59	26.56	26.51	26.34	26.38
V_4	26.28	26.39	26.34	26.1	26.32	26.21	26.39	26.35	26.26	26.3
$|V_1-V_2|$	0.07	0.20	0.19	0.16	0.23	0.28	0.14	0.11	0.23	0.21
$|V_1-V_3|$	0.01	0.01	0.19	0.31	0.01	0.23	0.17	0.08	0.31	0.22
$|V_1-V_4|$	0.20	0.01	0.03	0.34	0.10	0.15	0.00	0.08	0.23	0.14
$|V_2-V_3|$	0.06	0.21	0.00	0.47	0.24	0.05	0.03	0.03	0.08	0.01
$|V_2-V_4|$	0.27	0.20	0.22	0.50	0.33	0.43	0.14	0.19	0.00	0.07
$|V_3-V_4|$	0.21	0.01	0.22	0.03	0.09	0.38	0.17	0.16	0.08	0.08

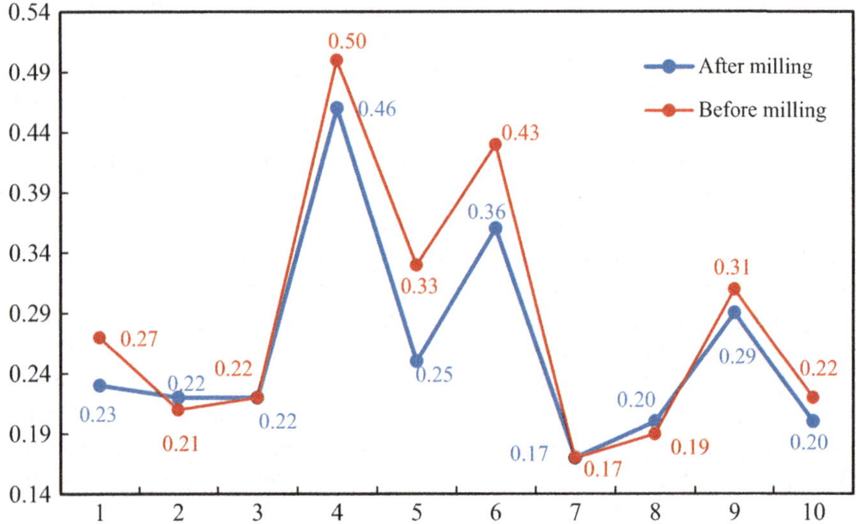

Fig. 8.29 Maximum volume variation of combustion chambers of each cylinder head before and after milling. Reprinted from Ref. [37], copyright 2017, with permission from ASME

of each cylinder before milling, $V_{\text{after milling}}$ is maximum volume variation of each cylinder after milling, and 0.8 is the sum of designed tolerance requirement ($\pm$0.4 ml). The decrease ratio after milling is calculated by Decrease Ratio (DR) = ($V_{\text{before milling}} - V_{\text{after milling}}$)/$V_{\text{before milling}}$.

It can be seen from Fig. 8.29 that the proposed method exhibits good performance on decreasing maximum volume variation of combustion chambers. Table 8.3 shows the explicit and notable decrease of maximum volume variation of combustion chambers of each cylinder head before and after milling. The maximum decrease is 24.24% and other positive decreases are very large, which can

Table 8.3 Control performance of maximum volume variation of each cylinder head after milling with the proposed method

		V_{before} milling	V_{after} milling	DR (%)	MVRR (%)
Maximum volume variation (ml)	1	0.27	0.23	14.81	5.00
	2	0.21	0.22	−4.76	−1.25
	3	0.22	0.22	0	0
	4	0.5	0.46	8.00	5.00
	5	0.33	0.25	24.24	10.00
	6	0.43	0.36	16.28	8.75
	7	0.17	0.17	0	0
	8	0.19	0.2	−5.26	−1.25
	9	0.31	0.29	6.45	2.50
	10	0.22	0.2	9.09	2.50
	Average	0.285	0.26	6.89	3.13

effectively enhance the compression ratio consistency. It also can be seen that the negative decreases occur when the maximum volume variation of combustion chambers before milling is small. The reason is that the cylinder head with small volume differences of any two ones of all chambers (the maximum volume variation is less than 0.22 ml) is more sensitive to the machining errors, such as the machine tools accuracy and positioning accuracy. Although there exist several negative and zero decreases of the maximum volume variation by the proposed method, most of the decreases are positive, and values of positive decreases are larger than values of negative decreases.

In order to better certify the performance of the proposed method, the normal production process is also conducted on the ten cylinder heads. In the normal production process of OP140, the milling parameter is 0.5 mm and the parameter remains unchanged. Every cylinder head is milled twice with 0.5 mm (nonoptimal strategy) and the optimal parameter h_0 (optimal strategy). The milling sequence depends on the values of 0.5 and h_0. For instance, if $0.5 < h_0$, the cylinder head is milled with 0.5 mm and measurement of the chambers is conducted. Then cylinder head is milled with $(h_0\text{-}0.5)$ mm (optimal strategy), and measurement of the chambers is conducted again. Table 8.4 shows the volumes of each cylinder head chamber and volume differences of any two ones of all chambers of each cylinder head after milling with nonoptimal strategy.

Since the milling process is on the same cylinder head, the volumes of chambers before milling are the same with Table 8.2. Control performance of maximum volume variation of combustion chambers of each cylinder head before and after milling with nonoptimal strategy is shown in Table 8.5.

The comparison of volume variation control performance of optimal strategy and nonoptimal strategy is shown in Table 8.6, Figs. 8.30 and 8.31.

Table 8.4 Volume of each cylinder head chamber and volume differences of any two ones of all chambers after milling with nonoptimal strategy

Number	1	2	3	4	5	6	7	8	9	10
v_1	24.42	24.43	24.33	24.52	24.39	24.35	24.36	24.42	24.16	24.24
v_2	24.53	24.60	24.47	24.68	24.64	24.66	24.52	24.52	24.36	24.45
v_3	24.48	24.38	24.47	24.22	24.42	24.63	24.53	24.50	24.47	24.43
v_4	24.27	24.4	24.26	24.20	24.35	24.27	24.34	24.33	24.41	24.36
$|v_1-v_2|$	0.11	0.17	0.14	0.16	0.25	0.31	0.16	0.10	0.20	0.21
$|v_1-v_3|$	0.06	0.05	0.14	0.30	0.03	0.28	0.17	0.08	0.31	0.19
$|v_1-v_4|$	0.15	0.03	0.07	0.32	0.04	0.08	0.02	0.09	0.25	0.12
$|v_2-v_3|$	0.05	0.22	0.00	0.46	0.22	0.03	0.01	0.02	0.11	0.02
$|v_2-v_4|$	0.26	0.20	0.21	0.48	0.29	0.39	0.18	0.19	0.05	0.09
$|v_3-v_4|$	0.21	0.02	0.21	0.02	0.07	0.36	0.19	0.17	0.06	0.07

Table 8.5 Control performance of maximum volume variation of each cylinder head before and after milling with nonoptimal strategy

		V_{before} milling	V_{after} milling	DR (%)	MVRR (%)
Maximum volume variation (ml)	1	0.27	0.26	3.70	1.25
	2	0.21	0.22	−4.76	−1.25
	3	0.22	0.21	4.55	1.25
	4	0.5	0.48	4.00	2.50
	5	0.33	0.29	12.12	5.00
	6	0.43	0.39	9.30	5.00
	7	0.17	0.19	−11.77	−2.50
	8	0.19	0.19	0	0
	9	0.31	0.31	0	0
	10	0.22	0.21	4.55	1.25
	Average	0.285	0.275	2.17	1.25

It can be seen from Table 8.6 that the average DR of optimal strategy is 6.89%, which is 4.72% higher than nonoptimal strategy (2.17%). The average MVRR of optimal strategy is also 1.88% higher than nonoptimal strategy. Besides, Figs. 8.30 and 8.31 show that the optimal strategy exhibits good performance on decreasing maximum volume variation of combustion chambers and outperforms the nonoptimal strategy. Since each cylinder head is milled twice, there exist processing errors (repeat positioning and clamping errors) which may influence the volume variation control performance of the optimal strategy. Although there are several worse decreases in the maximum volume variation by the optimal strategy, the total performance is better than the nonoptimal strategy.

Table 8.6 Comparison of volume variation control performance of optimal strategy and nonoptimal strategy

	DR		MVRR	
	Optimal strategy (%)	Nonoptimal strategy (%)	Optimal strategy (%)	Nonoptimal strategy (%)
1	14.81	3.70	5.00	1.25
2	−4.76	−4.76	−1.25	−1.25
3	0	4.55	0	1.25
4	8.00	4.00	5.00	2.50
5	24.24	12.12	10.00	5.00
6	16.28	9.30	8.75	5.00
7	0	−11.77	0	−2.50
8	−5.26	0	−1.25	0
9	6.45	0	2.50	0
10	9.09	4.55	2.50	1.25
Average	6.89	2.17	3.13	1.25

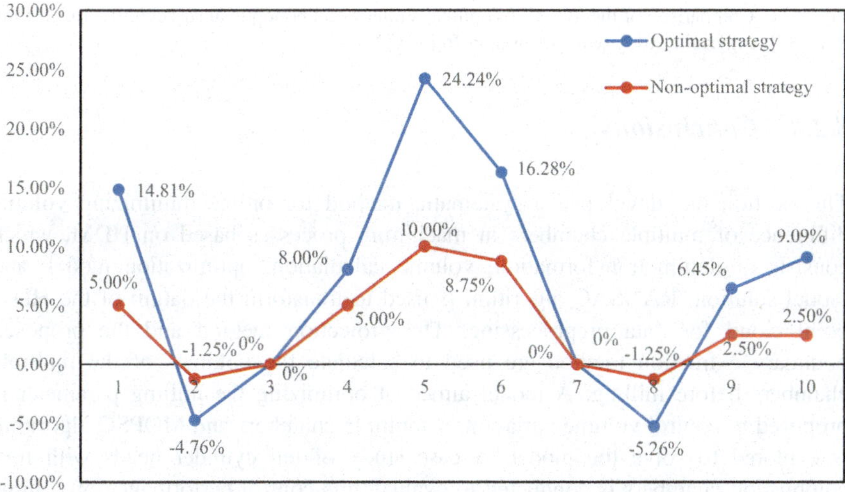

Fig. 8.30 Comparison of the DR of optimal strategy and nonoptimal strategy. Reprinted from Ref. [37], copyright 2017, with permission from ASME

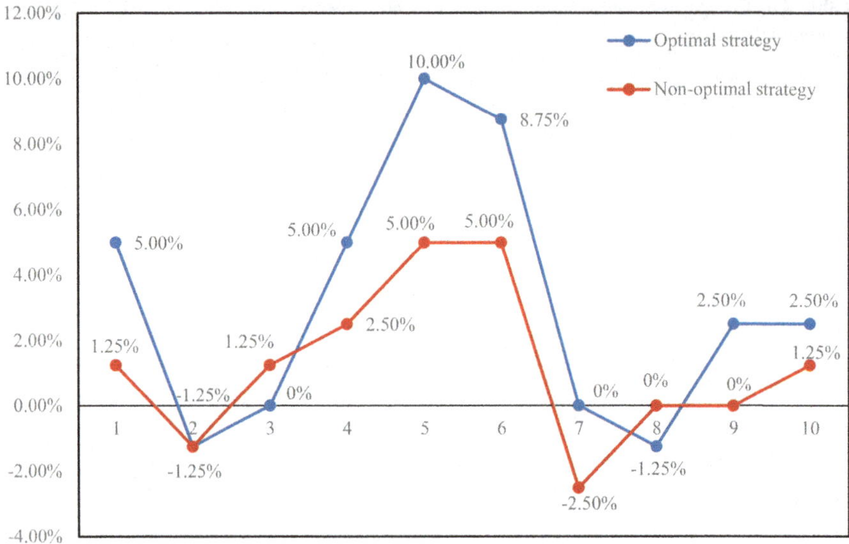

Fig. 8.31 Comparison of the MVRR of optimal strategy and nonoptimal strategy. Reprinted from Ref. [37], copyright 2017, with permission from ASME

8.2.4   Conclusions

This section has developed a systematic method for online minimizing volume difference of multiple chambers in machining processes based on HDM, which consists of datum transformation, volume calculation, optimization model, and model solution. RANSAC algorithm is used to transform the datum of the HDM point cloud for data preprocessing. The projection method and the proposed boundary extraction method are used to calculate the volumes of the multiple chambers before milling. A model aimed at optimizing the milling parameter is proposed to control volume variation of multiple chambers and MOPSO algorithm is explored to solve the model. A case study of ten cylinder heads with four combustion chambers is conducted to evaluate the control performance of volume variation. The results demonstrate that the proposed method can decrease the maximum volume variation of multiple chambers (decrease ratio ranges from 6.45 to 24.24% and maximum variation reduction ratio ranges from 2.50 to 10.00%) and outperforms the normal production process (nonoptimal strategy). In summary, the proposed method can be well applied online for controlling volume variation of multiple chambers.

The proposed method is applicable to minimizing the volume difference of multiple chambers of in-line engine, but cannot be directly used in V-type and W-type engines. For future work, control of volume difference of multiple chambers can be extended in three aspects.

(1) Spatial correlation of three-dimensional point cloud of multiple chambers can be applied to identify the mold failure modes of multichamber workpiece. Then, the mold of multichamber workpiece can be better maintained and casts blank workpieces with less volume variation.

(2) New machining technologies (i.e., machining the interior surfaces of the chambers) can be adopted. There are two options for controlling the volume variation of the multiple chambers with the ball milling cutter milling the interior surfaces of the chambers. The first option is shown in Fig. 8.32. Since the milling depths of the n chambers are the same, the control performance of volume variation is the same as the proposed method.

   The second option is shown in Fig. 8.33. In this way, the milling depths of the n chambers are different, which is possible to eliminate the volume variation of multiple chambers. But it is quite difficult to dynamically adjust the milling depth of each combustion chamber. For the second option, the new methodology needs to be developed.

(3) Further improvement is possible by using the software for the analysis of the alleged data and automated creation of a program for CNC machine.

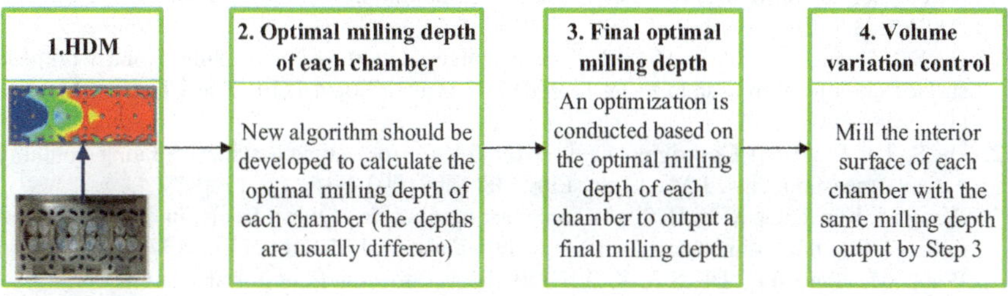

Fig. 8.32 Procedure of milling interior surfaces with a milling depth. Reprinted from Ref. [37], copyright 2017, with permission from ASME

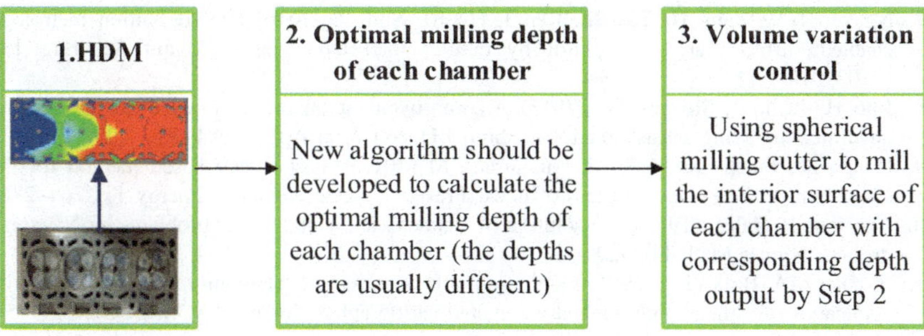

Fig. 8.33 Procedure of milling interior surfaces with n different milling depths. Reprinted from Ref. [37], copyright 2017, with permission from ASME

References

1. Yang MY, Choi JG (1998) Tool deflection compensation system for end milling accuracy improvement. J Manuf Sci Eng 120(2):222–229
2. Johnstone S, Peyton AJ (2001) The application of parametric 3D finite element modeling techniques to evaluate the performance of a magnetic sensor system. Sensors Actuat 93 (2):109–116
3. Lee DM, Choi SG (2004) Application of on-line adaptable Neural Network for the rolling force set-up of a plate mill. Eng Appl Artif Intell 17(5):557–565
4. Watanabe T, Iwai S (2006) A control system to improve the accuracy of finished surfaces in milling. J Dyn Syst Meas Contr 105(3):192–199
5. Liang HB, Li X (2009) A 5-axis milling system based on a New G code for NURBS surface. In: IEEE international conference on intelligent computing and intelligent systems, pp 600–603
6. Yuan G (2010) Online detecting system of roller wear based on laser-linear array CCD technology. Int Soc Opt Eng 76(1):467–479
7. Chen LX (2018) Online real-time control method for product manufacturing process. U.S. Patent No. 9891614
8. Available: introducing the latest in high-definition, non-contact metrology Shapix 1500 series. http://www.coherix.com
9. Du S, Liu C, Huang D (2015) A shearlet-based separation method of 3D engineering surface using high definition metrology. Precis Eng 40:55–73
10. Du S, Liu C, Xi L (2015) A selective multiclass support vector machine ensemble classifier for engineering surface classification using high definition metrology. J Manuf Sci Eng 137 (1):011003-1-15
11. Du SC, Huang DL, Wang H (2015) An adaptive support vector machine-based workpiece surface classification system using high-definition metrology. IEEE Trans Instrum Meas 64 (10):2590–2604
12. Du S, Fei L (2016) Co-kriging method for form error estimation incorporating condition variable measurements. J Manuf Sci Eng 138(4):041003-1-16
13. Wang M, Ken T, Du S, Xi L (2015) Tool wear monitoring of wiper inserts in multi-insert face milling using three-dimensional surface form indicators. J Manuf Sci Eng 137(3):031006-1-8
14. Wang M, Shao YP, Du SC, Xi LF (2015) A diffusion filter for discontinuous surface measured by high definition metrology. Int J Prec Eng Manuf 16(10):2057–2062
15. Wang M, Xi L, Du S (2014) 3D surface form error evaluation using high definition metrology. Prec Eng 38(1):230–236
16. Suriano S, Wang H, Shao C, Hu SJ, Sekhar P (2015) Progressive measurement and monitoring for multi-resolution data in surface manufacturing considering spatial and cross correlations. IIE Trans (ahead-of-print), pp 1–20
17. Nguyen HT, Wang H, Tai BL, Ren J, Hu SJ, Shih A (2016) High-definition metrology enabled surface variation control by cutting load balancing. J Manuf Sci Eng 138 (2):021010-1-11
18. Cho H, Luck R, Stevens JW (2015) An improvement on the standard linear uncertainty quantification using a least-squares method. J Uncert Anal Appl 3(1):1–13
19. Hongn M, Larsen SF, Gea M, Altamirano M (2015) Least square based method for the estimation of the optical end loss of linear Fresnel concentrators. Sol Energy 111:264–276
20. Anselone P, Rall L (1968) The solution of characteristic value-vector problems by Newton's method. Numer Math 11(1):38–45
21. Fischler MA, Bolles RC (1981) Random sample consensus: a paradigm for model fitting with applications to image analysis and automated cartography. Commun ACM 24(6):381–395

22. Kim T, Im YJ (2003) Automatic satellite image registration by combination of matching and random sample consensus. IEEE Trans Geosci Remote Sens 41(5):1111–1117
23. Yaniv Z (2010) Random sample consensus (RANSAC) algorithm, a generic implementation. Imaging
24. Raguram R, Chum O, Pollefeys M, Matas J, Frahm J (2013) Usac: a universal framework for random sample consensus. IEEE Trans Pattern Anal Mach Intell 35(8):2022–2038
25. Leon SJ, Björck Å, Gander W (2013) Gram-schmidt orthogonalization: 100 years and more. Numer Linear Algebra Appl 20(3):492–532
26. Pomerleau F, Colas F, Siegwart R, Magnenat S (2013) Comparing ICP variants on real-world data sets. Auton Robots 34(3):133–148
27. Di Maio F, Bandini A, Zio E, Alfonsi A, Rabiti C (2016) An approach based on support vector machines and a K-D Tree search algorithm for identification of the failure domain and safest operating conditions in nuclear systems. Prog Nucl Energy 88:297–309
28. Schauer J, Nüchter A (2014) Efficient point cloud collision detection and analysis in a tunnel environment using kinematic laser scanning and KD tree search. Int Arch Photogramm Remote Sens Spatial Inf Sci 40(3):289–295
29. Besl PJ, McKay HD (1992) A method for registration of 3-D shapes. IEEE Trans Pattern Anal Mach Intell 14(2):239–256
30. Du S, Xi L (2011) Fault diagnosis in assembly processes based on engineering-driven rules and PSOSAEN algorithm. Comput Ind Eng 60(1):77–88
31. Poli R, Kennedy J, Blackwell T (2007) Particle swarm optimization. Swarm Intelligence 1(1):33–57
32. Coello CAC, LechugaMS (2002) MOPSO: a proposal for multiple objective particle Sswarm optimization. In: IEEE Proceedings of the 2002 congress on evolutionary computation, pp 1051–1056
33. Reyes-Sierra M, Coello CC (2006) Multi-objective particle swarm optimizers: a survey of the state-of-the-art. Int J Comput Intell Res 2(3):287–308
34. Zhang Y, Gong D, Zhang J (2013) Robot path planning in uncertain environment using multi-objective particle swarm optimization. Neurocomputing 103:172–185
35. Halcon Solution Guide III-C 3D Vision. http://download.mvtec.com/halcon-9.0-solution-guide-iii-c-3d-vision.pdf
36. Dorsch R, Häusler G, Herrmann J (1994) Laser triangulation: fundamental uncertainty in distance measurement. Appl Opt 33(7):1306–1314
37. Huang DL, Du SC, Li GL, Wu ZQ (2017) A Systemic approach for on-line minimizing volume difference of multiple chambers with casting surfaces in machining processes based on high definition metrology. J Manuf Sci Eng 139(8):081003-1-17

Made in the USA
Monee, IL
07 July 2026